Regulation of Genome Editing in Plant Biotechnology

Hans-Georg Dederer • David Hamburger
Editors

Regulation of Genome Editing in Plant Biotechnology

A Comparative Analysis of Regulatory Frameworks of Selected Countries and the EU

Springer

Editors
Hans-Georg Dederer
Faculty of Law
University of Passau
Passau, Germany

David Hamburger
Faculty of Law
University of Passau
Passau, Germany

ISBN 978-3-030-17118-6 ISBN 978-3-030-17119-3 (eBook)
https://doi.org/10.1007/978-3-030-17119-3

This Springer imprint is published by the registered company Springer Nature Switzerland AG.
The registered company address is: Gewerbestrasse 11, 6330 Cham, Switzerland

Preface

The publication of a contributed volume is an undertaking that, by its very nature, is not possible without the fruitful collaboration of various parties.

That said, our gratitude goes first and foremost to our esteemed country rapporteurs Margaret Rosso Grossman, Tetsuya Ishii, Martin Lema, Karinne Ludlow, Ansgar Münichsdorfer, Stuart Smyth, Brigitte Voigt and Agustina Whelan. Without their commitment and dedication to the project, we and our readers would not have the opportunity to learn from, and be captivated by the accumulation of, their vast knowledge that is abundantly visible in their country reports.

The country rapporteurs presented their reports at a workshop in Munich on 22nd and 23rd of March 2018 which was attended by academics, regulators, practitioners and stakeholders both from Germany and abroad. The stimulating and intriguing discussions during the workshop undoubtedly left their mark on the final versions of the country reports.

Besides, it was the generous support from the German Federal Ministry of Education and Research (Bundesministerium für Bildung und Forschung, BMBF) which has created the financial underpinnings for this project to thrive. This contributed volume is published as part of the research project "Genome editing in plant biotechnology – a science-based legal analysis of regulatory problems" which is in its entirety funded by the BMBF (project no. 01GP1615). While this research project is mainly concerned with the regulation of genome edited plants in the European Union, this edited volume constitutes the project's contribution to the comparative law aspect of this field of study.

The funding by the BMBF was complemented by outstanding administrative support through the German Aerospace Center (Deutsches Zentrum für Luft- und Raumfahrt e.V., DLR) throughout the entire duration of the research programme.

In order for a book project to finally see the light of day, it is in the end the backing from a publisher that is indispensable. In that regard, we were fortunate to have received from early on the trust of such a renowned and experienced publishing house as Springer. As a result, we have been accompanied with the highest level of expertise and know-how during all stages of our research endeavour.

Finally, special thanks are due to the student assistants Sabrina Brzezinski, Clemens Dienstbier, Sebastian Graup and Katharina Schreiber, who provided with their exceptional work effort invaluable support in the completion of the final manuscript.

Since the country reports are for the most part based on the presentations made at the workshop in March 2018, changes after this date could only be partially taken into account.

Passau, Germany Hans-Georg Dederer
Passau, Germany David Hamburger
May 2019

Contents

Contributors

Hans-Georg Dederer is full professor of Constitutional and Administrative Law, Public International Law and European and International Economic Law at the University of Passau (Germany). His fields of research include international environmental law and agricultural biotechnology law. He is also a member of the Permanent Senate Commission on Genetic Research of the German Research Foundation (Deutsche Forschungsgemeinschaft—DFG).

Margaret Rosso Grossman Ph.D., J.D. is Bock Chair Emerita and Professor of Agricultural Law Emerita in the Department of Agricultural and Consumer Economics, University of Illinois (USA). Her research focuses on agricultural and environmental law in the United States and Europe. Grossman is the author of 130 law review articles and book chapters published in the United States, the United Kingdom, Canada, Australia, the Netherlands, Germany and Italy. She is author or editor of books published in Europe and the United States, and she has presented papers and lectures in numerous European countries as well as in Australia, China, Taiwan and Costa Rica. Professor Grossman received three Fulbright Senior Scholar Awards and a German Marshall Fund Research Fellowship. The American Agricultural Law Association awarded her the Distinguished Service Award, the Professional Scholarship Award (twice) and the Excellence in Agricultural Law Award. She received the Silver Medal of the European Council for Agricultural Law. She is a fellow of the Wageningen School of Social Sciences, Wageningen University, the Netherlands.

David Hamburger is a Ph.D. candidate at the Chair of Constitutional and Administrative Law, Public International Law and European and International Economic Law at the University of Passau (Germany) and holds an LL.M. degree from the University of Glasgow. As part of a project funded by the German Federal Ministry of Education and Research (BMBF), he is conducting research on the international aspects of the regulation of plants derived from genome editing.

Tetsuya Ishii Professor, Office of Health and Safety, Hokkaido University, earned his Ph.D. in bioscience at Hokkaido University (Japan). He joined the Center for iPS Cell Research and Application, Kyoto University, after working at Japan Science and Technology Agency, while completing a program officer program in US NIH. Currently, he is studying bioethics in the field of bioscience, particularly genetically modified food and gene therapy. He received the Hokkaido University President Prize, Encouragement Award 2017. He is also an associate member of the Science Council of Japan and a member of three committees regarding agricultural and medical uses of gene editing.

Martin Lema Biotechnologist, is Adjunct Professor at the Biotechnology School of the National University of Quilmes, Argentina. He also holds the position of Director of Biotechnology in the Secretariat of Foodstuff and Bioeconomy of the Ministry of Agroindustry of the Argentine Republic. He is the Chair of the National Advisory Commission of Agricultural Biotechnology (CONABIA), the Advisory Committee on the use of Bio-inputs for Agriculture and the National Advisory Commission of Biomaterials. He has 19 years of academic and entrepreneurial experience in biotechnology and 13 years' experience in the field of policymaking in agricultural biotechnology and biosafety. He has been appointed repeatedly as a Delegate representing the Argentine government in several bilateral and multilateral fora on agricultural biotechnology. He has published several publications on diverse biotechnology activities including research, teaching, biosafety and policymaking.

Karinne Ludlow B.Sc., LL.B., Ph.D. is an Associate Professor in the Faculty of Law, Monash University (Australia). Her research focuses on the integration of science, commercialization challenges and law, particularly around biotechnology and nanotechnology in all their applications including health, agriculture and industry. Ludlow is the author of 44 refereed articles and book chapters published in Australia, the United States, the United Kingdom and the Netherlands. She is author or editor of books published in Australia, the United States and the United Kingdom, and she has presented papers and lectures in numerous European countries as well as in Canada and the United States. Ludlow was lead author of a report commissioned by the Australian Government on preparing Australian regulatory frameworks for nanotechnology and a later report on preparing those frameworks for innovation more generally. Current projects include a project on regulating genome editing in food and agriculture as well as government-funded research on inheritable genetic modification of human embryos, including the use of mitochondrial donation and genome editing.

Ansgar Münichsdorfer is a Student Research Assistant at the Chair of Constitutional and Administrative Law, Public International Law and European and International Economic Law at the University of Passau (Germany) where he participates in a project on regulatory questions of genome editing in plant biotechnology funded by the German Federal Ministry of Education and Research (BMBF). He focuses on

Public International and European Law and holds a Certificate of Political Sciences from Sciences Po Lyon.

Stuart Smyth B.A., Ph.D. is an Associate Professor in the Department of Agriculture and Resource Economics at the University of Saskatchewan, where he holds the Industry Research Chair in Agri-Food Innovation. His research focuses on sustainability, agriculture, innovation and food. Dr. Smyth publishes a weekly blog on these topics at: www.SAIFood.ca. Recent publications include an authored book with William Kerr and Peter Phillips, *Biotechnology Regulation and Trade*, published by Springer (2017); a co-edited book with Nicholas Kalaitzandonakes, Peter Phillips and Justus Wesseler, *The Coexistence of Genetically Modified, Organic and Conventional Foods*, published by Springer (2016); and a co-edited book with Peter Phillips and David Castle of the *Handbook on Agriculture, Biotechnology and Development*, a 51-chapter book published by Edward Elgar (2014). An additional 29 journal articles have been published in the past 4 years.

Brigitte Voigt is a Ph.D. candidate at the Chair of Constitutional and Administrative Law, Public International Law and European and International Economic Law at the University of Passau (Germany). Her doctoral dissertation deals with the conceptualization and design of a novel EU regulatory framework for plants derived from genome editing, taking into account best regulatory practices of other states and requirements under world trade law.

Agustina Whelan provides technical and operational support to the Biotechnology Directorate in the Secretariat of Foodstuff and Bioeconomy of the Ministry of Agroindustry of the Argentine Republic and to the National Advisory Committee on Agricultural Biotechnology (CONABIA). These activities include the management of applications, participation in internal and external meetings and drafting safety assessment and other reports pertaining to scientific advice in biotechnology and biosafety.

In the academic field, she is Visiting Scholar at the National University of Quilmes, Argentina, for teaching graduate and postgraduate courses on biotechnology and biosafety. In addition, she has authored publications pertaining to biotechnology, biosafety and policymaking.

Agustina Whelan completed a Bachelor's Degree in Biotechnology at the University of Quilmes in 2009. She also holds Magister postgraduate studies in Policy and Management of Science and Technology at the University of Buenos Aires (UBA).

Chapter 1
Introduction: Regulation of Plants Derived from Genome Editing—What Lessons To Be Learned from Other Countries?

Hans-Georg Dederer and David Hamburger

Abstract The advent of genome editing in plant breeding and the resulting blurring of the boundaries between natural and artificially induced genetic modifications present regulators worldwide with new challenges. In such a time of regulatory uncertainty, or dispute over how to regulate genome edited plants, legislators are well advised to seek external guidance on how this issue could be addressed appropriately. Since genome edited organisms pose similar challenges to regulatory systems around the world, it seems sensible to study the practices of other jurisdictions in order to draw lessons for one's own regulatory efforts. To be able to choose from a diverse selection of regulatory approaches, countries with differing attitudes towards genetically modified plants were chosen as research objects. Broadly speaking the studied jurisdictions can be divided into those which embrace the cultivation of GMOs (Argentina, Canada and the USA), those which are reluctant adopters of GMOs (Australia and Europe) and a de facto absolute abstainer from GM crop cultivation (Japan). Based on a comparative analysis of the regulatory frameworks and an identification of possible best practices, the conclusion is made that a consistent regulatory regime should be product-based, i.e. the risk regulation should be triggered by a plant's traits. From a procedural point of view, an obligatory upstream procedure should be used for channelling the respective plant into the relevant regulatory framework. This process can be further catalysed by a voluntary early consultation procedure. Within such a framework the one-door-one-key principle should apply, which means that all relevant authorizations are granted upon a single application.

1.1 Introduction

The advent of so called new breeding techniques (NBTs) and the resulting blurring of the boundaries between natural and artificially induced genetic modifications present regulators worldwide with new challenges.

H.-G. Dederer (✉) · D. Hamburger
Faculty of Law, University of Passau, Passau, Germany
e-mail: hans-georg.dederer@uni-passau.de; david.hamburger@uni-passau.de

H.-G. Dederer, D. Hamburger (eds.), *Regulation of Genome Editing in Plant Biotechnology*, https://doi.org/10.1007/978-3-030-17119-3_1

The rapidly adopted and constantly improving genome editing technology, in particular the CRISPR-Cas technology, makes it possible to develop new genetically modified plant varieties that are indistinguishable from conventionally bred plants or naturally occurring mutants.[1] This development calls into question the established regulatory differentiation between genetically modified organisms (GMOs) and non-GMOs.[2]

Depending on the applicable domestic regulation, it may be unclear to what extent plants derived from new breeding techniques are subject to the relevant legal provisions. As a consequence, national legal frameworks for the regulation of GMOs are faced with the challenge that their scope of application has to be redefined or at least clarified with respect to genome edited organisms (GEOs). Even if the regulatory status of GEOs has, at least in significant part, been determined either by a governmental authority or by a court,[3] there may be doubts among regulators, or debates among politicians, seed developers, farmers, environmentalists and other actors of civil society, concerning the appropriateness of the applicable rules.

In such a time of regulatory uncertainty, or dispute over how to regulate GEOs adequately, legislators are well advised to seek external guidance on how this issue could be addressed appropriately. Such orientation can be provided *inter alia* by scientific gain of knowledge, public opinion, economic considerations, political necessities or ethical convictions. However, since GEOs pose similar challenges to regulatory systems around the world, it seems sensible to study the practices of other jurisdictions in order to draw lessons for one's own regulatory efforts. This way, a legislative endeavor can adopt advantages of different regulatory approaches as well as regulatory solutions already found in other legal frameworks while at the same time avoiding the repetition of their mistakes.

Just as there is a plethora of different jurisdictions, there is also a wide variety of regulatory approaches towards GMOs. In order to be able to filter out and make use of the best practices a specific regulatory regime has to offer, it is imperative that the objectives of the foreign regulatory system are identified as well. Legal provisions regarding GMOs can pursue different purposes like ensuring safety, promoting research and development, or facilitating the adoption of GMOs. Regulatory means are, in turn, aligned with regulatory objectives, *i.e.* they are intended, and hence specifically designed, to achieve a particular regulatory purpose. To ensure the compatibility of an identified foreign best practice with the domestic regulatory approach it is, therefore, decisive to ensure that the respective foreign regulatory measure, or mechanism, fits into the domestic overall regulatory structure and its object and purpose.

---

[1]Cf. Sprink et al. (2016), p. 1497; Voigt and Klima (2017), p. 321; Schenkel and Leggewie (2015), p. 265.

[2]Cf. Sprink et al. (2016), pp. 1494–1495; Globus and Qimron (2018), pp. 1293–1294.

[3]As is the case in the European Union (EU) with regard to the Court of Justice of the European Union's judgment of 25 July 2018 (CJEU, C-528/16, *Confédération paysanne et al.*). Cf. European Court of Justice (2018).

In order to be able to choose from a diverse selection of regulatory approaches with different objectives, countries with differing attitudes towards genetically modified plants were chosen as research objects. Very broadly speaking, the studied jurisdictions can be divided into three groups regarding their attitude towards GM crop cultivation: those which embrace and support the cultivation of GMOs (Argentina, Canada and the USA), those which are reluctant adopters of GMOs with regions that opted-out of GMO cultivation (Australia and Europe) and *de facto* absolute abstainers from GM crop cultivation (Japan).[4]

Consequently, the selection of these countries promises the identification of a wide variety of regulatory tools, which can be used to achieve a similarly manifold range of objectives.

1.2 Specific Characteristics of the Regulatory Approaches

A first step in determining transferable regulatory tools is the identification of the regulatory regimes' specific characteristics. This is crucial because the features that distinguish regulatory systems from each other facilitate the detection of a best practice by means of comparison.

1.2.1 Argentina

Argentina has been the first country in the world to adopt, in 2015, a new regulation[5] specifically addressed to NBTs including genome editing. However, this is not a substantive regulation (*i.e.* not laying down rules on, *e.g.*, risk assessment, authorization or labelling), but a regulation of a procedural nature only.

The decree lays down the process that is used to determine whether a plant derived from an NBT constitutes a GMO as defined by the Argentine regulatory system. The outlined administrative procedure, therefore, precedes the application of the regulatory framework for GMOs that will follow if the plant in question is found to constitute a GMO. Such an upstream procedure has the advantage that the original GMO regime can remain unchanged, while this newly established procedure ensures that it remains applicable whenever a new breeding technique emerges. Importantly, this novel procedure is not tied to specific technologies, *i.e.* it is technology-neutral.

At the same time, the upstream procedure serves the purpose of consulting with plant breeders. Plant breeders can ask the competent authority for an assessment

[4]For a detailed illustration of those countries' varying attitude towards GM cultivation see Chap. 8, Sect. 8.2.1.

[5]Ministerio de Agricultura, Ganadería y Pesca (2015). For detailed information on the content and working of this decree see Chap. 2, Sect. 2.3.2.

of an individual plant variety and they are entitled to an answer, whether the crop is considered to be a GMO or a non-GMO, within 60 days. A special feature of this procedure is the possibility to ask for a preliminary classification of a plant variety while it is still at the design stage. This early consultation procedure provides a developer with the opportunity to receive a regulatory classification of the envisaged product at an early point of his research and development efforts. As soon as the new plant variety has been developed, molecular biology studies on the relevant genetic alteration must be submitted to the authorities. If the final plant variety corresponds to the earlier product that was the subject of the early consultation procedure, the preliminary assessment regarding its regulatory status retains its validity.

If the plant, or plant variety, does not fit into the category of GMOs, it is subject to those rules which are applicable to 'conventional' plants, or plant varieties respectively. Interestingly, however, if the competent authority in charge of the (early) consultation procedure identifies (possible) risks arising from such a NBT-derived plant or plant variety, it may issue safety-related recommendations to the plant breeder. In addition, the authority has to notify the agency, which is in charge of plant variety registration, of such (possible) risks. The registration may be rejected, in the end, if commercialization of the variety poses unacceptable sanitary or phytosanitary risks. Thus, there is no, or at least no significant, regulatory gap between regulation of GM varieties and non-GM varieties as regards risks.

1.2.2 Australia

What specifically distinguishes the Australian regulatory framework for GMOs from the other examined approaches is the existence of a bi-national authority, the Food Standards Australia New Zealand (FSANZ).[6] FSANZ is responsible for the assessment and market approval of GM food prior to the commercial release into Australia and New Zealand. FSANZ does not make such an assessment separately for Australia and New Zealand, but issues a single market approval for both countries. Therefore, exactly the same rules and procedures are applied to both jurisdictions.

The existence of a common, in fact supranational, approval authority is especially interesting when taking into account that no cultivation of GMOs takes place in New Zealand while Australian farmers cultivate GMOs. Although Australia and New Zealand have a completely different approach towards the cultivation of GMOs, both countries were able to find common ground with regard to the consumption of GM food.

Besides the regulatory framework for GM foods implemented by FSANZ, there is a separate solely Australian regulatory framework for GMOs which applies to

---

[6]The situation seems at a first glance similar to the European approach. However, since the EU has the competence to shape the legal framework for marketing of GMO based on its own volition, the European situation is in this concrete instance more comparable to that of a federal state—even though the EU is not a state entity in legal terms.

"dealings" with GMOs such as GMO cultivation. Although both regulatory regimes are ultimately triggered by the use of "gene technology", the two frameworks define the decisive term "gene technology" differently. Due to those differing legal definitions, the Australian regulatory status of GEOs currently depends on whether they are used for, e.g., cultivation or food production which means, in the end, that GEOs may be covered, e.g., by the general GMO framework but not by the GM food framework.

1.2.3 Canada

What stands out in the case of Canada is the use of a solely product-based regulatory approach when it comes to the approval of GMOs for cultivation. Here, the decisive trigger for a stricter approval process is not the use of gene technology or any other, traditional or modern, breeding technique but the existence of a novel trait in a new plant variety. Such a variety with a novel trait is defined as "a plant containing a trait not present in plants of the same species already existing as stable, cultivated populations in Canada, or is present at a level significantly outside the range of that trait in stable, cultivated populations of that plant species in Canada".[7] Consequently, exactly the same rules apply to genome edited plants, crops derived from "classic" genetic engineering and conventionally bred varieties.

Such a purely product-based approach concerning the cultivation approval of new plant varieties is unique among the examined countries. To avoid confusion—and a common misconception—it should be stated, though, that it is only the approval process for cultivation, which is solely product-based. When it comes to GM food, a process-based component comes into play. However, that does not imply that all GM foods are reviewed simply because of the use of modern biotechnological techniques of genetic modification. Rather GM foods must also display a feature of novelty. GM foods are not considered to be sufficiently novel (so as to require administrative review) if there is a history of safe use abroad or if no "major change" concerning the food's composition has occurred.

1.2.4 European Union

Among the examined regulatory frameworks, the European regime is currently the only one that allows an ex ante determination of the regulatory classification of GEOs with regard to cultivation as well as concerning the marketing of food derived from genome edited plant varieties, even without knowing the specific product characteristics.

[7]Directive 94-08, Assessment Criteria for Determining Environmental Safety of Plants with Novel Traits, Sec. 1.

The Court of Justice of the European Court of Justice (CJEU) ruled that mutagenesis induced by genome editing leads to organisms which are fully covered by the current European GMO regulatory framework.[8] It can be safely derived from the Court's reasoning that directed mutagenesis through genome editing covers both SDN and ODM techniques. The CJEU did not make a determination on genome editing techniques that are not mutagenesis techniques. However, its interpretation of the European GMO definition allows for conclusions about other forms of genome editing as well. Since the Court ruled—at least implicitly, and insofar in line with the wording of the GMO definition—that the classification as GMO depends on whether the breeding technique used was natural,[9] it seems safe to assume that the use of all techniques of genome editing results in GMOs within the meaning of the European regulatory framework.[10]

Consequently, the European regulatory regime is characterized by two distinctive features: (1) the legal status of all forms of genome edited plant varieties and products derived from them can be considered to be settled, and (2) the legal framework is based on a process-emphasizing approach requiring minimal artificially induced genomic changes (such as point mutations) only to suffice to lead to a GMO.

Even though the restrictive interpretation of the European GMO framework by the ECJ has been widely criticized,[11] it cannot be denied that it provides a high degree of legal certainty.

1.2.5 Japan

Having a closer look at the Japanese approach, it is especially the overall regulatory outcome that stands out. While Japan has no ban for GMOs in place and even a comparable high number of GM varieties are approved for cultivation, no actual cultivation takes place.[12]

This peculiar situation can be partly explained by non-regulation-related factors. Japanese companies give a comparable high amount of attention to consumer's satisfaction and are, therefore, very sensitive with regard to voiced concerns or preferences. Due to the prevailing negative attitude towards GMOs within the Japanese society, local farmers and manufactures are reluctant to use GMOs (at least in food production). Moreover, it is argued that the traits of currently existing GM crop varieties are less compatible with Japanese farm sizes and agricultural habits. However, an adoption of GM crops is also hampered by local

---

[8]European Court of Justice (2018).

[9]European Court of Justice (2018), para. 29.

[10]Cf. Chap. 5, Sect. 5.3.2. The same view is held by Seitz (2018), p. 763.

[11]See for example Lappin (2018), Neslen (2018), Callaway (2018), Stokstad (2018) and Science Media Centre (2018).

[12]See Chap. 8, Fig. 8.5.

administrative provisions. Local governments impose burdensome application requirements, require coexistence measures that are difficult to adhere to or charge application fees that render the cultivation of GM crops economically less feasible.[13] This makes abundantly clear that not just the legal provisions should be taken into account when assessing a regulatory regime for GMOs but also the social, political, and economic environment.

Concerning the regulatory regime, the decisive trigger for its application, and for the classification of an organisms as GMO, is the presence of foreign "nucleic acids" (which includes both RNA and DNA) within an organism. In this regard, referring to "nucleic acids" as "products", the regulatory approach might be called "product-based". Nucleic acids are not foreign if they are present in organisms of the same species or in organisms of other species which exchange nucleic acids with the species of the genetically altered organism. Under this framework, the classification of GEOs as GMOs or non-GMOs depends decisively on whether guide RNA used for purposes of SDN techniques is a foreign "nucleic acid" and whether the guide RNA is stably integrated into the genome or for other reasons continuously present in the organism. In case of ODM, on the other hand, the classification of the resultant organisms depends on whether oligonucleotide sequence is a foreign DNA sequence.

1.2.6 United States

The US approach differs from others by the use of unique, but selective triggers for the regulation of GMOs. Accordingly, at least in theory, GMOs could escape the regulatory regime completely, if they do not meet any of the criteria which trigger GMO regulation.

With regard to the cultivation of plants, it is the existence of a plant pest, or plant pest risk respectively, or a plant incorporated protectant that subjects the plant to regulatory requirements. When it comes to the marketing of food, it is decisive for the applicability of the regulatory framework whether the food contains residues from plant incorporated protectants or a food additive or whether the food is adulterated or misbranded.

Another distinctive feature of the US regulatory regime is that it resorts to a purely product-based trigger with regard to regulation of food under an informal consultation procedure. At least among the examined countries, this is a unique feature of the United States regulatory framework.

Concerning the regulatory status of GEOs under the current regulatory framework, there is a remarkable degree of legal uncertainty. This is due to the "product-based" approach according to which only the presence of certain "products", or "products" with certain characteristics, (e.g. plant pests, noxious weeds, food additives, plants with pesticidal properties, pesticidal residues in foods, adulterated foods

[13]Cf. for this paragraph Sato (2015), 6, 15–16.

etc.) triggers existing regulations. Accordingly, GEOs escape the regulatory framework if they are not covered by one of the categories of regulated "products". However, this deficiency is mitigated, albeit to a limited extent only, by voluntary consultation procedures according to which developers of GEOs may request for a determination of the regulatory status of the respective organism.

1.3 Identification of Best Practices

"Best practices" is a rather frequently used term in science, management and politics. However, as it is typical for vogue expressions, they lose their clear-cut substantive meaning with the increasing frequency of their use and progressively deteriorate into mere buzzwords. To avoid confusion how the term is used subsequently, it is therefore necessary to define the underlying understanding of "best practices".

In general, from an abstract-methodological point of view, "best practices" are processes, methods or concepts that achieve the envisaged outcome more (1) efficiently, (2) effectively and (3) comprehensively than other practices (*i.e.* processes, methods or concepts).[14] A practice is most efficient if it is applied in such a way that the results achieved and the resources used are in the best possible cost-benefit ratio. Effectiveness describes the degree to which a practice is able to realize its objectives. The higher the efficacy and the level of attainment, the more efficient a measure is. A practice is comprehensive when it is able to take into account all concerns designated as its direct or indirect objectives. To qualify as a best practice, these three elements must be brought to bear in a manner that ensures that every single one of them is able to unfold its maximal potential.

Beyond that general concept, individual best practices can range from empirically well-established or scientifically evidence-based best practices over promising best practices to just emerging and not yet solidified best practices.[15] The subsequently

[14]There does not exist a uniform definition of "best practices" which is agreed upon. For different definitions see for example Bretschneider et al. (2005), p. 309; Bendixsen and de Guchteneire (2003), pp. 678–679.

[15]"Emerging best practices" describes a process or method for which there is only a low degree of scientific evidence to qualify as a best practice. In the case of a "promising best practice" the existing quantitative and qualitative data is elevated to a moderate level. An "evidence-based best practice" is supported by a convincing and strong set of scientific evidence regarding its general effectiveness and efficiency. For a more detailed illustration of different best practice categories and sources of best practice evidence see Spencer et al. (2013); Bhatta (2002), p. 102; Moore and Browne (2017), p. 385; Canadian Homelessness Research Network (2013), p. 7; Myers et al. (2006), p. 374.

An example for a widely adopted and well-regarded best practice in the realm of GMO regulation are the international frameworks for risk assessment. Cf. Cartagena Protocol on Biosafety to the Convention on Biological Diversity, 2226 U.N.T.S. 208, 39 ILM 1027, UN Doc. UNEP/CBD/ExCOP/1/3 42; UNEP International Technical Guidelines for Safety in Biotechnology; OECD Safety Considerations for Biotechnology 1992. See also the OECD Consensus Documents on Safety Assessment of Transgenic Organisms.

discussed concepts lack yet the scientific evidence to qualify already as best practices. Since regulation of GEOs has started only recently and is still heavily debated in most jurisdictions, there is—at least with regard to genome edited plants—insufficient available data and evidence to support such a conclusion. However, there are, in our opinion, certain indications which allow the assumption of emerging or promising best practices of how to regulate GEOs adequately.

After having outlined the terminology, it remains to be clarified how best practices can be identified. Best practice research is described as "the selective observation of a set of exemplars across different contexts in order to derive more generalizable principles and theories of management".[16] In a more instructive manner this process could be described as looking "for [solutions tried in other jurisdictions] that appear to have worked pretty well, [trying] to understand exactly how and why they might have worked, and evaluat[ing] their applicability to [one's] own situation".[17]

In order to perform such an inductive evaluation drawing on different practices, i.e., in our case, on different regulatory approaches and concepts, the objectives of the looked-for best practices must be clearly defined beforehand. With regard to the regulation of genome edited crops many different regulatory interests and factors are at play.[18] This raises the question for what objectives and purposes best practices should aim for. Due to the manifold and in part contradictory regulatory objectives, a generally valid assumption cannot be made. Therefore, the further examination will have a limited scope by concentrating on possible best practices that (1) take into account the science-based risk potential of genome edited crops, (2) facilitate a transparent, even-handed and appropriate approval process or other administrative oversight mechanism, (3) comply with international law obligations and (4) are not more restrictive than necessary.

Applying this methodology to the regulation of genome edited crops and products derived from them as they are discussed in the subsequent country reports, certain approaches and concepts can be identified as emerging best practices that could over time evolve into evidence-based, or empirically-established, best practices.

1.3.1 Voluntary Early Consultation Procedure

A voluntary early consultation procedure presents a plant breeder with the opportunity to ask the competent administrative authority for an early decision on the regulatory classification of a prospective new plant variety, i.e. whether it would constitute a GMO pursuant to the respective legal framework.

[16]Overman and Boyd (1994), p. 69.

[17]Bardach and Patashnik (2016), p. 125.

[18]For an overview see Hamburger (2018).

Such an early consultation on the regulatory classification of a genome edited plant or plant variety, which is in its design phase, enables the developer to decide on the breeding technique to be applied in the coming breeding process. Should the originally envisaged breeding technique lead to the classification of the plant or plant variety as a GMO, the developer might change to another breeding method to ensure the desired non-GMO status.

An early consultation procedure does not only improve legal certainty, but also reduces the economic risks which are associated with a costly GMO approval procedure. At the same time, this facilitates research and development since a developer can assess relatively early, whether a designed plant or plant variety is likely to be profitable.

While such a procedure ensures transparency, legal certainty and a timely administrative decision, it is also in line with safety interests. Due to the early involvement of the authorities, they can discourage the developer from pursuing the breeding of high-risk varieties or steer them into the direction of less risk-prone genetic alterations. That way, risks can be mitigated before they even materialize.

Decisive for the effectiveness of such an early consultation procedure is, on the one hand, a specified and rather short timeframe within which the competent authority has to arrive at a decision. Otherwise, a non-transparent and lengthy process would make such a procedure less attractive. On the other hand, it is important that the preliminary classification remains its validity if the marketable final plant or plant variety is congruent with the one that has been discussed during the early consultation process. This way it can be ensured that legal certainty continues beyond the early research and development phase.

An early consultation procedure can, however, not mitigate the problem that non-GMOs might remain unregulated even if the competent authority has identified potential risks of the novel plant or plant variety on health or the environment. This holds particularly true if the trigger of the regulatory framework for GMOs is the process, *i.e.* the breeding technique. In this case, the early consultation procedure would not be able to channel the plant or plant variety into the GMO framework. If no other risk assessment and risk management mechanisms are in place, the non-GM novel plant or plant variety could enter the market unregulated.

1.3.2 Single Point or Multiple Point of Entry

While the legal frameworks of Argentina, Japan and the EU use a single trigger for cultivation and marketing (i.e. a particular GMO definition), Australia, Canada and the USA use multiple triggers.[19]

[19]See Chap. 8, Table 8.5.

Different triggers within the same jurisdiction can lead to a situation where a plant variety is able to escape the strict GMO rules with regard to cultivation, but the distribution of its products are at the same time subject to them—or vice versa. This is especially apparent if GMO definitions trigger two different regulatory frameworks, e.g. for cultivation on the one side and foods on the other side, but these two definitions differ significantly from one another. In any case, jurisdictions applying multiple triggers for different regimes have a clear tendency to be fragmented and complex and may result in uncoordinated administrative procedures imposing undue burdens and most likely higher costs on the developer.

However, taking into account that Japan and the EU make use of a single point of entry regime, while the USA and Canada use different triggers, it becomes clear that a single point of entry is not in itself an indicator for a more permissive or flexible regulatory framework. This is the case, because a single-point of entry can be used to cover as wide a spectrum of GM varieties and uses as possible. This may be done in order to apply the same burdensome risk assessment and risk management tools to all covered new plants or plant varieties, independent of different risk levels. However, it may also be an unintended consequence since a single trigger does not allow for a higher degree in specificity.

Consequently, a multiple point of entry regime can make up for its disadvantages by its more specific, even-handed and less restrictive nature. That way only those plant varieties can be targeted which the regulator deems necessary with respect to a certain use (e.g. environmental release, food or feed).

If a single or a multiple point of entry approach qualifies as best practice, depends, therefore, on the concrete design in the individual case.

1.3.3 One-Door-One-Key Principle

The one-door-one-key principle describes a regulatory approach that requires only a single application to obtain both the approval for cultivation and the authorization of marketing for consumption as or in food and feed.

As a result, only one approval procedure has to be completed and the synergy effects usually save time and therefore financial resources. A further advantage of having both approvals at the same time is that the liability risk for unintentional presence of GM material, which is authorized for cultivation but not for consumption (or vice versa), can be minimized.[20] Furthermore, this approach creates legal certainty not just for the developer but also for the farmer who can be sure from the beginning that the harvest of the GM crops can be sold on the domestic market.

If, however, the procedure is designed in such a way that an approval is only issued when there are no objections to both cultivation and consumption, the risk

[20]Purnhagen and Wesseler (2016), p. 151.

that neither one is approved is elevated ("all or nothing approach").[21] Therefore, the one-door-one key principle should be implemented so that if only one envisaged usage cannot be approved the other one will still receive approval. Otherwise, it could be more attractive for developers to file two separate applications.

The one-door-one-key principle as described above is reflected, e.g., in the EU's GMO framework.[22] A more complete, but probably also more complex, variant of the one-door-one-key principle would be that the single application relates not only to the authorization of cultivation and consumption as or in food and feed but also, e.g., to plant variety registration.

1.3.4 Mandatory Upstream Procedure

For regulatory frameworks that make use of the dichotomy GMO and non-GMO, it is a common hurdle to establish a clear demarcation line between GMOs and non-GMOs with regard to NBTs. This issue can be tackled by designing the general regulatory scheme as a two-tiered one, which means that the application of the regulatory framework (e.g. for GMOs) is preceded by a classification procedure. Accordingly, the design of a new plant or plant variety is, at first, subject to a procedure that clarifies whether it is legally classified as a GMO. Depending on its outcome, e.g. if the plant or plant variety is classified as a GMO, the new plant or plant variety would be subject to the substantive and procedural requirements of the regulatory framework applicable to GMOs.

An upstream procedure to determine whether a new plant or plant variety falls within the scope of the GM regime has the advantage that the current regulatory framework can stay in place without any amendments. Since a change of an already existing legal regime is often a burdensome and lengthy political process, an upstream procedure is a comparably efficient way to ensure legal certainty regarding the classification of genome edited plants.

Furthermore, the well-established dichotomy of GMOs and non-GMOs can be maintained. Without recourse to an upstream procedure, it could become necessary to introduce a third category into the current framework to ensure its compatibility with genome edited crop varieties.

Another advantage of an obligatory upstream procedure is that the competent authorities are enabled to review all novel plants or plant varieties without exception no matter whether they will be classified later on as GMOs or non-GMOs. This way the frequently existing regulatory gap[23] caused by a tight regulation of GMOs on the one side and the lack of oversight with regard to non-GMOs one the other side can be

[21] Purnhagen and Wesseler (2016), p. 151.

[22] For a detailed illustration of that principle's application in the EU see van der Meulen and Yusuf (2014).

[23] For more on this issue see Voigt and Klima (2017), p. 335.

mitigated to a certain extent. However, an upstream procedure is only assisting in closing this regulatory gap if a risk assessment and risk management mechanisms for non-GMOs is in place.

1.3.5 Product-Based Approach

Compared to a process-based regime, a product-based approach, which takes into account only the traits of a new plant variety and does not consider the breeding technique used, has the advantage that it is more science-based. It is scientifically sound to assume that a risk potential is not inherent in genetic modification techniques as such but only linked to the traits of the resulting organisms in question.[24]

At the same time, therefore, it seems to be easier to ensure compatibility with international legal obligations stipulated, e.g., by WTO law or free trade agreements, since a product-based approach is more likely to avoid unjustifiable discrimination or unnecessary trade restrictions than a regulatory approach based on certain breeding techniques. This is because what may cause risks to human health and the environment are not breeding techniques as such but rather the resultant traits of the genetically altered organisms. Accordingly, it may be considered inconsistent to subject, e.g., herbicide tolerant plants or plant varieties to differently burdensome authorization procedures depending on the breeding technique used to provoke herbicide tolerance.

Additionally, a product-based approach has the advantage that it is not necessary to consider whether the criticized[25] dichotomy of GMO and non-GMO has to be supplemented by a third category.

This leads to a further benefit of a product-based regulatory regime: There is no regulatory gap between GMOs and non-GMOs. Since process-based GMO frameworks make a clear-cut differentiation between GMOs and non-GMOs not based on their actual risk potential but solely based on the breeding method used, the same risks arising from a particular trait may be treated differently in individual cases. Consequently, new plants or plant varieties are subject to either the strict GMO regulation or the far more permissive non-GMO regulation—but nothing in between. Even if a non-GMO plant variety poses a high risk potential, no stricter rules apply than for other conventionally bred crops. This results in a gap with regard to the risk assessment and the approval requirements between GMOs and non-GMOs including conventionally bred plants. A product-based approach, however, allows the approval requirements to be defined individually based on the specific product in question.

[24]Dederer (1998), pp. 32–49.

[25]Herring (2008), p. 459; Herring and Paarlberg (2016), p. 398.

Accordingly, a product-based approach is preferable only if it allows for a thoroughly tiered risk regulation. If all plants or plant varieties with novel traits are subjected to equally strict risk governance, this may lead to undesirable economic consequences in the case that, e.g., only the global players have the means and resources to cope with the onerous regulatory framework.

1.4 Conclusions

A comparison of the different national regimes' special features showed that there are indicators for emerging and promising best practices regarding the regulation of genome edited plant varieties.

However, it remains to be seen to what extent legislators are willing to adopt best practices. Since national legislators usually voice strong belief in the superiority of their own legal approach, a widespread dissemination of best practices must be viewed with scepticism. However, this view is often based not only on personal convictions but also on purely practical and political considerations. On the one hand, it is difficult for a legislator to acknowledge the inferiority of one's own concept. On the other hand, there is a strong incentive to promote its own regulatory regime, because the more countries that follow a similar approach, the easier it will be to trade products between these countries. Therefore, especially export-oriented countries have an interest to export not just their agricultural products but also their own regulatory approach to other countries to prevent trade barriers before they even arise.

This interest in establishing one's own approach as standard, however, can also promote the spread of best practices. Against this background, different national regulatory regimes are in a competitive relationship with each other. Therefore, it stands to reason that the regulatory approach will prevail, which suits the interest of the majority of parties best—i.e. which constitutes a best practice. Consequently, legislators, who are interested in disseminating their regulatory approach, are inclined to either adopt best practices or to make sure that the own approach constitutes a promising best practice.

With regard to the future, it can therefore be presumed that legislators might be more drawn towards best practices in an effort to prevail in this realm of regulatory competition and to shape an emerging international framework. If these or other best practices become widely accepted, the agricultural sector might move more closely towards a global regulatory standard.

In sum, based on the comparative analysis of the regulatory frameworks in Argentina, Australia, Canada, the EU, Japan and the US and the identification of possible best practices, our impression is that the purpose of a regulatory framework should be primarily aimed at preventing or, at least, minimizing risks to health and the environment. Such risks arise from plants and their traits. Of course, any such traits are gene-based. Accordingly, any genetic alteration may produce traits which cause the plant posing a risk to human health and the environment. However, it does not logically follow that it is techniques of genetic alteration, *i.e.* breeding techniques, as such which are inherently risk-prone. In fact, "classic" GMOs, especially GM crops, have been cultivated and consumed as feed or in foods on a global scale without any hint to risks to human health and the environment. This is in line with the continuous results of safety research aimed especially at the identification of GMO specific health and environmental risks. Hence, novel combinations of genetic material as such, even if brought about by transgenesis, should no longer be considered as a relevant trigger for risk assessment and risk management and, therefore, not for risk regulation related to genetically altered plants.

A consistent regulatory approach, therefore, should be product-based, *i.e.* the risk regulation should be triggered by a plant's traits. The regulatory problem then is to define those traits which deserve a closer look by administrative authorities. We think that the product-based trigger should be the "novelty" of the trait. Hence, "novelty" would be the single point of entry into the regulatory framework. "Novelty", in turn, should be defined in terms of "familiarity". That means that, independent of the breeding technique, only plants with "unfamiliar" traits should be considered "novel" and, therefore, subjected to the regulatory framework. We are fully aware, of course, that the term "familiarity" is vague and needs further specification. Factors to be considered within the concept of "familiarity" could be the long history of the trait in the crop plant species, the long history of safe use and consumption of plants with the respective trait, the substantial equivalence of the composition of the plant etc.

From a procedural point of view, an obligatory upstream procedure should be the initial step channelling the respective plant into the relevant regulatory framework. This process can be further catalysed from the outset by a voluntary early consultation procedure. Within that framework the one-door-one-key principle should apply, which means, *e.g.*, that all relevant authorizations (*e.g.* for cultivation as wells use as or in food and feed) including variety registration are granted upon a single application.

References

Bardach E, Patashnik EM (2016) A practical guide for policy analysis: the eightfold path to more effective problem solving, 5th edn. CQ Press/SAGE, Los Angeles/London

Bendixsen S, de Guchteneire P (2003) Best practices in immigration services planning. J Policy Anal Manage 22:677–682. https://doi.org/10.1002/pam.10163

Bhatta G (2002) Evidence-based analysis and the work of policy shops. Aust J Public Adm 61:98–105. https://doi.org/10.1111/1467-8500.00288

Bretschneider S, Marc-Aurele FJ, Wu J (2005) "Best Practices" research: a methodological guide for the perplexed. J Public Adm Res Theory 15:307–323. https://doi.org/10.1093/jopart/mui017

Callaway E (2018) CRISPR plants now subject to tough GM laws in European Union. Nature 560:16. https://doi.org/10.1038/d41586-018-05814-6

Canadian Homelessness Research Network (2013) What works and for whom? A hierarchy of evidence for promises practices research. Canadian Homelessness Research Network Press, Toronto

Dederer H-G (1998) Gentechnikrecht im Wettbewerb der Systeme: Freisetzung im deutschen und US-amerikanischen Recht. Springer, Heidelberg

European Court of Justice (2018) Case C-528/16, Confédération paysanne, Réseau Semences Paysannes, Les Amis de la Terre France, Collectif Vigilance OGM et Pesticides 16, Vigilance OG2M, CSFV 49, OGM: dangers, Vigilance OGM 33, Fédération Nature et Progrès v. Premier ministre, Ministre de l'agriculture, de l'agroalimentaire et de la forêt, ECLI:EU:C:2018:583

Globus R, Qimron U (2018) A technological and regulatory outlook on CRISPR crop editing. J Cell Biochem 119:1291–1298. https://doi.org/10.1002/jcb.26303

Hamburger D (2018) Normative criteria and their inclusion in a regulatory framework for new plant varieties derived from genome editing. Front Bioeng Biotechnol 6:176

Herring RJ (2008) Opposition to transgenic technologies: ideology, interests and collective action frames. Nat Rev Genet 9:458–463. https://doi.org/10.1038/nrg2338

Herring R, Paarlberg R (2016) The political economy of biotechnology. Annu Rev Resour Econ 8:397–416. https://doi.org/10.1146/annurev-resource-100815-095506

Lappin J (2018) EU Court extends GMO Directive to new plant breeding techniques_Brussels. https://gain.fas.usda.gov/Recent%20GAIN%20Publications/EU%20Court%20Extends%20GMO%20Directive%20to%20New%20Plant%20Breeding%20Techniques_Brussels%20USEU_Belgium%20EU-28_7-27-2018.pdf. Accessed 19 Sept 2018

Ministerio de Agricultura, Ganadería y Pesca (2015) Resolución 173/2015, Boletin Oficial N° 33.131:40–41

Moore C, Browne C (2017) Emerging innovations, best practices, and evidence-based practices in elder abuse and neglect: a review of recent developments in the field. J Fam Violence 32:383–397. https://doi.org/10.1007/s10896-016-9812-4

Myers SM, Smith HP, Martin LL (2006) Conducting best practices research in public affairs. Int J Public Policy 1:367–378. https://doi.org/10.1504/IJPP.2006.010842

Neslen A (2018) Gene-edited plants and animals are GM foods, EU court rules. https://www.theguardian.com/environment/2018/jul/25/gene-editing-is-gm-europes-highest-court-rules. Accessed 19 Sept 2018

OECD Consensus Documents on Safety Assessment of Transgenic Organisms. https://www.oecd-ilibrary.org/environment/harmonisation-of-regulatory-oversight-in-biotechnology_23114622. Accessed 9 Sept 2018

Overman ES, Boyd KJ (1994) Best practice research and postbureaucratic reform. J Public Adm Res Theory 4:67–83. https://doi.org/10.1093/oxfordjournals.jpart.a037195

Purnhagen K, Wesseler J (2016) The "Honey" Judgment of Bablok and Others Versus Freistaat Bayern in the Court of Justice of the European Union: implications for co-existence. In: Kalaitzandonakes NG, Phillips PWB, Wesseler J, Smyth SJ (eds) The coexistence of genetically modified, organic and conventional foods: government policies and market practices. Springer, New York, pp 149–165

Sato S (2015) Japan's regulatory system for GE crops continues to improve. https://gain.fas.usda.gov/Recent%20GAIN%20Publications/Agricultural%20Biotechnology%20Annual_Tokyo_Japan_7-13-2015.pdf. Accessed 20 May 2019

Schenkel W, Leggewie G (2015) New techniques in molecular biology challenge the assessment of modified organisms. J Consum Protect Food Saf 10:263–268. https://doi.org/10.1007/s00003-015-0958-4

Science Media Centre (2018) Expert reaction to Court of Justice of the European Union ruling that GMO rules should cover plant genome editing techniques. http://www.sciencemediacentre.org/expert-reaction-to-court-of-justice-of-the-european-union-ruling-that-gmo-rules-should-cover-plant-genome-editing-techniques/. Accessed 20 Sept 2018

Seitz C (2018) Modifiziert oder nicht? – Regulatorische Rechtsfragen zur Genoptimierung durch neue biotechnologische Verfahren. Europäische Zeitschrift für Wirtschaftsrecht (EuZW) 21:757–764

Spencer LM, Schooley MW, Anderson LA, Kochtitzky CS, DeGroff AS, Devlin HM, Mercer SL (2013) Seeking best practices: a conceptual framework for planning and improving evidence-based practices. Prev Chronic Dis 10. https://doi.org/10.5888/pcd10.130186

Sprink T, Eriksson D, Schiemann J, Hartung F (2016) Regulatory hurdles for genome editing: process- vs. product-based approaches in different regulatory contexts. Plant Cell Rep 35:1493–1506. https://doi.org/10.1007/s00299-016-1990-2
Stokstad E (2018) European court ruling raises hurdles for CRISPR crops. http://www.sciencemag.org/news/2018/07/european-court-ruling-raises-hurdles-crispr-crops. Accessed 20 Sept 2018
van der Meulen B, Yusuf N (2014) One-Door-One-Key Principle: observations regarding integration of GM authorization procedures in the EU. Penn State Law Rev 118:877–891. https://doi.org/10.2139/ssrn.2605766
Voigt B, Klima J (2017) CRISPR-Plants & Co. – the Quest for Adequate Risk Regulation: Modern plant breeding techniques and the current legal framework for risk regulation in the European Union. Zeitschrift für Europäisches Umwelt- und Planungsrecht:319–333

Chapter 2
Regulation of Genome Editing in Plant Biotechnology: Argentina

Agustina I. Whelan and Martin A. Lema

Abstract Argentina is a world leader in regards to regulation and adoption of GM crops. As a consequence, the regulatory aspects of gene editing applied to agriculture were considered proactively, and a simple but sound pioneer regulation was developed.

At present, the Argentine regulatory system is fully able to establish if a gene-edited crop should be classified (and handled) either as a GM crop or a conventional new variety. To this end, the concept of "novel combination of genetic material" derived from the Cartagena Protocol is of paramount importance.

After some pilot cases that have been handled under the new regulation, applicants appreciate the ease, speed and predictability of this regulation. Moreover, it has been considered by other countries in developing their own regulations, thus acting also as a harmonization factor for the safe and effective insertion of these technologies in the global market.

The information and views are those of the authors as individuals and experts in the field, and do not necessarily represent those of the organizations where they work.

A. I. Whelan
Ministry of AgroIndustry, Buenos Aires, Argentina

University of Buenos Aires, Politics and Management of Science and Technology, Buenos Aires, Argentina
e-mail: awhelan@magyp.gob.ar

M. A. Lema (✉)
Ministry of AgroIndustry, Buenos Aires, Argentina

National University of Quilmes, Bernal, Argentina
e-mail: mlema@unq.edu.ar

H.-G. Dederer, D. Hamburger (eds.), *Regulation of Genome Editing in Plant Biotechnology*, https://doi.org/10.1007/978-3-030-17119-3_2

2.1 Introduction

Argentina belongs to a group of six countries which were the first in the world to simultaneously allow genetically modified (GM) crops to be marketed. This happened in 1996, and particularly in Argentina it began with the introduction of herbicide (glyphosate) tolerant soy. Ever since then, Argentina has increased its production of GM crops, and it is currently the third largest grower of biotech crops in the world, after the United States and Brazil. During the 2015–2016 season the country produced 13% of the world's total biotech crop harvest; this included soybeans (18.7 million planted hectares), corn (4.74 million hectares) and cotton (400,000 hectares).[1] Regarding the degree of adoption of GM varieties, in the case of soy and cotton GM seeds make up 99% of total trade with these crops; in the case of maize it is slightly lower, at 94%.[2]

From a world trade perspective, Argentina is currently the main world exporter of soya oil and meal, and the third exporter of soy grains; it is also the second main exporter of corn grain, according to INDEC[3] and COMTRADE.[4]

In total, Argentina has issued 48 commercial authorizations for GM crops (which in some cases include more than one event or stacked events),[5] and it displays the highest number of events approved in recent years.[6]

There are some studies available that estimate the productive, social and economic impacts derived from the introduction of GM crops.[7] One of them, which spans the first 20 years of commercialization, has estimated that the gross benefit derived exclusively from the introduction of genetic engineering (i.e. the difference between the actual economic figures and the estimated incomes of a modelled scenario without GM crops) was close to US$127 billion. This GDP surplus, according to the authors, might account for the creation of two million jobs during that period.[8]

The introduction of GM crops has contributed with the sustainability of agriculture in two ways. On one hand, through a reduction in the use of chemical insecticides, in the case of insect-resistant "BT" crops. On the other side, through the synergy between herbicide-resistant crops and no-till farming practices where the latter enables better conservation of soils, through reduced erosion and reduced oxidation of organic matter. Intensification of no-till, greatly facilitated by the use of GM crops, also reduces the emission of greenhouse gases from exposed (plough) soil organic matter and from fuel consumption of agricultural machinery, as well as

[1] ISAAA (2016).

[2] ASA (2018).

[3] INDEC (2018).

[4] COMTRADE (2018).

[5] MINAGRO (2018).

[6] ISAAA (2016).

[7] Barfoot and Brookes (2014).

[8] Trigo (2016).

improved carbon sequestration. Additionally, these practices also facilitate having multiple cropping in one season (e.g. second crop soybeans after wheat in the same growing season).[9]

In the last couple of years, Argentina has also allowed the market release of GM varieties with innovative traits (including cases of added value), and in different crops. This includes, for instance, high-oleic and drought-resistant soy, virus-resistant potato and safflower expressing bovine chymosin for the cheese-making industry.[10] However, these are very recent and (for commercial reasons) slow-paced innovations whose presence in the market is still negligible.

In regards to Gene Editing Techniques (GETs), although the regulators of a few other countries took earlier decisions on the regulatory standing of specific products, Argentina was the first in the world to incorporate specific provisions on its regulatory framework for dealing with products derived from New Breeding Techniques (NBTs) based on innovative biotechnology approaches. This was the outcome of a 3-year science-based policymaking work, which reviewed national and international legislation, the state of the art and parallel discussions overseas.[11]

2.2 The Regulatory Framework for Genetically Modified Organisms (GMOs): An Overview

2.2.1 Overview, Applicable Laws and Regulations

The Argentine GMO regulatory framework has been described *in extenso* elsewhere.[12] It is one of the pioneers in the world and the second-oldest in Latin America after the Mexican regulatory system. It has been active uninterruptedly since 1991, when the National Commission on Agricultural Biotechnology (CONABIA) was created.

Since its inception, applicable laws and implementing regulations have been updated frequently. At present, the activities involving GM crops are regulated in Argentina under several laws including the National Law 20.247 on Seeds and Phytogenetic creations,[13] the National Law 27.233 on Animal and Plant Health,[14] the National Law 22.520 on the Ministries of the Executive Branch (the latter, in turn, combined with its implementation Decrees 1940/2008 13/2015 and 32/2016).[15]

[9]Penna and Lema (2003).

[10]Bustamante (2018).

[11]Whelan and Lema (2015).

[12]Burachik (2012) and Burachik and Traynor (2002).

[13]INFOLEG (1973).

[14]INFOLEG (2015).

[15]INFOLEG (1992). Noteworthy, there is also a National Law 20.270 on the Promotion of the Development and Production of Modern Biotechnology—see INFOLEG (2007). However, it does

A Ministerial Resolution 763/11 sets forth the specifics for the regulation of GMOs.[16] Then, 22 subsidiary regulations, including Resolutions from State Secretaries, the National Seeds Institute (INASE) and the National Agrifood Health and Quality Service (SENASA), rule different detailed aspects for diverse activities involving GM plants, animals and microorganisms.[17]

In regards to international law or treaties, the Argentine Republic is a member of the World Trade Organization, including its treaty on Sanitary and Phytosanitary measures (SPS). Therefore, its GMO regulatory framework is fully compliant with the relevant standards of the organizations recognized as reference in the SPS Agreement: the *Codex Alimentarius* Commission, the International Plant Protection Convention (IPPC), and the World Organization for Animal Health (OIE). In addition, Argentina abides by the WTO Dispute Settlement Understanding, and has applied it specifically to GM crops when challenging the past functioning of the European Communities' regulatory system.[18]

In regards to the Cartagena Protocol on Biosafety (CPB), which regulates transboundary movements of GMOs,[19] Argentina was among the founder signatories in 2000 but the country has not ratified the Protocol yet.[20] Nevertheless, the current Argentine regulatory system is fully compatible with the CPB text (particularly regarding definitions, as discussed later) and the decisions adopted by its COP-MOP (Conference of the Parties to the Convention on Biological Diversity serving as the Meeting of the Parties to the CPB).[21]

By the time the CPB was drafted, Argentina was already a member of the parent treaty, the Convention on Biological Diversity, and one of the few countries in the world with a full functioning regulatory system for GMO biosafety assessment. Therefore, Argentina participated actively in the negotiation of the CPB text, and its experience was taken into account when devising the Protocol. As a consequence, not only Argentina but also most of its partners for transboundary movements of GMOs have a CBP-compliant regulation.

Finally, the senior GMO regulatory body in Argentina, which is the National Advisory Commission for Agricultural Biotechnology (CONABIA) has been granted the role of Center of Reference in GMO Biosafety by the United Nations Organization for Food and Agriculture (FAO). FAO Reference Centers are institutions selected to provide specific, independent, technical or scientific advice on issues related to its mandate. The Centers are chosen on the basis of their high level scientific expertise, as well as their commitment and demonstrated ability to

---

not play any role in regulatory aspects, since it is focused on granting funding and tax incentives to biotechnology-Based Innovative Enterprises—see Rosada (2018).

[16]INFOLEG (2011a).

[17]INFOLEG (2011b).

[18]WTO (2018).

[19]CBD (2000).

[20]CBD (2018a).

[21]CBD (2018b).

perform capacity building in their areas of expertise. This status has been mutually agreed by a treaty signed by FAO and the Argentine Government.[22]

2.2.2 Authorities and Responsible Agencies

The Ministry of Agroindustry is the competent authority in regards to the authorization of activities involving environmental and market release of GM organisms of agricultural use and/or belonging to species of agricultural use. For GM crops in particular, the decision upon each individual authorization is currently delegated on its Secretariat of Foodstuff and Bioeconomy.

The Directorate of Biotechnology, within that Secretariat, is the leading regulatory bureau. It coordinates activities and collects documentation from CONABIA (of which it also acts as the Executive Secretariat) and other relevant agencies whose intervention in different aspects is required prior to the decisions by the Secretariat of Foodstuff and Bioeconomy.

CONABIA is a multi-disciplinary, inter-institutional commission of experts. Its main role pertains to biosafety assessment and/or the evaluation of confinement/containment measures for every application pertaining activities with GMOs. It also has a broad role as advisor in scientific and technical issues pertaining agroindustrial biotechnology.

Prior to market release, in addition, GM crops are also subject to food safety assessment. It is performed by the Technical Advisory Committee on the food use of GMO (CTAUOGM). This Committee is hosted by SENASA.

Also for market approval, there is an additional requirement to perform an analysis of the impacts on the production and commercialization that may arise from the commercial release of a GM crop. It is carried out by the Undersecretariat of Agricultural Markets.

For the control of biosafety measures in confined releases such as field trials, as well as for investigating or responding to any deviation of the regulation, inspectors from the National Seeds Institute (INASE) and SENASA exert the role of sanitary police.[23] INASE is competent on seed or any viable plant propagation material, while SENASA is competent on grains and plant-material derived products, including foodstuff, as well as microorganisms and animals. In addition, INASE is the entry point for applications pertaining GM crops (Table 2.1).

[22]FAO (2014a).

[23]SENASA (2018).

Table 2.1 Authorities and responsible agencies

Authorities and responsible agencies	Responsibility	Pertinent laws and regulations (only chief instruments are mentioned here)
Ministry of Agroindustry: Secretariat of Foodstuff and Bioeconomy	– Decision making (Permits, administrative sanctions) – Enacting of main Administrative Regulations	Law 22.520 on the Ministries of the Executive Branch
Biotechnology Directorate	– Coordination of the regulatory framework – CONABIA (Biosafety Assessment) Chair	– Decrees 1940/2008 13/2015 and 32/2016 – Ministerial Resolution 763/11 on the structure of the regulatory system, and several subsidiary regulations
Undersecretariat of agricultural markets	Market assessment for commercial release	Resolution 510/11 for the assessment of impacts on production and commercialization
SENASA	– CTAUOGM (Food Safety Assessment) Chair – Food and plant health police	– Law 27.233 on Animal and Plant Health – Resolution 412/02 on Food and Feed Safety Assessment (domestication of Codex guidelines)
INASE	Seed (i.e. any plant propagative material) police	– Law on Seeds and Phytogenetic creations – Resolution 46/04. GM crops Operators' register

2.3 Regulatory Status of Genome Edited Plants

2.3.1 Applicability of the Regulatory Framework for GMOs

For the reasons stated before, Argentina follows the definitions under CPB in its regulatory system. In particular, Resolution 701/11[24] defines "*Genetically Modified Plant Organism*" as "*any vegetable organism that possesses a novel combination of genetic material obtained through the use of modern biotechnology*" which is akin with the CPB definition of "*Living Modified Organism*" which is "*any living organism that possesses a novel combination of genetic material obtained through the use of modern biotechnology*".

Both the same Argentine regulation and the CBP define biotechnology identically as "*the application of: a. In vitro nucleic acid techniques, including recombinant deoxyribonucleic acid (DNA) and direct injection of nucleic acid into cells or organelles, or b. Fusion of cells beyond the taxonomic family, that overcome natural physiological reproductive or recombination barriers and that are not techniques used in traditional breeding and selection*".

[24]INFOLEG (2011c).

For practical purposes, most relevant products correspond to the item (a) because in their development there is an application of recombinant DNA (rDNA) and direct injection of nucleic acid in cells.

This definition is the touchstone to decide if a product is under the Argentine regulation for GMOs (and under the CPB). Since the beginning of the regulatory system till recent years, there have been no controversies or even the need for developers to consult the regulators about a particular product being GMO or not.

This has been classified as a "product-based" trigger for regulation[25] because the definition requires the crop to have "a novel combination of genetic material". However, it is also a process-based regulation, because such novel combination must have been "obtained through the use of modern biotechnology" which, in turn, means the use of r-DNA techniques.

The assessment and authorizations processes are "case by case", which in practice means "transformation event by transformation event". Besides, the stacking of transformation events that have been previously approved requires a new separate assessment and authorization process. In order to implement this, the Argentine regulation found it useful to incorporate two other definitions, which follows:

Individual transformation event, also referred to as "*event*": the insertion in the plant genome in a stable and joint arrangement, of one or more genes or DNA sequences that are part of a defined gene construct.

Stacked events: introduction of two or more events in the same genome.

2.3.2 *Regulatory Classification of Genome Editing/Genome Edited Plants*

In 2015, Resolution 173 regarding new breeding techniques in plants was issued; its unofficial translation is hereby included as supplementary material in the Annex 1. This regulation does not alter the preexisting regulatory framework applicable to GM plants (in particular, it respects the operational definitions mentioned previously). Instead, it rather clarifies the procedure to determine if a crop obtained by new breeding techniques is subject to the preexisting GMO rules and regulations. This procedure is outlined in Fig. 2.1 and described next.

The determination has to be made in a case by case basis, which means that each line carrying a certain genotype will be assessed separately. This is because a determination of the GM/Non GM status at the level of "techniques" was considered to be inviable, since the same molecular biology tool can be used in different ways to produce very dissimilar results, from no change in DNA sequences to small deletions to insertions of foreign DNA. In addition, new techniques or variants of older ones are developed continuously; therefore any reference to a closed list could unnecessarily hamper the applicability of the regulation in the near future.

[25]Ishii and Araki (2017).

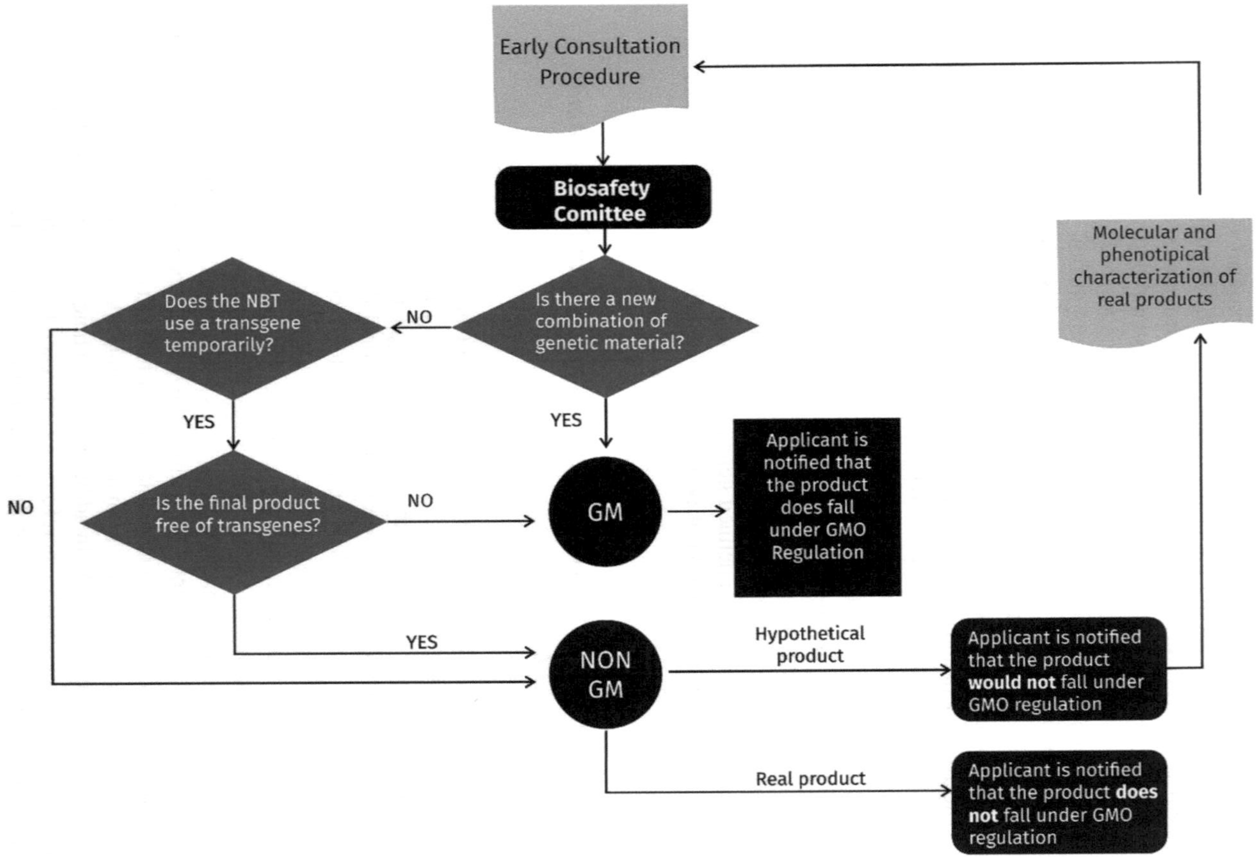

Fig. 2.1 Roadmap for the regulatory classification of new breeding techniques, including gene-edited crops, in Argentina. See text for details

The process begins when the interested party submits information regarding a product (NBTs-derived crop) to CONABIA. This information shall cover the breeding methodology used to obtain and develop the crop, the new trait and the genetic changes present in the final product.

The applicant will be compelled to perform the consultation usually because of the use of r-DNA as part of the NBT (since the regulation establishes that this makes the crop to be presumed GM until CONABIA establishes otherwise). Therefore, and taking into account the GMO definition, the main task of the Commission is to establish whether the result of the breeding process is a novel combination of genetic material or not. In this regard, a genetic change shall inescapably be considered a novel combination of genetic material when a genetic construct has been introduced permanently into the plant genome.

The regulation clarifies that GM-plant offspring shall be presumed also a GM-plant until scientific information proves otherwise. This is relevant for those cases where GETs resort to the transient use of a gene construct, for instance to express a nuclease. In this context, by "transient use" it is understood that a stable integration of the gene construct in the plant genome has either not occurred, or alternatively it has occurred but lasted one (or a few) generations and then the transgene was outbred of the final line to be introduced in the market; this latter case is known as "null segregant". Since null segregants are not considered GM crops, if the applicant shows evidence that the gene construct was not integrated and/or inherited by the line introduced in the market, that transient use by itself does not lead to the product being considered a GM crop.

An interesting feature of this regulation is that it also allows developers to anticipate the regulatory standing of a gene-edited crop before obtaining it. For projects still in the design stage, CONABIA may issue a preliminary assessment that has a putative standing. When the new crops are finally obtained, the applicant must return and submit molecular biology studies on the genetic modification actually generated. If the product correspond to what was described in the preliminary inquiry, the earlier assessment regarding its regulatory status would remain firm. Developers have found that the option of requesting a preliminary determination when the product is on design stage is particularly useful to choose between the novel techniques and a "more traditional" transgenesis approach in certain cases, as well as to estimate the investment required by the projects and taking a decision on its viability.

The regulation includes a 60 days' time-limit for delivering the assessment.

Finally, the regulation also states that even in those cases where the product is not found to be a GM crop, CONABIA may recommend the adoption of follow-up measures on a case-by-case basis, if the crop has specific features that could warrant it on scientific and technical grounds. In practice, this means that in those cases where the crop is non-GM but nevertheless the GMO regulator recognizes a risk hypothesis, this hypothesis must be also reported to the appropriate regulator of varieties obtained by "conventional" breeding for consideration (see below). Somehow, this also addresses a peculiarity of almost all regulatory systems worldwide (exception made perhaps of Canadian's regulation) where products having the same traits and therefore comparable potential risks, are handled with very different regulatory approaches—and likely different level of protection—if they are conventional or GM plants, with the latter being under more stringent regulation thus higher level of protection.

As explained, the determination has to be made on a case by case basis. Nevertheless, for the deliberations that the Commission maintained in order to draft the current regulation, a representative case of a real product derived from each NBT under consideration was obtained from the scientific literature and assessed to test the regulatory criteria. In addition, from the onset of Resolution 173/2015, developers have made formal consultations on some cases that have been evaluated already. In Fig. 2.2, this caseload for products derived from GETs as well as other NBTs is represented.

Specifically in regards to gene-editing, the following rationale sustains the criteria displayed in Fig. 2.2:

Site-Directed-Nucleases or "SDN": the terminology and scope of the term (including SDN1, SDN2 and SDN3) is according to what was originally proposed by the European Food Safety Authority[26] (which can be considered to include techniques that appeared later, such as CRISPR-CAS9). In regards to Oligonucleotide-Directed-Mutagenesis or "ODM", the understanding from the Joint Research Centre of the European Union is adopted.[27]

SDN1: Where the technique involves the introduction of the nuclease gene in the cell, in the cases assessed so far the scope of the genetic intervention (small deletions or base-change mutations) was not considered to constitute a novel combination of genetic material. In all cases, the nuclease gene or other helper transgenes were proven to be absent or removed from the final plant line intended to be introduced in the market. The same conclusion applies if the technique involves introducing the nuclease protein and RNA guide instead; in this case, since no foreign DNA is introduced in the cell, the technique is not even considered to be reached by the definition of "modern biotechnology".

ODM: The same conclusion as for SDN1 under an analogous rationale. In this case, moreover, the indel or base-change mutation can be of a very small size, and the nucleic acid inserted in the cell is not a r-DNA, and it is not incorporated in the genome.

SDN2: Rationale was the same as for SDN1, the fact that the DNA repair process was guided by a template was not considered to introduce a relevant change in the criteria adopted.

SDN3: Since SDN3 does involve the insertion of a DNA in the cell, it is almost certain to be considered a GMO. The only exception could be a "perfect allelic replacement", where a gene allele already present in another plant line of the same species is inserted in exactly the same position and without any further insertion

[26]EFSA (2012).

[27]Lusser et al. (2011).

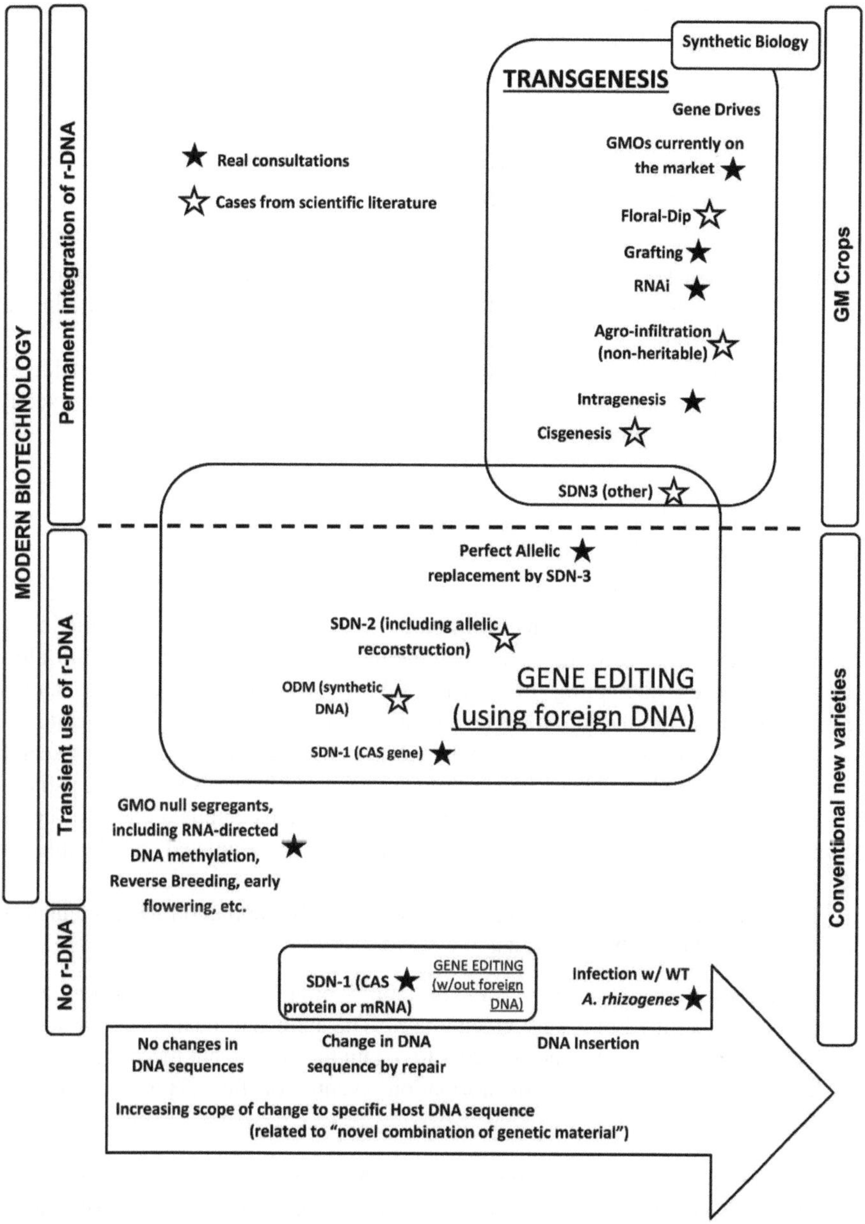

Fig. 2.2 Likely classification of crops obtained by new breeding techniques, including gene-edited crops, in Argentina. Definite results will be dependent upon each crop on a case by case basis. This is an indicative projection after analyzing cases taken from scientific literature and actual consultations from interested developers. The vertical axis separates techniques that do not use r-DNA from those that use it transiently or permanently; and the horizontal axis indicates an increasing intervention in plant DNA sequences by each technology. Therefore, each axis relates to one of the two main elements of the GMO definition: (a) application of modern biotechnology—i.e. r-DNA—

of other DNA sequence; therefore rendering the exact same result as a conventional crossing with the other plant line.

2.4 Regulatory Prerequisites for Activities Relating to Genome Edited Plants

According to what has been described earlier, depending on the case the regulatory regime for GMOs may be applicable (scenario 1) or not (scenario 2) to genome edited plants or products derived from them.

In case of scenario 1: After the applicant is notified that the product is considered to be a GMO, the gene edited plant and its derived products will be subject to exactly the same regulatory procedures for any other GMO, described next:

2.4.1 Contained/Confined Use of GM Plants

Research laboratories developing GM plants should have a general accreditation of Biosafety Level 2, in accordance to the guidelines of the WHO Laboratory biosafety manual.[28] When attempting to transfer GM plants into a greenhouse and/or performing a field trial, there must be an earlier permit for an *ad hoc* application.

In order to perform any kind of contained or confined activity beyond a research laboratory, the applicant must be previously registered in the National Registry of Operators of Genetically Modified Plant Organisms. This entails a check of applicant's credentials and other formal aspects; this is regulated by Resolution 46/2004.[29]

There are three categories of permits enabled in the regulation for contained or confined activities with GM plants:

- Biosafety greenhouses: There is the possibility of applying for performing activities under certified biosafety greenhouses. The applicant must specify a defined list of species and transgenes in order to obtain the permit, but there is no need of having a defined list of transformation events for this. This regulation is

Fig. 2.2 (continued) and (b) the generation of a novel combination of genetic material. The line incorporated in the figure attempts to mark the approximate separation between techniques more likely to result in plants considered either GM or conventional new varieties

[28] WHO (2018).

[29] INFOLEG (2004).

particularly useful for local developers of GM plants in order to have an *ex ante* flexible authorization to move a substantial number of novel transformation events from the laboratory to the greenhouse for initial characterization. This is regulated, *inter alia*, by Resolution 241/2012.[30]
- Field trials: open field experimental releases also require applying for a permit where, in addition to the proposed confinement measures and other aspects, the characteristics of all events involved is required to be analyzed. This is regulated, *inter alia*, by Res. 763/11 already mentioned.
- Production of counter-season seeds or biomass: this applies whenever an event has not yet been approved for commercialization but there is intent to use it for productive purposes. Two main situations are considered: (a) counter-season production of seeds to be later exported to another country where such seeds have been approved for commercialization; and (b) the production of valuable biomass (e.g. for biofuel generation, or for purifying a valuable recombinant product) in a process that will end up in the destruction of the GM plant viability. This is regulated, *inter alia*, by Res. 17/2013.[31]

In all cases, the application for performing such activities is filed in INASE, which performs a formal completeness check, and then it is sent to the Biotechnology Directorate for a technical completeness check, and finally the CONABIA performs an evaluation of the safety assessment. This assessment is used as the main base for a final decision on the application, which is taken by the Secretary on Foodstuff and Bioeconomy under the Ministry of Agroindustry.

In addition, the importation of GM plants or propagation material for the purposes described above requires a special clearance by SENASA. It is granted after corroborating that such material will be managed under laboratory containment or a previously-approved activity. This is regulated, *inter alia* by Res. 498/2013.[32]

2.4.2 Marketing of GM Plants and Foodstuffs

The authorization for commercial release of GM crops and derived products is regulated, *inter alia*, by Resolution 763/11. In this case, the applicant files for three independent assessments. One of them is the biosafety assessment, performed by CONABIA (according to Res.701/11), the other one is the food safety assessment (performed by CTAOUGM-SENASA according to Resolution 412/02[33]), and the third one is an assessment of productive and commercial impacts (performed by

[30]INFOLEG (2012a).
[31]INFOLEG (2013a).
[32]INFOLEG (2013b).
[33]INFOLEG (2002a).

SSMA-MINAGRO according to Resolution 510/11).[34] The three assessments are later taken into consideration by the Secretary on Foodstuff and Bioeconomy for a final decision on the application.

Until 2018 all market authorizations covered every possible use of the GM crops and derived products: cultivation, any food use, feed uses, importation and exportation. By the time this chapter is being finished, a new Resolution 26/18[35] has just established a procedure that allows for the alternative of issuing approvals with a restricted scope usually known as "food, feed and processing" (i.e. excluding cultivation), in particular cases where the developer or an importer requires such restricted scope.

2.4.3 Conventional (Non-GM) New Varieties

In the case of scenario 2, gene-edited plants would be regarded just as ordinary new varieties, like any new mutant (spontaneous or chemically/radiation-induced) from a conventional breeding process. In order to introduce new varieties in the market, there must be a registration in the National Register of Seed Control and Trade (a.k.a. National Register of Cultivars).[36]

This registration system derives from the above-mentioned National Law on Seeds and Phytogenetic creations. It is summarized next but the interested reader can find it fully described elsewhere.[37]

The registration is based on a crop-species *ad hoc* questionnaire, which once filled is assessed by a committee of specialists in that particular crop. Then, the assessment from the specialists is further reviewed by the National Commission on Seeds (CONASE). This assessment includes many agronomic indicators, but it can also include characterization of sanitary issues like harmful metabolites or pest susceptibility. If the crop has a novel trait that may impact any issue of relevance to the general interest, including sanitary aspects, CONASE is enabled to take this into account, to require further specific information from the applicant, and to demand supplementary analysis by other organizations such as SENASA. On the basis of the final conclusions by the National Seed Commission, the President of INASE will decide upon granting or not the permit (registration) for commercialization.

As it has been mentioned before, if during the analysis of a gene-edited plant CONABIA concludes that the crop is not a GMO (scenario II), but nevertheless a specific risk hypothesis connected to the plant phenotype or genotype is identified,

---

[34]INFOLEG (2011d).

[35]INFOLEG (2018).

[36]Not to be confused with the National Register of Property of Plant varieties, which is a different register also under the National Seed Institute but devoted to the protection of intellectual property rights on a plant variety.

[37]Cascardo et al. (1998).

Res 173/2015 mandates that CONABIA conveys the risk hypothesis to the relevant regulator, in this case CONASE, for its further assessment.

2.5 Status Quo of Genome Edited Plants and Products Derived from Them

Until now, only a few cases of gene edited plants have been presented for consultation regarding their regulatory status. This includes annual crops as well as ornamental plants and fruit trees. The introduced traits range from agronomic improvements like herbicide resistance, to enhancements that are valuable to the consumer and/or industry, therefore adding value to the harvested products. Most of the cases (which are mapped in Fig. 2.2) have been found not to fulfill the GMO definition, and therefore to be treated as new varieties of conventional crops.

About half of the cases involved plants that have been already obtained, and the remainder consisted in products which are still on the design stage (where consultation is preliminary). In regards to hypothetical products, if they are obtained in the future, the developer must return to CONABIA and present a molecular characterization, which will be used to determine if the preliminary findings remain valid.

In the case of crops that have been already obtained, the next step for commercialization would be the registering of such varieties, as described, but this has not happened yet for any of them. It seems that early consultations were made merely in order to find out what would be the stance of Argentine regulatory framework regarding each particular product.

2.6 Reform Efforts

In general, the regulation of GM plants in Argentina is kept constantly updated. In regards to the specific regulation aimed at establishing if a particular product is GM or not, it is quite recent (2015). It is relevant to recall, among other strengths of this regulation, that it has been framed in such a way (and proven to be) flexible enough to accommodate for technologies that were not considered initially. Therefore, it is not expected that a reform will be introduced in the short-medium term.

The main issue reported from users is the lack of a detailed questionnaire for the applicant, who has to present information covering the general issues described in Resolution 173/2015. However, the regulator is enabled to perform clarification questions and therefore this shortcoming is compensated adequately.

In summary, there is no perceived need of updating the mechanism for determining if a plant is GM or not. In 2017, this regulatory approach has been extended also to gene-edited animals,[38] and for the sake of coherency it will likely be applied to gene-edited microorganisms as well in the near future.

2.7 Low Level Presence

There is no general provision in the regulation for low level or adventitious presence of non-approved GMO in seeds or foodstuff, including grain. Therefore it is implicitly interpreted that there is a zero tolerance in general.[39]

In the scenario that a gene edited plant is considered to be a GMO, therefore, a zero tolerance for non-approved GMO in seeds or foodstuff, including grain, should be assumed. Since the regulation presumes that plants obtained with the use of recombinant DNA tools are GM until there is a determination on the contrary from the regulator, any product grown in third countries that has not been still submitted for consideration would be presumed GM. In regards to detectability, gene-edited plants that are GM or presumed GM are not considered to pose a different challenge for performing controls in comparison with other GM products.[40]

For the opposite scenario, where a gene edited plant has been submitted for consideration and found not to be a GMO, there would not be an issue of adventitious presence in the regulatory sense. In any case, specifically in regards to seed trade, the usual thresholds related to tolerance for off-type varieties may apply. These are derived from existing international standards of commercial purity, such as the standards for genetic purity and identity from the Association of Official Seed Certifying Agencies (AOSCA)[41] and the Organization for Economic Co-operation and Development (OECD) Seed Schemes.[42]

---

[38]INFOLEG (2017a).

[39]FAO (2014b).

[40]It has been argued elsewhere—see OECD (2015)—that detection of gene-edited products may be more challenging compared to GMOs currently on the market, based on reasons of both technical and human nature. However, from a technical viewpoint, even the slightest mutations obtained by gene editing can be easily detected by molecular techniques currently available, such as those applied in marker-assisted breeding programs to detect Single-Nucleotide Polymorphism (SNPs). Of course, this assertion is true as long as the mutation is known but the same is also valid for GMOs, where the introduced gene or protein must be known in order to design a detection method. In regards to "human nature", considerations relate to the fact that gene editing may lead to mutations that could be generated spontaneously (or using physical or chemical techniques for mutagenesis); this has been presented as a potential temptation for concealing the method used for obtaining the trait. However, such covering would be very difficult in reality since biotech developers employ several people to generate such products (who should be accomplice of the plot for an indefinite time), and they forcedly leave a trace of patents, papers, presentations to regulatory frameworks of third countries and other public information.

[41]AOSCA (2018).

[42]OECD (2018).

2.8 Labelling

There is no regulation for mandatory labelling of GMOs in Argentina. Since the products that are allowed to be in the market have been found to be equivalent to their conventional counterparts, such labelling has been considered unnecessary and potentially misleading by the regulatory authorities under the executive branch. Besides, national legislation cannot clash with the Mercosur Treaty, which includes an Agreement on Labelling of prepacked foods that does not differentiate between GMOs and non GMOs.[43]

Nevertheless, there have been some law projects in the National Congress to impose a mandatory labelling, mostly related with the concept of "consumer-right-to-know" or otherwise inspired by the EU regulations. However, they did not receive wide support and did not progress.[44]

Some legislation on mandatory GMO labelling has been actually passed by a few provincial or municipal governments, for instance the city of Bariloche,[45] a world-wide famous tourist destination. However, its sub-national nature clashes with the principle of free movement of goods within the country. For these reasons and others of practical nature, such legislation in the end proved impossible to be implemented.

In regards to voluntary "non-GM" labelling, there are very few examples of products in the Argentine market having small trade volumes. Such kind of labelling does not seem to be very effective in attracting the Argentine customer, since it has not been widely adopted.

Moreover, if such labelling is misused, this can risk the vendor to be accused of misleading advertising. For instance, by labelling a product "non-GM" for which GM counterparts have not been released to the market, such as "non-GM lemon juice"; this could mislead the consumer to suspect that some of the competing brands could be using GM lemons, when it cannot be the case.

Of course, if a GMO or a non-GM variety has a trait that causes food to be significantly different to the conventional counterpart for consumer purposes, such as high-oleic oil, it may warrant a special labelling on a case-by-case basis. Nevertheless, the trigger of the labelling and the content of the label would relate to the novel properties of the end product and not the method of obtaining the variety.

For instance, the labelling of "high oleic soybean oil" has been recently introduced in the Argentine legislation[46] after the approval of a GM soybean event having that trait by the technology of RNA interference. However, the right to use such denomination is defined according to the product composition, not the breeding process; therefore it also could be used by gene-edited or otherwise null mutant soybeans having the same trait.

---

[43]Schiavone et al. (2013).

[44]H. Cámara de Diputados de la Nación (2010, 2013).

[45]Berto (2018).

[46]INFOLEG (2017b).

It is also important to highlight that such criteria regarding misleading labelling as well as labelling triggered by special properties is fully in line with *Codex Alimentarius* standards.[47]

Given the rationale above, there would be no difference in regards to labelling of food products derived from gene-edited plants, regardless of these being considered GMO or not.

2.9 Identity Preservation System (Coexistence)

According to Phillips and Smyth,[48] a difference can be established between the concepts of *Identity Preservation* and *Segregation*: the first is a strategy applied to reach niches markets, while the second aims at ensuring safety. Where a product has been found to be safe, segregation is not needed. However, Identity Preservation may be used to reach premiums from the trade of a specialty. From this is derived the corollary that responsibility and costs of coexistence measures in the case of Identity Preservation are to be borne by the producer of the higher-value specialty, while on segregation they fall on the shoulders of the unsafe product's operator.

In Argentina there is no mandatory segregation of GM food and feed. Since products are allowed to be placed on the market after it has been established that they are as safe as their conventional counterparts, they are not considered to constitute a relevant difference for the ordinary consumer.[49] Therefore, non-GM identity preservation is a voluntary action of private operators (companies or cooperatives) targeting mostly external markets, based on standards that are agreed on contractual basis.[50]

Currently in Argentina, identity preservation is used only for products consisting of or derived from non-GM maize or non-GM soybean. This is because GM crops of other species have not been released to the market, with the exception of cotton where there seems to be no significant demand for non-GM products.

In addition, positive (or "Does contain..") identity preservation schemes and labeling are used for conventional varieties with special traits obtained by mutagenesis and conventional breeding, such as sunflower having high contents of stearic and oleic acids.[51] All GM crops effectively traded in the Argentinean market are still first generation products, this is, they only have traits of agronomic interest, and the advantage that they may represent to the consumer is mainly through reduction in the cost of derived food. Positive identity preservation for special GM products having

47FAO (2011).

48Phillips and Smyth (2002).

49FAO (2000).

50Pensel et al. (2007).

51INFOLEG (2012b).

benefits to the consumer, such as the high oleic GM soybeans is anticipated[52] but not effective yet.

Non-GM identity Preservation in Argentina is therefore an activity targeting the international market. There are basically two identity preservation systems in use.[53] The first one is based on the use of big bags where the harvest is packed on the farms. Then, these bags are gathered in containers and sent by vessel for export. The second is a bulk supply system, where information is traced along the bulk aggregation from the first batches in the farm to the silos and then to ships. In addition to traceability, the system is reinforced by instances of sampling and testing (strip tests or Polymerase Chain Reaction—PCR) at key steps.

It is important to note that identity preservation through management of batches is easier in countries such as Argentina and Brazil compared with other regions. Typically, these countries' farms and plots are of bigger size and handling facilities are greater and modern. These allow segregation systems for exporting different kinds of products to Europe to be implemented without many difficulties.

Probably the more representative systems for identity preservation currently in use are the ones applied for exports of non GM maize of the Flint varieties to Europe.[54] Details of the systems vary slightly according to each one of the four companies that handle most of the international shipments. However, the final traceability summary usually covers from farm to port. It typically includes seed supply, source plot report (including general description, location, owner, previous crops, and use of adjacent plots), a GM laboratory test report of samples taken in the farm and during loading of the shipment, and the list of transportation means, storage facilities, and port elevators used.

In contrast, the production of non-GMO soybean comes in two flavors: "organic" and "non-GMO conventional" soybean; where soybean milk beverages are perhaps the more important component of the latter.[55]

In recent years, production of non-GM soybean has had a marginal place in a country which is the third largest grower of GM soybean. Therefore, one of the paramount elements of coexistence is the cleaning of all the machinery used, from seeding to containers, since they are only temporarily used for this purpose, and most of the time they handle GM produce. An important element is the widespread use of "silo-bags" to store grains in the farm, which in this case allows for the storage of relatively small amounts of harvest clearly separated from the other plots' production. The other control element of paramount importance is the PCR laboratory analysis of samples taken when trucks arrive from the farm to the storing plants, and of samples taken later for each container once they have been filled.

---

[52]Valor Soja (2017).

[53]Co-Extra (2005).

[54]Ghezan and Tapia (2006).

[55]Ghezan et al. (2006).

In addition, organic production is ruled by law 25.127[56] (1999), which excludes the use of GMOs and establishes that SENASA has the responsibility to enact detailed regulations, including those related to coexistence, and to oversee the private certification control systems as well as to impose punitive damages in case of faults to the law.

PCR analyses for non-GMO products of maize and soybean are usually performed by the molecular biology laboratory of the National Institute of Agricultural Technology (INTA). However, this is a service of private nature, and the result is not a governmental certificate.

The National Seed Institute controls the purity of all seed varieties, including non-GM. This is one of the bases for the non-GM voluntary identity preservation schemes.

Conversely, for ordinary (without identity-preservation) international shipments of corn and soybean products, SENASA issues an accompanying certificate that states the potential presence of GMOs ("may contain" according to the latest decisions of the COP-MOP on implementing article 18 of the CBP). This reduces even further any uncertainty associated with shipments from Argentina in regards to their content of GMOs.

Concerning gene-edited crops, the above considerations will be equally applicable if they are classified as GMOs either in Argentina and/or in the export market. The Identity Preservation schemes described, from seed supply to shipment, will be equally implementable on the basis of working with defined varieties whose breeding history is known to be non-gene edited (whenever that is the consumer's preference), or conversely for gene edited crops improved for nutritional or health benefits, such as the gene-edited soybeans having a high content of oleic acid). In any case, of course these are specialty products and therefore it is up to the customer to pay extra due to a mix of costs including not only the identity preservation measures, but also intellectual property, yield penalties of the specific desired variety derived from suboptimal genetics, etc.

2.10 Liability

The duty of preserving the environment and the obligation to redress its damage is explicitly stated in the Argentine Constitution[57] (article 41). From that starting point, the liability regime for environmental damage in Argentina is established in law no. 25.675 (dated year 2002), which is known as the General Law on the Environment.[58] It defines environmental damage as any relevant alteration that negatively

[56]INFOLEG (1999).

[57]INFOLEG (1995).

[58]INFOLEG (2002b).

modifies the environment, its resources, the ecosystems' balance, as well as collective values or assets.

It is a civil liability regime, and it holds anyone causing current or future degrading effects responsible for the costs of preventive and reparation measures. It is a strict liability regime. However, in case of concurrent fault (violation of administrative regulations), it reverses the burden of proof against the defendant. Conversely, a standard of care approach exonerates the operators of the technology if all measures aimed at preventing the damage were taken and the damage was caused exclusively by the victim or a third party.

The right to appeal in court for expedited measures aiming at avoiding or mitigating environmental damage is explicitly acknowledged in the Argentine Constitution (article 43) to the affected person(s), the Ombudsman, or relevant NGOs. In continuation, this law also recognizes those actors as enabled to claim compensation for the costs of those measures taken for the restoration of the environment.

Concerning human health, among the offences against public health in the Argentine Penal Code[59] (Title VII, Chapter IV), articles 200 to 203 mandates prison and fines to those who knowingly or carelessly would introduce dangerous or adulterated foods in the market. These would be generally applicable in case a food derived from a GM plant would be introduced without the appropriate regulatory permit, as well as foods from conventional plants.

In regards to commercialization, Law 20.247 on Seeds and Phytogenetic creations restricts the commercialization of any plant variety that has not been registered. As mentioned, for GMOs there is a requirement of ex ante assessment by CONABIA and other technical bodies. For conventional varieties, the assessment is made by CONASE.

The commercialization of any plant variety (GM or not) without this registration process is a violation of the seed law which is penalized, according to the details of the misdemeanor, with fees (punitive damages), destruction of the plant materials and/or commercial proscription of the offender.

Also, if the introduction of a GMO causes an environmental, human health or commercial damage, the general provisions of the Civil and Commercial code (Title V, Chapter 1, on Civil Liability) are applicable in order to obtain restoration or monetary compensation. This Argentine code has quite standard provisions in continental legal systems including, for instance, the duty to avoid damage (and claim from compensation for legitimate measures taken by anyone with this purpose), objective and subjective attribution, *force majeure*, *res judicata,* etc.

These administrative, civil and penal responsibility rules are not tied to a specific technology. Therefore, they are fully and equally applicable to damages derived from conventional or GM plants, and therefore to any gene-edited crop regardless of it being considered a GMO or not.

[59]INFOLEG (2009).

2.11 Perception of Genome Editing

2.11.1 Position of Public Authorities

The regulatory authorities have shown to be in favor of clarifying the situation of genome edited crops, by giving a certain response in due time to developers regarding if their products are considered GM or not. This adds robustness to a sanitary system caring for human welfare and the environment, and at the same time shows an aspiration for providing legal certainty and predictability for commercial R&D activities as well as for agroindustrial business. Remarkably perhaps from a policymaking perspective, public authorities have found no need of giving a "special" third category or special treatment beyond what already exists for GM vs. conventional crops.

As a token of a public authority position, the Head of Staff of the Minister of Agroindustry has recently declared that "*there is a particular enthusiasm for the new gene editing techniques*".[60]

We have claimed elsewhere[61] that policymakers in any country, in addition to caring for the safety aspects and legal implications of the regulation applied to GETs, would also benefit from studies (none currently available, but some in process of elaboration) on the impact that over-regulation or decisions-delay (*de facto moratorium*) may have on innovation and productivity.

2.11.2 Public Opinion

Media journalists have welcomed the possibilities of genome edited crops, particularly in connection to developments by national companies or research institutes.[62] There have been some surveys and analysis of public perception,[63] but in general the public and consumers in Argentina are not occupied with the GMO issue. This could be interpreted either as an implicit acceptance or a lack of knowledge. In any case, this attitude can be considered to be also transposed to gene edited products for the time being.

We have analyzed elsewhere[64] that society and the general public worldwide are still in a stance of interpretative flexibility regarding products derived from GETs. This flexibility will still be molded for some time by dissimilar factors, including the discourse of interested actors, opinions or concepts in media coverage of agricultural and non-agricultural applications, references to these technologies in popular culture, etc.

---

[60]Rumi (2018).

[61]Whelan and Lema (2017).

[62]Preciado (2015), Biz (2017), Ingrassia (2017), Roman (2017) and Longoni (2017).

[63]MINCYT (2015) and Trigo et al. (2002).

[64]Whelan and Lema (2017).

2.12 Treatment of Other New Breeding Technologies

The Argentine regulation addresses NBTs in general, which embraces GETs. Examples of other NBTs include Cisgenesis/Intragenesis, Grafting, Agro-infiltration, RNA-dependent DNA methylation (RdDM) or Reverse Breeding, where this terms are understood in accordance with the Joint Research Centre of the European Union.[65] As it has been mentioned before, the status or treatment of products from these other techniques (or others to arise in the future) are the same as described for gene-edited crops.

As it has been mentioned, regulation was developed by considering real cases of different NBTs extracted from the scientific literature, and afterwards some cases have been brought to CONABIA for their assessment by applicants, as reflected in Fig. 2.2.

Although the determinations are made on a case by case (plant line by line) basis, some rationale can be offered regarding the outcome in relationship with the particular NBT involved.

For instance, products from cisgenesis and intragenesis are likely to be considered a GM because in most cases there is an introduction of a DNA construct. In regards to grafting, the GM part will certainly have to be regulated as such before allowing its market release, therefore for most practical purposes under the Argentine framework the grafted plant and derived products will be regulated as GM.

Now, turning to open field agro-infiltration (assuming a non-heritable modification in the plant), such activity is likely going to be regulated as the environmental release of a GM microorganism. Regarding RNA-dependent DNA methylation, it is unlikely to have commercial use in the short-medium term due to trait instability; however its products would likely be considered non-GM because in most cases there is no modification in the genome sequence. Finally, products derived from techniques like early flowering (accelerated breeding) and reverse breeding are likely to be considered non-GM if the line intended to be released is a null segregant of the helper transgene.

2.13 Other Regulations in the Region

Argentina is the first country having a GMO regulatory framework where a provision has been issued for dealing specifically with products derived from NBT (including GETs). However, soon after this, two other countries in the region, Chile and Brazil, issued their own regulatory guidance in the same regard. By the time this chapter is being finished, Colombia has just notified the WTO a Resolution project on NBTs. The translations of these regulations are included here as an Annex 1, and an overall analysis of similarities and differences is offered next.

[65]Lusser et al. (2011).

Chile[66] In this case, regulatory criteria are similar to the Argentine approach in the following: (a) it relies on the concept of NBTs, (b) it highlights the importance of the notion of "novel combination of genetic material" for the decision, (c) the decision process will end up classifying the product as GM or conventional, but does not create any third category or special regulatory treatment within those categories, (d) a null segregant from a GM plant is considered non-GM, (e) the same "event" definition is used for signaling those cases where definitely there is a novel combination of genetic material, (f) regulatory determinations are established "line by line" and no reference to specific techniques is made, and (g) a time limit for obtaining a determination is provided.

In regards to the differences (a) the criteria are only published as "guidance", but each determination is issued as a separate Resolution, (b) the regulatory status in other countries is taken into consideration and (c) there is no explicit option for an early and preliminary consultation in the design stage.

Brazil[67] In this case, for the main corpus of the regulation, the term "Precision Breeding Techniques" is employed. However, it is clarified that it is also meant to cover the same as the early European-borne term NBT, as well as "Precision Breeding Innovation" (PBI). The latter term looks like a mixing of "Plant Breeding Innovation" and "Precision Breeding Techniques", both introduced recently by biotech developers and the seed industry.[68]

References to "final lines", "null segregants" and "absence of r-DNA" are compatible with the Argentine and Chilean criteria, although expressed with a different language. Also, reference is made to "transgene insertion" which is comparable to the use of the "event" definition in the other Latin-American regulations introduced so far.

The Brazilian regulation does include an explicit, open list of NBTs, and a closed list of genetic interventions or characteristics that non-GM NBT products may have. This is compatible with the implicit criteria and early caseload analysis made in Argentina. Such explicit listings may provide higher clarity and predictability for existing technologies in the case of Brazil, but may result in complications to encompass future technologies.

The questionnaire included in this regulation is more detailed, although it covers the same main aspects as in the Argentine and Chilean cases. Therefore, it is anticipated that after a process of questions and answers, the three countries will work on dossiers of similar content and level of detail. As in the case of Chile, the Brazilian form inquires for information regarding the regulatory status in third countries.

In addition to the explicit list of NBTs, there are interesting references to the topical use of nucleic acids and gene drives, although it is not clear what this reference implies regarding chances of such products to be considered GM or not.

---

[66]SAG (2017).

[67]CTNBIO (2018).

[68]ISF (2018), Gepts and Hancock (2006) and Sprink et al. (2016).

Finally, as in the case of the other two Latin American countries, there is a time limit for issuing a determination.

Colombia[69] The content of this draft regulation is very similar to the Argentine legal text. For instance, it clearly refers to the Cartagena Protocol definition as the touchstone for deciding if a new variety is GMO or not. Also, it includes a definition of foreign genetic material that is identical to the Argentine definition of transformation event. Moreover, it describes the required information as in the Argentine case and it also includes a reasonable time limit for the whole process.

The Colombian regulation is more explicit in certain aspects, beginning with the title *"Resolution setting out the applicable procedure for crops where any stages over the plant-breeding process incorporate innovative phyto-improvement techniques through modern biotechnology and the final product does not contain any foreign genetic material.* In addition, it does explicitly state that marketing of non-GMO NBT products will be regulated as conventional varieties. In regard to the latter, it does also acknowledge the regulatory "gap"—as in the Argentine counterpart—by stating that the regulator may propose special follow-up measures for non-GMO varieties on a case by case basis.

Finally, Colombian draft regulation employs a slight variation of the term NBT: "New technologies of Plant Breeding derived from Modern Biotechnology" (fortunately, no new acronym is coined. . .). It also echoes the Brazilian use of the term "Plant Breeding innovation" as a synonym, but it is also defined here as the "*refinement of existing modern biotechnology methods*". Unlike Chile and Brazil (but alike Argentina), this regulation does not request information regarding the regulatory situation in third countries.

2.14 Conclusion

The Argentine Republic was the first country in the world to issue an official *ad hoc* explicit regulation to deal with new breeding techniques (including gene-edited crops) in 2015. It basically establishes a procedure to determine if a plant line intended to be introduced in the market is GM or not. In brief, the regulation requires all products obtained with the aid of recombinant DNA to be considered by the National Biosafety Commission, which will analyze the genetic intervention and decide if it merits to be considered a "novel combination of genetic material".

Remarkably, the rationale under the Argentine approach to regulate these products is based on the Cartagena Protocol definitions of "Living Modified Organism" and "Modern Biotechnology". Thus, third countries that are members of the Protocol may also follow the same rationale for transboundary movement of GM plants viable material, as well as for other purposes if the CPB language is also taken as a base for

[69]ICA (2018).

their national regulation. Maybe a token of this can be found in the resemblance of the Argentine approach with others that emerged later and are described here.

The early experience gained while applying this rationale has shown its robustness, since all queries have analyzed satisfactorily so far. Future developments will be highly dependent on the stance that trade counterparts and influential food importers may adopt, and for this reason Argentina includes the regulation of NBTs-GETs in every relevant bilateral dialogue or international fora.

The amount and variety of products from SMEs and public sector research brought into consideration so far evidences the enormous potential of these technologies for innovation and increase of agroindustrial productivity with higher sustainability. Many lessons have arisen from the long and sometimes painful experiences of developing national regulatory systems for GM crops over a quarter of a century, and such lessons must not be overlooked now. A timely, scientifically sound, internationally harmonized approach for the regulation and trade of these products would greatly benefit people of all latitudes in many ways in the following years.

Annex 1: First NBT Regulations from LA Countries

Resolution No. 173/2015 (Argentina)

Note: The following text is offered for illustrative purposes only and it is not an official translation (Spanish is the official language of the Argentine Republic).

In Considering:

Resolution No. 763 dated August 17, 2011 of the Ministry of Agriculture, Livestock and Fisheries (MAGYP) sets forth the guidelines for the activities involving Genetically Modified Organisms (GMO) in the Republic of Argentina.

Pursuant to article 3.A of the Resolution No. 763/11, risk assessment, design of biosafety measures and risk management during each stage of GMO assessment hereof shall be conducted by the National Advisory Commission on Agricultural Biotechnology (CONABIA), which Executive Secretariat is held by the Biotechnology Directorate of the National Directorate of Processes and Technologies of the Undersecretariat of Added Value and New Technologies under the Secretariat of Agriculture, Livestock and Fisheries (SAGYP) under the MAGYP.

Article 3 of Resolution (SAGYP) No. 437 dated August 06, 2012 sets forth as actions pertaining to CONABIA, among others, to advise the Secretary of Agriculture, Livestock and Fisheries on "risk assessment, design of biosafety measures and risk management in the various stages of assessment, authorization and release into the agro-ecosystem of genetically modified organisms" and "every issue to be submitted to its scientific evaluation".

Resolution (SAGYP) No. 701 dated October 27, 2011 sets forth the requirements and proceedings that must be met by biosafety assessments for the release of GM-plants into the agro-ecosystem.

Resolution No. 701/11 defines GM-plant as a plant organism bearing a combination of genetic material obtained through the application of modern biotechnology.

Such regulation defines event as "the combined and stable insertion into the plant genome of one or more genes or DNA sequences that are part of a defined genetic construct".

The development of agricultural biotechnology is a key tool for the addition of value in the agribusiness value chain in the Argentine Republic.

In the Argentina Republic, as in the rest of the world, major advances are being produced in the development of new breeding techniques in plants (NPBT).

The characteristics of the crops derived from these techniques are of such heterogeneity that demand a prior scientific assessment in order to determine whether any such crop falls under the rules and regulations applicable to GM-plants or, on the contrary, are not subject to such regulations.

That this decision does not alter the regulatory framework applicable to GMO but rather sets forth proceedings to determine the cases in which a crop obtained by NBT that use modern biotechnology to generate genetic modifications are subject to GMO rules and regulations.

CONABIA, after an extended debate in several meetings during 2013 and 2014 has rendered its agreement to this regulation during its ninth meeting of the year 2014, which took place on November 25, 2014.

The General Directorate of Legal Affairs of the Ministry of Agriculture, Livestock and Fisheries has expressed its legal opinion.

The Secretary of Agriculture, Livestock and Fishery is competent to issue this resolution pursuant to Decree No. 357 dated February 21, 2002 as amended.

Therefore, the Secretary of Agriculture, Livestock and Fisheries Hereby Orders as Follows:

Article 1 The procedure to determine in which cases a crop obtained by new breeding plant techniques (NBPT) using modern biotechnology, does not fall under GMO rules and regulations pursuant to Resolution (MAGYP) No. 763 dated August 17, 2011 and its complementary regulations, is hereby enacted.

Article 2 In regards to the situations mentioned in Article 1, the interested party shall submit its case for the assessment of CONABIA through a Previous Consultation Stage ("ICP") pursuant to Resolution No. 701/11. During the ICP the Applicant shall submit data on the breeding methodology used to obtain and select the crop, on the new trait or characteristic introduced, and on evidence of the genetic changes present in the final product. Within the framework of the ICP, the applicant shall request CONABIA to establish whether the result of the breeding process is a novel combination of genetic material.

A genetic change shall be regarded as a novel combination of genetic material when the assessment establishes the occurrence of a stable and joint insertion in the plant genome of one or more genes or DNA sequences being part of a defined genetic construct.

Article 3 Any GM-plant offspring shall be presumed also a GM-plant until scientific information proves otherwise. Therefore, in addition to the provisions contained in Article 2 herein, applicants shall inform if any transformation event was used during the breeding process, even when it is no longer present in the crop to be introduced into the agro-ecosystem, and include evidence of its absence.

Article 4 The Biotechnology Directorate will conduct a preliminary assessment on the data provided by applicants in a period not exceeding 60 calendar days, and proceed to list the matter for debate in the following CONABIA meeting. On the basis of the information filed during the ICP, CONABIA will establish whether a novel combination of genetic material has been created. Also, if appropriate, CONABIA will determine if there exists enough scientific evidence to support the absence of the event(s) used transiently during the crop breeding process. Both the Biotechnology Directorate and CONABIA may request the Applicants to file additional data and information in order to complete their assessments.

Article 5 Upon CONABIA finding that a novel combination of genetic material has not been created and, if applicable, that no unauthorized events subsist in the crop, the SAGYP, through the Biotechnology Directorate shall notify the Applicant that the product does not fall under the scope of Resolution No. 763/11 and its complementary regulations.

Notwithstanding the aforementioned, CONABIA may also recommend the Secretary of Agriculture, Livestock and Fisheries, the adoption of follow-up measures on a case-by-case basis, taking into account the crop features and/or novelty, based on scientific and technical grounds.

Article 6 Applicants must be previously registered under the National Registry of Genetically Modified Plants Organisms Operators (RNOOVGM) created by Resolution (ex-SAGPYA) No. 46 dated January 7, 2004 before filing for the ICP. Otherwise, applicants will be subject to register with the Biotechnology Directorate in order to prove their legal standing. If the product is considered a GM-Plant, applicants must register under the RNOOVGM before continue filing their first application for GM-Plant environmental release.

Article 7 Alternatively, applicants may file for a preliminary inquiry aiming at anticipating whether a hypothetical product from projects still in the design stage would fall under the scope of Resolution No. 763/11 and its complementary regulations. In these cases, no registration under the RNOOVGM or equivalent documentation shall be required and CONABIA shall perform a preliminary assessment and provide an indicative answer that the Biotechnology Directorate will notify to applicants. If such new crops are obtained later, they shall be subjected to the provisions hereinabove in order to establish whether they have the features anticipated in the preliminary inquiry.

Article 8 This resolution shall come into effect the day after its publication in the Official Gazette.

Article 9 Be it communicated, published, given to the National Directorate of the Official Registry and filed.-.Sgd.: G DELGADO. Secretary of Agriculture, Livestock and Fisheries.

Resolution No. 16/2018 (Brazil)

Note: The following text is offered for illustrative purposes only and it is not an official translation (Portuguese is the official language of the Federative Republic of Brazil).

Ministry of Science, Technology, Innovations and Communications

National Biosafety Technical Committee

The National Technical Commission on Biosafety—CTNBio, in the use of its legal and regulatory attributions and in compliance with the provisions contained in items XV and XVI of art. 14 of Law 11,105 of March 24, 2005;

Considering

The need to consider the Innovative Precision Improvement Techniques (TIMP), from the English Precision Breeding Innovation (PBI) which also encompasses the so-called New Breeding Technologies-NBTs, in the light of the precepts provided in the Law No. 11,105 of March 24, 2005;

Considering that Law No. 11,105 of 2005 defines recombinant DNA/RNA molecules, genetic engineering and genetically modified organisms—GMOs in items III, IV and V of its art. 3, respectively;

Considering that TIMP encompass a set of new methodologies and approaches that differ from the genetic engineering strategy by transgene, as it results in the absence of recombinant DNA/RNA in the final product;

Considering that TIMP can introduce innovative uses of molecular biology tools, which can result in:

1. The precise editing of genomes, by induction of specific mutations, generating or modifying wild and/or mutated alleles without transgene insertion (s);
2. Genetic transformation and/or control of gene expression (activation/inactivation);
3. Epigenetic regulation of the expression of genes by natural mechanisms without genetic modification of the individual;
4. Genetic transformation and/or control of gene expression with genes from sexually compatible species;
5. Temporary and non-inheritable genetic transformation of cells and tissues;
6. Permanent or non-permanent host infection of genetically modified viral elements;
7. The creation of alleles with autonomous inheritance and potential of recombination with the possibility of altering a whole population (gene drive); and
8. The construction of heterologous genes or new copies of homologous genes

Resolves

Article 1 The technologies described in Annex 2 which are part of this Normative Resolution, may originate a product not considered as a Genetically Modified Organism (GMO) and its derivatives, as defined by Law No. 11,105 of March 24, 2005, are considered examples of Innovative Precision Improvement Techniques (TIMP), but not limited to them.

§ 1—The product referred to in the section of this article is defined as the offspring, lineage or end product of a process that uses Innovative Precision Improvement Techniques in one of its phases of development.

§ 2—The cases to be classified are not limited to the technologies described in Annex 2, considering that different technologies are rapidly and continuously advancing and may provide new products, to which the provisions of this Normative Resolution will also apply.

§ 3—The products referred to in the main paragraph of this article imply at least one of the following characteristics:

I. product with proven absence of recombinant DNA/RNA, obtained by a technique employing GMOs as a parent;
II. product obtained by a technique using DNA/RNA that will not multiply in living cells;
III. product obtained by a technique that introduces targeted site mutations, causing gain or loss of gene function, with the proven absence of recombinant DNA/RNA in the product;
IV. a product obtained by a technique where there is a temporary or permanent expression of recombinant DNA/RNA molecules, without the presence or introgression of these molecules in the product; and
V. a product where techniques employing DNA/RNA molecules are used which, whether absorbed or not systemically, do not cause permanent modification of the genome.

In the case of a product obtained from a GMO with favorable opinion from CTNBio for commercial release, the conditions described will apply only to the characteristic introduced by TIMP.

Article 2 In order to determine whether the product obtained by TIMP will be considered as a GMO and its derivatives, pursuant to article 3 of Law 11,105 of 2005, the applicant must submit a consultation to CTNBio.

§ 1—The consultation shall be accompanied by the information contained in Annex 3 of this Normative Resolution.

§ 2—Once the consultation is registered at CTNBio, its view will be published in the Official Journal of the Union and distributed to one of the members, titular or alternate, for reporting and drawing up a final opinion.

§ 3—The final opinion of the member shall be based on an analysis, on a case-by-case basis, of proof of compliance with at least one of the conditions described in § 3° of article 1 of this Normative Resolution.

§ 4—For the products and technologies obtained using the techniques exemplified in Annex 2, CTNBio's decision shall observe compliance with one or more of the conditions described in § 3 of article 1 of this Normative Resolution and will be conclusive regarding the application of the definitions of arts. 3 and 4 of Law 11,105 of 2005.

Article 3 The final opinion referred to in paragraphs 2nd of art. 2nd of this Normative Resolution shall be submitted to at least one of the Standing Sectoral Subcommittees, in agreement with the parental organism and the proposed use of the technique submitted for consultation and, after its approval, shall be referred to the CTNBio plenary for deliberation.

The Subcommittees shall within 90 days analyze and elaborate the opinions, and this term may be extended for the same period by decision at the CTNBio plenary.

Article 4 CTNBio may, as a result of the consultation and with due scientific justifications, request additional information or studies.

Article 5 Any situations not foreseen herein will be assessed and decided, case by case, by CTNBio.

Article 6 This Normative Resolution enters into force on the date of its publication.

EDIVALDO DOMINGUES VELINI
President of the Commission

Annex 2

Examples of "Innovative Precision Improvement Techniques (TIMP)".

1. Technique: Induced Early Flowering.
1.1 Summary of the Technique: Silencing and/or overexpression of genes related to flowering by insertion of the genetic modification into the genome and subsequent segregation or through temporary expression by viral vector.
2. Technique: Technology for Seed Production.
2.1 Summary of the Technique: Insertion of the genetic modification for restoration of fertility in naturally male-sterile lines in order to multiply these lines maintaining the male-sterility condition, without, however, transmitting the genetic modification to the offspring.
3. Technique: Reverse Breeding improvement.
3.1 Summary of the Technique: Inhibition of meiotic recombination in selected heterozygous plants for the characteristic of interest in order to produce homozygous parental lines.
4. Technique: Methylation of RNA-Dependent DNA.(RNA-directed DNA methylation)
4.1 Summary of the Technique: Methylation directed by interfering RNAs ("RNAi") in promoter regions homologous to RNAi with the objective of inhibiting the transcription of the target gene in living beings.
5. Technique: Site Directed Mutagenesis.

5.1 Summary of the Technique: Protein or riboprotein complexes capable of causing site-directed mutagenesis in microorganisms, plants, animals and human cells.
6. Technique: Oligonucleotide Directed Mutagenesis.
6.1 Summary of the Technique: Introduction into the cell of an oligonucleotide synthesized complementary to the target sequence, containing one or a few nucleotide changes, which may cause substitution, insertion or deletion in the target sequence through the cell repair mechanism (microorganisms, plants, animals and human cells).
7. Technique: Agroinfiltration/Agroinfection.
7.1 Summary of the Technique: Leaves (or other somatic tissue) infiltrated with *Agrobacterium sp.* or gene constructs containing the gene of interest to obtain temporary expression at high levels located in the infiltrated area or with viral vector for systemic expression, without the modification being transmitted to subsequent generations.
8. Technique: Topical application of RNAi/ systemic use.
8.1 Summary of the Technique: Use of double stranded RNA ("dsRNA") sequence homologous to the target gene (s) for specific silencing of such gene (s). The engineered dsRNA molecules can be introduced/absorbed by the cell from the environment.
9. Technique: Viral Vector.
9.1 Summary of the Technique: Inoculation of living organisms with recombinant virus (DNA or RNA) expressing the genetic modification and amplification of the gene of interest through the mechanisms of viral replication, without modification of the host genome.

Annex 3

1. Regarding to the original organism (parental organisms), indicate:
1.1. the identification of the genetic technology, purpose and intended use of the resulting organism and its derivatives;
1.2. the taxonomic classification, from family, to the most detailed level of the organism to be released, including, where appropriate, subspecies, cultivar, pathovar, strain and serotype;
1.3. the risk classification of the genetically modified organism in accordance with Normative Resolution No. 2 of November 27, 2006 4. the gene(s) and/or genetic element(s), body(ies) of origin and their specific functions, where applicable;
1.4. the genetic strategy (s) used to produce the desired modification (s); the genetic map (s) of the building (s) used in the process indicating, with all genetic elements present;
1.5. Molecular characterization of the result of manipulation in the recipient organism (parent and end product), where applicable, providing information related

to: (1) number of manipulated copies (e.g. number of genomic sequences, number of alleles, etc.); (2) location in the genome of the manipulated region, where possible; (3) identify the presence of unintentional genetic modifications (off-target), when applicable.

1.6. the product of expression of the manipulated genomic region (s), described in detail, where applicable.

2. With regard to the product (offspring, lineage or final product) inform:

2.1. proof of the absence of recombinant DNA/RNA molecules, through the use of molecular methods.

2.2. whether the product containing DNA/RNA molecules for topical/systemic use has the recombinant ability to enter into target species and/or non-target species.

2.3. whether the product covered by the application is commercially approved in other countries.

2.4. If the product uses the gene drive principle that may allow the phenotypic change conferred to have the potential to spread throughout the recipient organism population, explain the care to monitor the organism using at least two strategies.

2.5. how the possibility of possible unintentional (off-target) effects of the technology that may be present on the product has been assessed.

Guidance on the Applicability de Resolution no. 1.523/2001 (Chile)

Note: The following text is offered for illustrative purposes only and it is not an official translation (Spanish is the official language of the Republic of Chile).

Applicability of Resolution No. 1.523/2001 on propagation material developed by new plant breeding techniques.

Individuals, natural or legal who want to internalize and introduce to the environment live modified plant propagation organisms, must strictly comply with the provisions of Exempt Resolution No. 1523 of 2001, of the Agricultural and Livestock Service (SAG). [1]

Considering that scientific progress has allowed the development of a new generation of biotechnological techniques of plant genetic improvement other than transgenics, the Agricultural and Livestock Service has considered it necessary to solve case by case if the propagative material developed by any of these techniques is within or outside the scope of Resolution No. 1523 of 2001.

In this context the SAG, based on the information presented by the interested party that intends to introduce to the national environment, a propagation material developed by any of these techniques, whether imported or national, will be pronounced by resolution, with respect to if said material is within or beyond the scope of Resolution No. 1523 of 2001. To this end, the SAG will evaluate the background information presented on the technique and verify whether the propagation material in question has a novel combination of genetic material.

For these purposes, a novel combination of genetic material will be understood as a stable insertion of one or more genes or DNA sequences that encode proteins, interfering RNA, double-stranded RNA, signaling peptides or regulatory sequences.

This procedure allows to obtain an official statement from the SAG, through a Resolution, that will indicate whether the propagation material developed by any of the new biotechnological techniques of plant breeding, which is intended to be introduced into the national environment, is within or outside the scope of Resolution No. 1523 of 2001, which for this purpose means that the material is considered or not LMOs, respectively.

In the event that the SAG determines that the propagation material submitted for the Service's consideration is outside the scope of Resolution No. 1523 of 2001, the interested party may carry out activities for the purposes of agricultural production and use, without restrictions and therefore both without having to comply with the biosecurity measures established by the SAG, in accordance with the aforementioned Resolution in force.

The pronouncement will have an indefinite validity or until the Service determines otherwise based on new scientific background.

To whom is addressed:

Natural or legal persons, research centers or universities that intend to introduce to the national environment a propagative material produced from a new plant breeding technique different from transgenics.

Documentation to present

The presentation made by the interested party in introducing propagation material produced from new plant breeding techniques other than transgenics into the environment, should be addressed to the Chief of the Division of Agricultural and Forestry Protection, consulting as to whether the material is outside or within the scope of Resolution No. 1523 of 2001, in consideration of the information that is presented and exists about said material and its development process.

For these purposes, the application is submitted through the form established by the SAG, containing and accompanying the information in this required, at least [2]:

All information must be presented in Spanish and attach all reference articles, analytical results and documentation from official agencies that support its content.

The sole intention of introducing another type of propagation material, imported or national, to the environment in the country, in different stages of development, will require the presentation of a new application.

All the information delivered by the user will be protected according to the current regulations.

Response time: 20 business days, from the receipt of the request by the Division of Agricultural and Forestry Protection of the SAG.

Cost of the procedure: The evaluation of the background will be governed by the Supreme Decree No. 142 of 1990 of the Ministry of Agriculture.

[1] Breaches of current regulations on the subject will be sanctioned according to the Law of Agricultural Protection (Decree Law No. 3,557) and the procedure established in Law No. 18.755 Organic Agriculture and Livestock Service.

[2] The veracity of the information provided will be the sole responsibility of the applicant and in no case exempts it from compliance with other regulations that apply to the same material.

Form

The undersigned that is identified below, comes to present to you for processing an official statement on whether the propagation material developed by a new biotechnological techniques of plant breeding, which is intended for introduction into the national environment, is within or outside the scope of Resolution No. 1523 of 2001, which establishes standards for the confinement and introduction into the environment of living modified plant propagation organisms.

Section II: Technical Information Presented (**)

Section I: Identification of the Applicant

1. Applicant's background:

Name or company name:
Genre:
Natural person or Legal person
Nationality (only natural person case):
Home address:
Commune: Region: Country:
Email: Phone number:

2. Background of the legal representative (only in case of legal entity):

Name of the legal representative (*):
Nationality:
Gender: F M
Address in Chile:
Commune: Region:

3. Background of the technical counterpart before the SAG:

Name of the technical counterpart before the SAG:
E-mail technical counterpart:
Phone number:

(*) Accompany document stating the power conferred in accordance with Law No. 19,880.

Section II: Technical Information Presented (**)

1. Individualization of the propagation material to be introduced into the environment:

a. Species.
b. Variety/Line.
c. Description of obtained phenotype.
d. Company or institution that developed the material.

2. Regarding the biotechnological process used:
2.1 Background of the biotechnological technique used, indicating the modified DNA sequences.
2.2 Include genetic scheme detailing the lines that will be introduced in Chile and the techniques used to rule out the insertion of genetic sequences that encode proteins, RNA interference, double-stranded RNA, signaling peptides or regulatory sequences.
3. Indicate if the propagation material has been authorized by the official agency of a country. If this is the case, you must indicate the type of authorization referring exclusively to the material that is requested to be entered into the national environment, providing all the written information that you have.

(**) The information must be presented in Spanish and attach all reference articles, analytical results and documentation from official agencies that support its content.

Along with the above, I declare under oath to be aware of the following:

1. The veracity of the information provided will be the sole responsibility of the applicant and in no case exempts it from compliance with other regulations that apply to the material.
2. The sole intention of introducing another type of propagation material, imported or national, to the environment in the country, at different stages of development, will require the submission of a new application.

Signature of the applicant.

Draft Resolution (Colombia)

Note: The following text is the translation provided by the Colombian Government to the WTO on its notification G/SPS/N/COL/282 presented to the Committee on Sanitary and Phytosanitary Measures (Spanish is the official language of the Colombia Republic).

Draft Resolution setting out the applicable procedure for crops where any stages over the plant-breeding process incorporate innovative phyto-improvement techniques through modern biotechnology and the final product does not contain any foreign genetic material.

Resolution No. (to be assigned)

The General Manager of the Colombian Agricultural & Farming Institute (ICA)

In use of his legal powers and particularly as conferred by Article 65 under Act 101 of 1993, Article 4, Decree 3761 of 2009 and Article 2.13.1.6.1, Decree 1071 of 2015 and

In Consideration That:

In accordance with Article 65, Act 101 of 1993 "General Agricultural, Farming & Fishing Development Act", it is the responsibility of the Ministry of Agriculture, through the Colombian Agricultural & Farming Institute—ICA, to develop policies and plans intended for protection of nationwide agricultural and farming health, production and productivity; therefore, it shall be responsible for undertaking agricultural and farming health actions and exerting technical control over imports, exports, manufacturing, commercialization and use of agricultural and farming feedstock intended for protecting domestic agricultural and farming production and minimizing the risks for foods and the environment arising from use thereof and facilitating access of domestic products to the international market.

By means of Act 740 of 2002 Colombia ratified the Cartagena Protocol on Biosafety to the Convention on Biological Diversity, the purpose of which, in accordance with the precaution approach, is to contribute to guaranteeing an appropriate level of protection within the realm of safe transfer, manipulation and use of Living Modified Organisms generated by modern Biotechnology that may have adverse effects on preservation and sustainable use of biological diversity and specifically focusing on cross-border movements.

Over development of such provisions, the Colombian National Government issued Decree 4525 on 6th December 2005 "Regulating Act 740 of 2002" and set out that the Ministry of Agriculture & Rural Development, by means of the Colombian Agricultural & Farming Institute—ICA, shall be competent for authorizing the activities described in Article 2 therein when dealing with Living Modified Organisms—LMO exclusively for agricultural, farming, fishing, commercial forest plantation and agroindustrial use, as may have adverse effects on preservation and sustainable use of biological diversity.

By means of Resolution 0946 of 2006, ICA established the procedure for processing at such bureau applications for Living Modified Organisms—LMO, approving the internal CTNBio LMO by-laws exclusively for agricultural, farming, fishing, commercial forest plantation and agroindustrial purposes issuing other related provisions.

The Colombian Agricultural & Farming Institute—ICA is responsible for exerting technical control over production and commercialization of agricultural and farming feedstock, animal genetic material and seeds for sowing, with the purpose to prevent risks that may affect agricultural and farming health, food safety and domestic agricultural and farming production.

ICA is the agency in charge of granting, suspending or revoking licenses, registrations and permits for operation, commercialization, mobilization, importation or exportation of animals, plants, feedstock, seeds and agricultural and farming products and by-products, as well as imposing any applicable penalties, pursuant to current legal standards.

Dynamic innovation in phyto (vegetable) improvement/plant breeding allows for obtaining heterogeneous products, which makes it necessary to conduct prior technical analysis with the purpose to determine whether the regulation on Living Modified Organisms (LMO) shall be applied thereto.

On the basis of the above, the present Resolution sets forth the procedure that must be applied to crops obtained by means of using phyto-improvement innovation techniques through modern biotechnology where the final product does not contain any foreign genetic material in order to determine if it is LMO or not and consequently decide whether said regulation on Living Modified Organisms (LMO) shall be applied thereto or not.

By virtue of the above,

It Is Hereby Resolved As Follows:

Article One—Purpose: Set out the procedure applicable to crops where any stages over the plant-breeding process incorporate innovative phyto-improvement techniques through modern biotechnology and the final product does not contain any foreign genetic material, for which it shall not be considered as LMO.

Article Two—Scope of Application: The present Resolution shall be applicable to all individuals or companies committed to genetic improvement research, assessing, producing, conditioning, importing, exporting, storing and/or marketing crops that have been materials for sowing.

Article Three—Definitions: For the purposes of this resolution, the following definitions are adopted:

3.1 Crop: Generic name used to refer indistinctly to varieties, lines, hybrids and clones that are being used as commercial planting materials.
3.2 Phyto-Improvement: This is the art and science of altering or modifying the heredity of plants to obtain genetically improved crops (varieties or hybrids), adapting to specific conditions and resources of the producer, industry and consumers.
3.3 Plant Breeding Innovation: This corresponds to scientific progress over the last few years, which has allowed for developing a new generation of methods/techniques based on refinement of existing modern biotechnology methods designed for increasing phyto-improvement speed, accuracy and efficiency.
3.4 Foreign Genetic Material: This refers to exogenous stable and joint insertion in a genome of one (1) or more genes or DNA sequences forming part of a specific genetic construction.
3.5 Unintentional (Off-Target) Modifications: These occur when over the phyto-improvement process of a crop innovation techniques through modern biotechnology have been used and the final product—the crop—contains unplanned modifications in its genome.
3.6 Living Modified Organism (LMO): Any live organism having a new genetic material combination that has been obtained by means of applying modern biotechnology.

Article Four—Application: Individuals or companies interested in processing an application for having used innovation techniques over phyto-improvement of a crop through modern biotechnology with the final product not containing any foreign genetic material, shall be previously registered with ICA as seed producers or importers or as phyto-improvement research units and shall submit to ICA the respective application meeting the following requirements:

4.1 Full name or corporate purpose, address, telephone number, electronic mail address, name and identification of legal representative and proxy, if applicable, and full name or corporate purpose of developing entity.
4.2 For companies, certificate of existence and legal representation issued by the respective Chamber of Commerce with date no older than thirty (30) calendar days before submission of the application; and for individuals, commercial registration, RUT (Single Taxation Registration) or Colombian Citizenship Card.
4.3 The application shall indicate the number of the resolution whereby ICA granted registration of the business activity (seed producer or importer or phyto-improvement research unit).
4.4 Provide information related to:

- Crop: Taxonomic species classification, description of phenotype obtained and use.
- The improvement machinery used for obtaining the desired result, genetic map (s) of construction(s) used over the improvement process, including all present genetic elements, protein sequences and RNA used in the free DNA edition process.
- Details of any new characteristics or modifications of existing characteristics.
- Evidence of genetic changes present in the final product—molecular characterization, describing the number of genes, sites, loci or DNA sequences manipulated, location in the genome and, where applicable, identification of any unintentional (off-target) modifications.
- Analytical evidence showing that the improved crop (final product) does not contain any foreign genetic material.
- Evidence (by means of DNA sequences) that off-target sites, those that could have predictably been intentionally modified, did not sustain any changes in the improved crop.

Proviso One: The above information shall be submitted in Spanish attaching all reference articles/papers, as well as analytical results.

Proviso Two: Individuals or companies interested in conducting phyto-improvement research shall be previously registered with ICA as research units. In addition, if over the phyto-improvement (plant breeding) process any foreign genetic material was introduced, it shall be reported to ICA irrespective of the final product containing or not any such foreign genetic material. Furthermore, they shall count on authorization from ICA to perform any research activities on a confined basis in meeting the applicable regulations and the established biosafety plan.

Article Five—Processing Applications: Upon submitting an application, within a maximum period of thirty (30) working days counted as from the filing date thereof,

ICA shall review the information and documents described in Article Four of the present Resolution, as applicable, and shall require the interested party, where applicable, to clarify any of the information provided or attach any additional documents, for which a maximum period shall be granted up to thirty (30) working days counted as from the date when the communication is received.

Upon expiry of such period, if the interested party has not clarified the information or sent the documents required, it shall be considered that it desists from the application and ICA shall proceed with return thereof complete with the respective supporting documents within the next fifteen (15) working days, without prejudice to the interested party's right to submit a new application in meeting all the requirements set out in the present Resolution.

Article Six—Response To Application: Once the requirements set out in the above Article have been complied with, ICA shall carry out an assessment of the information received within a period no longer than sixty (60) working days, determining whether the new crop contains any foreign genetic material inserted in its genome due to the use of modern biotechnology techniques.

For a genetic change to be considered as foreign genetic material, it shall be analyzed whether a stable and joint exogenous insertion has been produced in one (1) or more genes or DNA sequences forming part of a specific genetic construction.

Upon determining the above or upon expiry of the period initially indicated, ICA shall inform the applicant in writing if the submitted crop is considered an LMO or not and consequently whether it is or not within the scope of regulation of Living Modified Organisms.

Proviso One: When dealing with a crop obtained by innovative phyto-improvement techniques through modern biotechnology and the final product does not contain any foreign genetic material, for marketing seeds from such crop the applicant shall comply with the provisions of ICA Resolution 3168 of 2015 or any other amending o replacing it or adding thereto, and when dealing with LMO it shall comply with the provisions in Decree 4525 of 2005 and regulatory resolutions thereto and ICA Resolution 3168 or 2015 or any standards amending or replacing it or adding thereto.

Proviso Two: An analysis conducted on one variety/hybrid shall be applicable to another such variety/hybrid of the same species as long as the second variety or both varieties are derived from the same parent as initial source of the new characteristic obtained by means of new phyto-improvement technologies derived from modern biotechnology. The Authority reserves the right to request more information in case it deems so relevant or convenient, on the basis of scientific criteria.

Proviso Three: ICA may request special follow up of any crop analyzed when the characteristics and/or novelty thereof merit so, on the basis of scientific and technical criteria.

Article Seven—Validity: The present Resolution shall govern as from the date of its publication in the Official Journal.

References

AOSCA (2018) Seed certification. https://www.aosca.org/programs-and-services/seed-certification/. Accessed 02 May 2018

ASA (2018) Cultivos Bt y manejo de resistencia de insectos. http://www.asa.org.ar/material-de-interes/biotecnologia/. Accessed 02 May 2018

Barfoot P, Brookes G (2014) Global socio-economic and environmental impacts 1996–2014. GM Crops & Food. http://www.pgeconomics.co.uk/pdf/2016globalimpactstudymay2016.pdf. Accessed 02 May 2018

Berto G (2018) Defienden los alimentos genéticamente modificados. Congreso de Bromatología analizó el tema. Afirman que los controles son los adecuados. https://www.rionegro.com.ar/sociedad/defienden-los-alimentos-geneticamente-modificados-AAHRN20071123102002. Accessed 02 May 2018

Biz P (2017) Biotecnología rosarina desarrollará una super soja resistente a herbicidas. https://puntobiz.com.ar/noticias/val/113299/val_s/129/biotecnologica-rosarina-desarrollara-una-super-soja-resistente-a-herbicidas.html. Accessed 02 May 2018

Burachik M (2012) Regulation of GM crops in Argentina. GM Crops Food 3(1):48–51

Burachik M, Traynor PL (2002) Analysis of a national biosafety system: regulatory policies and procedures in Argentina. ISNAR, The Hague. (No. 660.6/B945)

Bustamante E (2018) La Argentina tiene un plan para la biotecnología. CLARIN Rural. https://www.clarin.com/rural/argentina-plan-biotecnologia_0_HkIhhaOBf.html. Accessed 07 Feb 2018

Cascardo R, Gianni C, Piana JA (1998) Variedades vegetales en Argentina: El comercio de semillas y el derecho de obtentor (No. D50/39). Latín Gráfica, Buenos Aires

CBD (2000) Text of the Cartagena protocol on biosafety. http://bch.cbd.int/protocol/text/. Accessed 02 May 2018

CBD (2018a) Parties to the Cartagena protocol and its supplementary protocol on liability and redress. https://bch.cbd.int/protocol/parties/. Accessed 3 Mar 2018

CBD (2018b) COP-MOP decisions. https://www.cbd.int/decisions/mop. Accessed 02 May 2018

Co-Extra (2005) Comparative review of coexistence and traceability systems in selected countries of Europe, North and South America. Deliverable D7.27 of the GM and non-GM supply chains: their CO-EXistence and TRAceability. http://www.coextra.eu/deliverables/deliverable1348.pdf. Accessed 02 May 2018

COMTRADE (2018) UN Comtrade Database 2014-2017. https://comtrade.un.org/. Accessed 02 May 2018

CTNBIO (2018) Resolução Normativa N° 16, de 15 de janeiro de 2018. http://ctnbio.mcti.gov.br/inicio/-/asset_publisher/58KNi0CuF68J/content/resolucao-normativa-n%C2%BA-16-de-15-de-janeiro-de-2018. Accessed 02 May 2018

EFSA (2012) Scientific opinion ad-dressing the safety assessment of plants developed using Zinc Finger Nuclease 3 and other Site-Directed Nucle-ases with similar function. http://onlinelibrary.wiley.com/doi/10.2903/j.efsa.2012.2943/epdf. Accessed 02 May 2018

FAO (2000) Codex Alimentarius Commission. Comments submitted by Argentine Governments regarding FOODS DERIVED FROM BIOTECHNOLOGY, page 4 comments on GMO Labeling. http://www.fao.org/tempref/codex/Meetings/CCFBT/ccfbt1/bt0041ce.pdf. Accessed 02 May 2018

FAO (2011) Compilation of Codex texts relevant to the labelling of foods derived from modern biotechnology. http://www.fao.org/fao-who codexalimentarius/thematic-areas/biotechnology/en/. Accessed 02 May 2018

FAO (2014a) La FAO y Argentina refuerzan la seguridad en biotecnología. http://www.fao.org/director-general/newsroom/news/detail/es/c/264211/. Accessed 02 May 2018

FAO (2014b) Low levels of genetically modified crops in International Food and Feed Trade: FAO International Survey and Economic Analysis. www.fao.org/docrep/019/i3734e/i3734e.pdf. Accessed 02 May 2018

Gepts P, Hancock J (2006) The future of plant breeding. Crop Sci 46(4):1630–1634
Ghezan G, Tapia C (2006) Technical visit report - experiences of traceability: Non-GM Maize in Argentina', INTA. CO-ExTra Project
Ghezan G, Iriarte L, Tapia C, D'Annunzio J (2006) Argentine non-GMO Soybean Chain, INTA
H. Cámara de Diputados de la Nación (2010) Proyecto de Ley. "Etiquetado de alimentos que sean o contengan Organismos Genéticamente Modificados o sus derivados". http://www1.hcdn.gov.ar/proyxml/expediente.asp?fundamentos=si&numexp=8307-D-2010. Accessed 02 May 2018
H. Cámara de Diputados de la Nación (2013) Proyecto de Ley. Informacion al consumidor sobre etiquetado de alimentos y bebidas transgenicos o que contengan Organismos Geneticamente Modificados. http://www1.hcdn.gov.ar/proyxml/expediente.asp?fundamentos=si&numexp=2324-D-2013. Accessed 02 May 2018
ICA (2018) SPS Notification G/SPS/N/COL/282. https://docs.wto.org/dol2fe/Pages/FE_Search/FE_S_S009-DP.aspx?language=E&CatalogueIdList=243401&CurrentCatalogueIdIndex=0&FullTextHash=1&HasEnglishRecord=True&HasFrenchRecord=True&HasSpanishRecord=True. Accessed 02 May 2018
INDEC (2018) Database of the National Institute of Statistics and Surveys. https://www.indec.gob.ar/. Accessed 02 May 2018
INFOLEG (1973) Ley de Semillas y creaciones Fitogenéticas. http://servicios.infoleg.gob.ar/infolegInternet/anexos/30000-34999/34822/texact.htm. Accessed 02 May 2018
INFOLEG (1992) Ley de Ministros 22.520. http://servicios.infoleg.gob.ar/infolegInternet/anexos/45000-49999/48853/texact.htm. Accessed 02 May 2018
INFOLEG (1995) Constitucion de la Nacion Argentina. Ley N° 24.430. Publicación del texto oficial de la Constitución Nacional (sancionada en 1853 con las reformas de los años 1860, 1866, 1898, 1957 y 1994). http://servicios.infoleg.gob.ar/infolegInternet/anexos/0-4999/804/norma.htm. Accessed 02 May 2018
INFOLEG (1999) Produccion Ecologica, Biologica u Organica. Ley 25.127. http://servicios.infoleg.gob.ar/infolegInternet/anexos/55000-59999/59885/norma.htm. Accessed 02 May 2018
INFOLEG (2002a) Resolución 412/2002. http://servicios.infoleg.gob.ar/infolegInternet/anexos/70000-74999/74376/norma.htm. Accessed 02 May 2018
INFOLEG (2002b) Política Ambiental Nacional Ley 25.675. http://www.infoleg.gob.ar/infolegInternet/anexos/75000-79999/79980/norma.htm. Accessed 02 May 2018
INFOLEG (2004) Organismos Vegetales Geneticamente Modificados. Resolución 46/2004. http://servicios.infoleg.gob.ar/infolegInternet/anexos/90000-94999/92241/norma.htm. Accessed 02 May 2018
INFOLEG (2007) Ley de Promoción del Desarrollo y Producción de la Biotecnología Moderna, Ley 20270. http://www.infoleg.gob.ar/infolegInternet/anexos/130000-134999/130522/norma.htm. Accessed 02 May 2018
INFOLEG (2009) CODIGO PENAL Ley 26.524 Modificación. http://servicios.infoleg.gob.ar/infolegInternet/anexos/155000-159999/159680/norma.htm. Accessed 02 May 2018
INFOLEG (2011a) Resolución 763/2011. http://servicios.infoleg.gob.ar/infolegInternet/anexos/185000-189999/185806/norma.htm. Accessed 02 May 2018
INFOLEG (2011b) Subsidiary regulations to Resolución 763/2011. http://servicios.infoleg.gob.ar/infolegInternet/verVinculos.do?modo=2&id=185806. Accessed 02 May 2018
INFOLEG (2011c) Resolución 701/2011. http://servicios.infoleg.gob.ar/infolegInternet/anexos/185000-189999/189067/norma.htm. Accessed 02 May 2018
INFOLEG (2011d) Resolución 510/2011. http://servicios.infoleg.gob.ar/infolegInternet/anexos/185000-189999/185853/norma.htm. Accessed 02 May 2018
INFOLEG (2012a) Resolución 241/2012. http://servicios.infoleg.gob.ar/infolegInternet/anexos/195000-199999/198153/norma.htm. Accessed 02 May 2018
INFOLEG (2012b) Resolución 49/2012. http://servicios.infoleg.gob.ar/infolegInternet/anexos/190000-194999/194394/norma.htm. Accessed 02 May 2018
INFOLEG (2013a) Resolución 17/2013. http://servicios.infoleg.gob.ar/infolegInternet/anexos/220000-224999/224874/norma.htm. Accessed 02 May 2018

INFOLEG (2013b) Resolución N° 498/2013. http://servicios.infoleg.gob.ar/infolegInternet/anexos/220000-224999/224698/norma.htm. Accessed 02 May 2018

INFOLEG (2015) Sanidad de los Animales y Vegetales Ley 27233. http://servicios.infoleg.gob.ar/infolegInternet/anexos/255000-259999/257451/norma.htm. Accessed 02 May 2018

INFOLEG (2017a) Resolución 79-E/2017. http://servicios.infoleg.gob.ar/infolegInternet/anexos/285000-289999/287337/norma.htm. Accessed 02 May 2018

INFOLEG (2017b) Resolución Conjunta 9-E/2017. http://servicios.infoleg.gob.ar/infolegInternet/anexos/280000-284999/280580/norma.htm. Accessed 02 May 2018

INFOLEG (2018) Resolución 26/2018. http://servicios.infoleg.gob.ar/infolegInternet/anexos/310000-314999/310441/norma.htm. Accessed 02 May 2018

Ingrassia V (2017) Avance científico mundial: investigadores argentinos lograron generar clones equinos con genomas mejorados. https://www.infobae.com/salud/ciencia/2017/12/04/avance-cientifico-mundial-investigadores-argentinos-logran-generar-clones-equinos-con-genomas-mejorados/. Accessed 02 May 2018

ISAAA (2016) International service for the acquisition of agri-biotech applications. Ithaca, NY. Global Status of Commercialized Biotech/GM Crops: 2016. ISAAA Brief No. 52. ISAAA. https://www.isaaa.org/resources/publications/briefs/52/download/isaaa-brief-52-2016.pdf. Accessed 02 May 2018

ISF (2018) Technological advances drive innovation in plant breeding to create new varieties. http://www.worldseed.org/our-work/plant-breeding/plant-breeding-innovation/. Accessed 02 May 2018

Ishii T, Araki M (2017) A future scenario of the global regulatory landscape regarding genome-edited crops. GM Crops Food 8(1):44–56

Longoni M (2017) La "edición genética", una nueva técnica que alborota a la industria semillera. Infobae. http://bichosdecampo.com/la-edicion-genetica-una-nueva-tecnica-que-alborota-a-la-industria-semillera/. Accessed 02 May 2018

Lusser MP, Parisi C, Plan D, Rodriguez Cerezo E (2011) New plant breeding techniques – state-of-the-art and prospects for commercial development. http://ftp.jrc.es/EURdoc/JRC63971.pdf. Accessed 02 May 2018

MINAGRO (2018) List of GM events approved in Argentina. https://www.agroindustria.gob.ar/sitio/areas/biotecnologia/ogm/. Accessed 02 May 2018

MINCYT (2015) Encuesta Exploratoria Sobre Percepción Pública de la Biotecnología Alimentaria en la Argentina. http://indicadorescti.mincyt.gob.ar/documentos/Informe_Percepcion_Biotecnologia_Alimentos.pdf. Accessed 02 May 2018

OECD (2015) Workshop on environmental risk assessment of products derived from novel plant breeding techniques. https://one.oecd.org/document/ENV/JM/MONO(2016)5/en/pdf. Accessed 02 May 2018

OECD (2018) OECD seed schemes: rules and regulations. http://www.oecd.org/tad/code/oecdseedschemesrulesandregulations.htm. Accessed 02 May 2018

Penna JA, Lema MA (2003) Adoption of herbicide tolerant soybeans in Argentina: an economic analysis in the economic and environmental impacts of Agbiotech. Springer, Berlin, pp 203–221

Pensel N, Gargiulo G, Barrios M (2007) Traceability, coexistence systems and practices in Argentina. INTA

Phillips PWB, Smyth S (2002) Product differentiation alternatives: identity preservation, segregation, and traceability. AgBioforum 5(2):30–42

Preciado J (2015) La biotecnología después de la transgenia. Infocampo. http://www.infocampo.com.ar/la-biotecnologia-despues-de-la-transgenia/. Accessed 02 May 2018

Roman V (2017) Edición genética en Argentina: cómo se usa para beneficiar a pacientes, mascotas y productores agropecuarios. Infobae. https://www.infobae.com/salud/ciencia/2017/08/17/edicion-genetica-en-argentina-como-se-usa-para-beneficiar-a-pacientes-mascotas-y-productores-agropecuarios/. Accessed 02 May 2018

Rosada C (2018) Ley de desarrollo y promoción de biotecnología. https://www.casarosada.gob.ar/informacion/eventos-destacados-presi/41676-se-reglamento-la-ley-de-desarrollo-y-promocion-de-biotecnologia. Accessed 18 Jan 2018

Rumi MJ (2018) La biotecnología argentina apunta al exterior. La Nación. https://www.lanacion.com.ar/2101488-la-biotecnologia-argentina-apunta-al-exterior. Accessed 02 May 2018

SAG (2017) Aplicabilidad de Resolución N° 1.523/2001 en material de propagación desarrollado por nuevas técnicas de fitomejoramiento. http://www.sag.gob.cl/ambitos-de-accion/aplicabilidad-de-resolucion-ndeg-15232001-en-material-de-propagacion-desarrollado-por-nuevas-tecnicas-de-fitomejoramiento. Accessed 02 May 2018

Schiavone E, Morón P, Lema M (2013) Normas locales sobre identificación de "alimentos transgénicos" y el "derecho a la información del consumidor" Revista Alimentos Argentinos N° 32. http://www.alimentosargentinos.gob.ar/HomeAlimentos/Publicaciones/revistas/nota.php?id=436. Accessed 02 May 2018

SENASA (2018) Se constató la destrucción de 147 hectáreas de alfalfa transgénica ilegal. http://www.senasa.gob.ar/senasa-comunica/noticias/se-constato-la-destruccion-de-147-hectareas-de-alfalfa-transgenica-ilegal. Accessed 02 May 2018

Sprink T, Metje J, Schiemann J, Hartung F (2016) Plant genome editing in the European Union to be or not to be a GMO. Plant Biotechnol Rep 10(6):345–351

Trigo EJ (2016) Veinte Años de Cultivos Genéticamente Modificados en la Agricultura Argentina. http://argenbio.org/index.php?action=novedades&note=747. Accessed 02 May 2018

Trigo E, Chudnovsky D, Cap E, López A (2002) Los transgénicos en la agricultura argentina. Una historia con final abierto. Libros Del Zorzal, Buenos Aires

Valor Soja (2017) Incorporaron al aceite de soja alto oleico al Código Alimentario Argentino. https://www.valorsoja.com/2017/10/09/incorporaron-al-aceite-de-soja-alto-oleico-al-codigo-alimentario-argentino-estaria-disponible-en-los-proximos-anos/. Accessed 02 May 2018

Whelan AI, Lema MA (2015) Regulatory framework for gene editing and other new breeding techniques (NBTs) in Argentina. GM Crops Food 6:253–265

Whelan AI, Lema MA (2017) A research program for the socioeconomic impacts of gene editing regulation. GM Crops Food 8(1):74–83

WHO (2018) Laboratory biosafety manual. http://www.who.int/csr/resources/publications/biosafety/WHO_CDS_CSR_LYO_2004_11/en/. Accessed 02 May 2018

WTO (2018) Measures affecting the approval and marketing of biotech products. https://www.wto.org/english/tratop_e/dispu_e/cases_e/ds293_e.htm. Accessed 02 May 2018

Chapter 3
Regulation of Genome Editing in Plant Biotechnology: Australia

Karinne Ludlow

Abstract Two Australian regulatory frameworks are fundamental to the cultivation of genome edited plants and marketing of their products. These are the gene technology and food regulatory frameworks. Both frameworks rely on process triggers—that process being the use of gene technology. Unfortunately, these frameworks use different definitions of gene technology, creating the likelihood of different responses to genome edited plants, particularly to plants produced using SDN-2 or ODM.

No genome edited plants are currently cultivated in Australia but the relevant regulators are each currently undertaking reviews to determine whether some or all genome editing techniques are gene technology and how their frameworks should respond to those techniques. Final decisions are expected during 2018. In the meantime, the regulators have each adopted interim approaches to genome edited plants or their products, summarised in the first table.

3.1 Introduction

Australia approved the commercial release of its first GM plant, GM carnations, in 1995.[1] In 1996, the approved commercial release of GM cotton made Australia one of the first six countries to commercialise GM field crops[2] and by 2016, Australia ranked eleventh in countries planting GM crops.[3]

As at November 2017, there were 171 accredited organisations and 26 'other' organisations authorised by the Australian national gene technology regulator to deal

This paper was correct at the time of writing, April 2018.

[1]Huttner (1997), p. 10.

[2]ISAAA (2016), p. 58.

[3]ISAAA (2016), pp. 58–59.

K. Ludlow (✉)
Faculty of Law, Monash University, Melbourne, VIC, Australia
e-mail: karinne.ludlow@monash.edu

H.-G. Dederer, D. Hamburger (eds.), *Regulation of Genome Editing in Plant Biotechnology*, https://doi.org/10.1007/978-3-030-17119-3_3

with GMOs. 119 GM events for food, feed, and/or cultivation are approved under the gene technology regulatory framework. These include alfalfa (3 events), Argentine canola (21), carnation (12), cotton (24), maize (27), potato (10), rice (1), rose (1), soybean (17), sugar beet (2), and wheat (1).[4] However, over 97% of authorisations for dealings with GMOs over the past five years have been for notifiable low risk dealings, a category imposing minimal regulatory burden and not permitting release of the GMO into the environment.[5]

Nevertheless, approvals to release GM crops into the environment are possible under the Australian gene technology framework, and such approvals can authorise limited field trials or general commercial release. Varieties of canola and cotton altered to be resistant to particular pests and/or herbicides are the only GM crops approved for commercial release to date.[6] An application for commercial release of safflower modified for high oleic acid composition is under evaluation.[7] Field trials are currently underway for GM banana, barley, perennial ryegrass, safflower, sugarcane, wheat and white clover.[8] Licenses have also been granted for field trials for Indian mustard, potato, and sorghum.[9] In addition, GM products can be used in Australia as ingredients in foods, including GM varieties of soybean, corn, potato, sugar beet, wheat and rice and for the production of stockfeed, including GM cottonseed meal, imported GM soybean and GM canola meal.[10]

GM crops are estimated to have enhanced Australian farm income by AUS$1.37 billion in the period 1996 to 2015 and the benefits for 2015 alone is estimated at AUS $64.1 million.[11] The adoption rate of GM crops in Australia continues to grow, that rate being 36% in 2016.[12] This comprised 852,000 hectares of GM cotton and canola; 405,000 hectares cotton and 447,000 hectares canola.[13]

GM cotton has a 98% adoption rate, meaning 405,000 hectares of the 413,000 total hectares planted to cotton in 2016, were planted to GM cotton.[14] The adoption rate for GM canola was 23% in that year, being 447,725 hectares of the total canola area of 1.95 million hectares. Herbicide tolerant canola was grown in three states: New South Wales (NSW), Victoria and Western Australia. Farmers in Western Australia grew 346,000 hectares (30% of total canola) of biotech canola, 46,582 hectares (16%) in Victoria, and 55,143 hectares (11%) in NSW. Biotech canola was

[4]ISAAA (2016), pp. 58–59.

[5]OGTR (2017a), p. 16.

[6]Productivity Commission (2016), p. 264.

[7]OGTR (2018), Licence No DIR 158.

[8]Productivity Commission (2016), p. 264.

[9]OGTR (2018), License Nos DIR 149, 150 and 153.

[10]Productivity Commission (2016), p. 264.

[11]Brookes (2016), p. 6.

[12]ISAAA (2016), pp. 58–59.

[13]ISAAA (2016), pp. 58–59.

[14]The data is this paragraph is from ISAAA (2016), pp. 59–60.

planted by more than 1000 Australian farmers in 2016 with more than 180 growers planting it for the first time. In recent research by Brookes, since 2008, the average yield gain from biotech canola technology has been 11%. This has resulted in an additional 226,000 tonnes of canola produced in the country.[15]

3.2 The Regulatory Framework for GMOs: An Overview (Table 3.1)

Regulatory responsibility around agricultural produce and gene technology in Australia is shared between the federal government and the eight state and territory governments.[16] Nationally consistent regulation therefore requires inter-governmental cooperation and sharing of responsibilities. Generally, a federal act is enacted, which is mirrored in counterpart legislation in each state and territory. This creates the structure for each of the regulatory frameworks of most relevance to genome edited plants: the regulatory frameworks for gene technology, food, and agricultural and veterinary (agvet) chemicals.[17]

In each case a national regulator is established to administer the scheme, that regulator being independent of government. However, each scheme is subject to broader policy guidance by ministerial forums, comprising ministers from federal and state/territory governments. The food and agvet regulators are required to consult with the gene technology regulator before making decisions regarding GMOs or their products.

Regulation around environmental protection is more complex. The regulatory frameworks for gene technology and for agvet chemicals both require that the environment be protected. Additional specific federal and state legislation is directed at environmental protection more generally.[18] However, the federal environmental protection legislation is triggered only where a matter of national environmental significance will be significantly impacted. The limited scope of these matters means the federal environmental protection legislation is unlikely to be relevant to genome edited plants.[19] The state-based environmental legislation is potentially legally

[15]Brookes (2016), p. 7.

[16]There are also over 600 local governments. They have limited authority on the issues relevant here and are not considered here.

[17]The frameworks differ in regards to responsibility for monitoring and enforcement. Monitoring and enforcement of the food regulatory regime is carried out by the states (and local governments in some states); with respect to the agvet chemicals framework, national responsibility continues until the point of sale, when responsibility reverts to the states/territories; federal officers are responsible for monitoring and enforcement of the gene technology regime.

[18]*Environment Protection and Biodiversity Conservation Act 1999* (Cth). An example of such state legislation is the *Environmental Protection Act 1970* (Vic).

[19]Ludlow (2005a).

Table 3.1 Regulatory framework for GMOs

Authority	Area of responsibility	Pertinent legislation
Gene Technology Regulator (GTR) [within the Australian Government Department of Health]	Protection of human health and safety and the environment re gene technology	*Gene Technology Act 2000* (Cth)
Food Standards Australia New Zealand (FSANZ) [within the Australian Government Department of Health]	Protection of human health and safety through food standards, relevantly including on: – Foods produced using gene technology – Novel foods	*Food Standards Australia New Zealand Act 1991* (Cth)
Australian Pesticides and Veterinary Medicines Authority (APVMA) [within the Australian Government Department of Agriculture and Water Resources]	Protection of human health and safety and the environment regarding: – agricultural chemical products (including fungicides, herbicides, plant nutrients) – veterinary chemical products (including veterinary biologics, vaccines)	*Agricultural and Veterinary Chemicals (Administration) Act 1992* (Cth) *Agricultural and Veterinary Chemicals Code Act 1994* (Cth)

relevant but given the past approach to GM crops generally, is unlikely to be applied with respect to genome edited plants.[20]

3.3 Regulatory Status of Genome Edited Plants

3.3.1 Applicability of the Regulatory Framework for GMOs

3.3.1.1 Gene Technology

The Australian gene technology regulatory framework commenced operation on 21 June, 2001.[21] The centrepiece of that framework is the national *Gene Technology Act 2000* (Cth) (GT Act) which establishes the Gene Technology Regulator (GTR). The GTR is an independent statutory office holder and heads the national Office of the GTR (OGTR), a federal regulatory agency within the federal government Department of Health. Each Australian state and territory then has mirroring legislation, in turn referring powers to the national GTR.[22] This approach was intended to create a nationally consistent approach to gene technology regulation across

[20]Ludlow (2004a).

[21]Australia had a voluntary self-regulatory system for GM research from the mid-1970s until the introduction of the gene technology regulatory framework.

[22]Although only in New South Wales and Northern Territory are amendments to the federal legislation automatically adopted into state mirroring legislation. In the other jurisdictions, periods

Australia although as discussed below, additional regulations imposed by some states means that this is not the case.[23]

All dealings with GMOs in Australia are prohibited *unless* authorised under the GT Act. Australian regulation also allows for strict controls on adopters even after the relevant crop has been approved for release into the environment.[24] Possible controls include prohibition on cultivation of GM crops in specific areas, information duties, and technical segregation measures such as imposition of buffer zones around GM crops, restrictions on the time of year or place where GM crops are grown, and ongoing monitoring of fields for volunteers or the spread of GM material from the authorised area.

GMO dealings are authorised only if they fall within one of four defined categories and relevant regulatory conditions are satisfied.[25] The term 'dealings' includes most uses of GMOs, including research, development, production of the GMO, import and commercial release. Approval conditions become more onerous as perceived risk increases and only the final two categories allow release into the environment. The four categories are:

1. *Exempt dealings*, assessed as posing negligible risks, and do not require approval from or notification to the GTR.
2. *Notifiable low risk dealings* (NLRDs), assessed as posing low risk provided regulatory conditions are met, and must be assessed by an Institutional Biosafety Committee (IBC) and notified to the GTR annually.
3. *Dealings on the GMO Register*, determined by the GTR to pose minimal risk and not requiring a licence to adequately protect human health and safety or the environment.
4. *Licensed Dealings.*

Two types of licences can be issued by the GTR. Assessment of applications for both types of licences requires case-by-case assessment by the GTR and licensed dealings must be carried out in accordance with tailored licence conditions. Licences for *Dealings Not Involving Intentional Release* (DNIRs) to the environment, are for

---

of legislative inconsistency (and uncertainty), occur as each jurisdiction arranges for the passage of new amendments to the federal legislation. Productivity Commission (2016), p. 281.

[23] *Gene Technology Act 2000* (Cth) s 5. The state and territory governments also agreed to maintain legislation that corresponds with the Commonwealth Gene Technology Act under the intergovernmental Gene Technology Agreement 2001. Nevertheless, there are differences in the corresponding mirroring legislation.

[24] *Gene Technology Act 2000* (Cth) s 62.

[25] There are two additional avenues to authorisation but these are exceptional. These are licenses for inadvertent dealings, pursuant to which the GTR can grant a licence to enable the disposal of a GMO inadvertently present on the applicant's land; and an Emergency Dealing Determination (EDD) which is a legislative instrument made by the Minister under section 72 of the Act to expedite approval of dealings with a GMO in an emergency. This was introduced during the 2007 outbreak of equine influenza, to enable the import of a vaccine. Four inadvertent dealings licences were issued in 2016–2017 following the unknowing importation of GM petunias which had not been authorised under the regulatory framework, to allow all plants and seeds to be disposed of.

contained dealings that 'do not meet the criteria for classification as exempt dealings, NLRDs or DIRs.'[26]

Dealings Involving Intentional Release (DIRs) licences authorise release into the environment.[27] There are two categories of DIRs: Limited and controlled release for field trials and large-scale applications for commercial release.

As noted above, while a nationally consistent framework for the regulation of gene technology was intended, this has not eventuated due to inconsistent state/territory government interventions.[28] The limited scope of the GTR's assessment process and the proposed commercial release of GM canola in 2003 were important in the decision by some states/territories to introduce their own additional legislation. Public consultation during the framework's creation showed the Australian public considered ethical issues equally important to safety and environmental concerns. However, the final national regulatory framework does not address ethical or social issues. Importantly, possible trade and marketing ramifications of GMO releases on other forms of agriculture, such as non-GM agriculture, are not considered.

State/territory legislation was introduced in some jurisdictions, purporting to address these other concerns. This legislation resulted in bans or 'moratoria' on certain GMO releases in every Australian state and territory, except Queensland and the Northern Territory where the trigger for these moves, canola, cannot be grown. The purpose of these moratoria was to preserve the identity of non-GMOs for marketing purposes and provide time to consider the socio-economic ramifications of GMO releases. It was not for safety reasons.[29]

The moratoria have now expired in all states and territories except Australian Capital Territory, South Australia and Tasmania, where the bans are expected to continue until at least 2025 in South Australia and 2019 in Tasmania. All three jurisdictions prohibit certain dealings with GMOs unless licensed by the GTR *and* authorised under state legislation. In Tasmania, this applies to all dealings including research and to all GMOs. The Australian Capital Territory and South Australian ban applies, in contrast, only to the *cultivation* of GM food crops, with food crop being defined to include crops intended for animal consumption.

Although the majority of moratoria have expired, the legislation allowing the introduction of additional moratoria remains in place in all states/territories except Western Australia, which repealed its legislation. The state/territory legislation therefore remains relevant if, and when, the commercial release of GM crops other than canola and cotton is authorised under the national scheme. Confusingly, the legislation differs between the states in terms of what GMOs are regulated, reflecting

---

[26]Australian Department of Health (2018), p. 98.

[27]The legislation explains that 'a dealing with a GMO involves the *intentional release of the GMO into the environment* if the GMO is intentionally released into the open environment, whether or not it is released with provision for limiting the dissemination or persistence of the GMO or its genetic material in the environment.' *Gene Technology Act 2000* (Cth) s 11.

[28]Croplife Australia (2017), p. 1.

[29]Ludlow (2004b).

the differing needs of the states/territories. Western Australia, for example, has a greater weed problem than other states leading to greater concern about the development of herbicide tolerance. States/territories also differ in the crops grown in their jurisdiction and their export markets.

The moratoria and moratoria legislation continue to raise concerns. They were the main technology-related regulatory issue raised by participants to the 2016 Australian Productivity Commission inquiry into agricultural regulation.[30] The Productivity Commission is the Australian Government's independent research and advisory body on a range of economic, social and environmental issues affecting the welfare of Australians. Its role, expressed most simply, is to help governments make better policies, in the long term interest of the Australian community. That inquiry recommended the repeal of all state/territory legislation imposing or giving power to impose moratoria.[31] Repeal was recommended to end the need for farm businesses wanting to use new GM-related biotechnologies to seek approval from two levels of government, namely the GTR and the relevant state/territory authorities. Further, the moratoria prevent the transport of GM crops and seeds through certain states, increasing transport costs. For example, the Australian Seed Federation told the Commission that because of the moratorium in South Australia, 'canola seed companies/producers in the Eastern states and Western Australia are now forced to ship GM canola seed by sea or move by road transport through Darwin, avoiding the natural transport route through South Australia'.[32] If the legislation is not repealed, the definitions relevant to the scope of the moratoria legislation (such as gene technology and GMO) will need to be reviewed to assess whether genome edited plants are effected.

3.3.1.1.1 Triggers for Gene Technology Regulatory Framework

Although the full name of the GT Act is 'An Act to regulate activities involving gene technology, and for related purposes', the GTR's powers relate to dealings with GMOs and not with gene technology. As the GTR has noted, the protections of the public and the environment by the scheme commences when gene technology is used to modify an organism[33] and the potential risks focused on are those posed by the organism itself.[34]

GMO is defined in the GT Act.[35] That definition is in the box below.

[30]Productivity Commission (2016), p. 265.

[31]Productivity Commission (2016), recommendation 6.1.

[32]Productivity Commission (2016), p. 281.

[33]OGTR (2017b), p. 23.

[34]OGTR (2017b), p. 24.

[35]'GMO' is itself defined in the legislation as 'genetically modified organism'. *Gene Technology Act 2000* (Cth) s 10(1).

Section 10(1), GT Act 2000 (Cth) definition of 'GMO'
Genetically modified organism means:

(a) an organism that has been modified by gene technology; or
(b) an organism that has inherited particular traits from an organism (the initial organism), being traits that occurred in the initial organism because of gene technology; or
(c) anything declared by the regulations to be a genetically modified organism, or that belongs to a class of things declared by the regulations to be genetically modified organisms;

but does not include:

(d) a human being, if the human being is covered by paragraph (a) only because the human being has undergone somatic cell gene therapy; or
(e) an organism declared by the regulations not to be a genetically modified organism, or that belongs to a class of organisms declared by the regulations not to be genetically modified organisms.

Paragraph (b) makes it clear that progeny of an initial founder GMO are included themselves as GMOs. However, to be included as GMOs, progeny must inherit the relevant GM trait. As will be discussed below, the food regulatory definitions do not distinguish between the initial and progeny organisms, all of these being treated as an organism modified by gene technology. This difference has possibly created or added to the perceived uncertainty around the meaning of paragraph (b) and whether it would exclude null segregants as GMOs. As discussed in section F, the GTR intends amending the GT Regulations to clarify that null segregants are not GMOs.

The definition of GMO also illustrates that the Australian gene technology framework is applicable to all species, including humans.[36] The GMO definition would include humans (under paragraph (d)) where they have undergone germline gene therapy. However, human germline gene therapy is prohibited in Australia by other legislation and so no such human being should be captured by the definition. As discussed in Sect. 3.6, this aspect of the legislation is part of the broader policy concerns which could be addressed in the GT Regulatory Scheme Review, around effective regulation of gene technology when it is used in the different fields of agriculture, medicine and environmental science.

There are no declarations for the purposes of paragraph (c) of the above definition, although declarations have been proposed as part of the reforms around genome

[36]'Organism' is defined in the GT Act as 'any biological entity that is: (a) viable; or (b) capable of reproduction; or (c) capable of transferring genetic material'. *Gene Technology Act 2000* (Cth) s 10 (1), definition of 'organism'.

editing discussed in Sect. 3.6. A list of organisms that are not GMOs for the purposes of paragraph (e) is contained in the GT Regulations.[37] Relevantly this includes:

> Schedule 1, GT Regulations 2001 (Cth)
> Relevant exceptions to s 10(1)(e) definition of 'GMO'
> Item 1—A mutant organism in which the mutational event did not involve the introduction of any foreign nucleic acid (that is, non-homologous DNA, usually from another species).
> ...

Gene technology is also defined in the GT Act. That definition is in the box below.

> Section 10(1), GT Act 2000 (Cth) definition of 'gene technology'
> ***Gene technology*** means any technique for the modification of genes or other genetic material, but does not include:
>
> (a) sexual reproduction; or
> (b) homologous recombination; or
> (c) any other technique specified in the regulations for the purposes of this paragraph.

The GT Regulations then list techniques that are not gene technology.[38] Relevantly, the Regulations exclude natural mutagenesis and mutagenesis induced by particular named methods: electromagnetic radiation (item 2), particle radiation (item 3) and chemical (item 4).[39]

> Schedule 1A, GT Regulations 2001 (Cth)
> Relevant exceptions to s 10(1) definition of 'gene technology'
> Item 2—Electromagnetic radiation-induced mutagenesis.
> Item 3—Particle radiation-induced mutagenesis.

(continued)

[37]Gene Technology Regulations 2001 (Cth) reg. 5 and Schedule 1.

[38]Gene Technology Regulations 2001 (Cth) reg. 4 and Schedule 1A.

[39]While the definitions in the legislation have not been amended since their introduction in 2000, the Regulations (and the Schedules that list the exclusions from those definitions) were amended in 2006. These changes resulted in the provisions described here. The original version of reg. 4 listed only somatic cell nuclear transfer (cloning) as being excluded from gene technology and there was no schedule of other exclusions. The 2006 amendments replaced reg. 4 with the current reference to techniques listed in Sch. 1A, that Schedule also being added at that time. The further amendments in 2011 to Sch. 1 are not relevant here.

Item 4—Chemical-induced mutagenesis.

. . .

Item 10—A natural process, if the process does not involve genetically modified material.

Examples: Examples of natural processes include conjugation, transduction, transformation and transposon mutagenesis.

The development of genome editing techniques has highlighted a number of ambiguities in the current GT regulatory framework important in the classification of genome edited plants. In particular, ambiguities arise in the definition of GMO and in one exception to that definition.

1. Definition of GMO, paragraphs (a) and (b)—The phrase 'an organism that has been *modified* by gene technology' in paragraph (a) of the definition of GMO is ambiguous because it is unclear whether that modification must be permanent. As discussed below, it seems that while the new DNA is present, the organism is a GMO; if the introduced DNA is later removed, the organism ceases to be a GMO.

   Paragraph (b) concerns progeny of these organisms. Genome editing may be undertaken on the initial organism to enable some other breeding process or objective to be achieved. At a later stage, the gene construct introduced during genome editing may be removed or progeny selected which have not inherited the construct. Accelerated breeding following induction of early flowering is an example of such a process which could be used to facilitate the introduction of a disease resistance gene using traditional cross breeding. It is arguable that such progeny do inherit a particular trait from the initial organism (being the disease resistance) that occurred in the initial organism because of gene technology given the shortened flowering time was enabled by genome editing.
2. Exception to the definition of 'GMO' for certain mutated organisms, Schedule 1—The exception for mutant organisms 'in which the mutational event did not involve the introduction of any foreign nucleic acid (that is, non-homologous DNA, usually from another species)' is particularly problematic. First, the reference to 'nucleic acid' could include DNA or RNA. However, because of the bracketed text following that term (namely, '(that is, non-homologous *DNA*,. . .') it is arguable that nucleic acid here is limited to DNA. Secondly, 'introduction' is ambiguous as to whether the DNA must be included in the plant's genome and whether that inclusion must be permanent. Finally, the term 'foreign DNA' has been acknowledged by the GTR as being ambiguous because it is unclear whether it includes the introduction of some ODM (whether RNA or DNA) or SDNs.[40] Further, it is unclear if foreign means foreign to that individual or to the species concerned. This is significant for cisgenesis and intragenesis, where genes (and functional elements for intragenesis) from the same species or species that

[40]OGTR (2017c), p. 3.

are crossable are used. The use of 'usually' arguably indicates that the non-homologous DNA does not have to be from another species.

3.3.1.2 Food

Non-living food products produced from GMOs (for example, canola oil from GM canola plants), are regulated by food standards which are developed by Food Standards Australia New Zealand (FSANZ). FSANZ is a bi-national statutory authority, within the Australian Government Department of Health. Since May 1999, GM foods must be assessed and approved by FSANZ prior to initial commercial release onto the Australian or New Zealand market.[41] There are also specific regulations concerning labelling of such food.

Standard 1.5.2 – Food Produced Using Gene Technology was included in the *Australia New Zealand Food Standards Code* in 1999 and revised in 2016. Food produced using gene technology cannot be sold unless expressly permitted by, and listed in, Schedule 26 of the Code. Over 70 approved foods are listed in the standard. It is an offence under Australian Commonwealth, state and territory (and New Zealand) food laws to not comply with the Code.[42]

Section 1.5.2-3 of Standard 1.5.2 concerns the sale of foods covered by the standard and section 1.5.2-4 provides for labelling requirements. A food for sale may consist of, or have as an ingredient, a food produced using gene technology if the food has been specifically approved by FSANZ and complies with any specific conditions. Approval requires a risk based case-by-case safety assessment by FSANZ and approval by the high level intergovernmental committee which oversights food regulation and sets policy, the Australia and New Zealand Ministerial Forum on Food Regulation. The FSANZ safety assessment is undertaken in accordance with procedures outlined in the FSANZ Application Handbook. These procedures are consistent with internationally established scientific principles and guidelines developed through the work of the Organisation for Economic Cooperation and Development (OECD), the Food and Agriculture Organisation (FAO) of the United Nations, the World Health Organisation (WHO) and the *Codex Alimentarius* Commission and is discussed in Sect. 3.4 below. As part of that safety approval, special conditions may be imposed on the sale of the food, such as special labelling. For example, approval for sale requires that the food have been determined to be at least as safe as its traditional counterpart. However, if it contains a factor known to cause an allergic reaction in some part of the population, it may need appropriate labelling. Labelling may also be required where the food has altered

[41]Despite finding there was no intrinsic difference between GM food and other food produced using random breeding practices, the Australian food regulator nevertheless recommended that GM food should be specifically authorised before use for food. At that time, there was no monitoring of any new food products regardless of how they were produced. That has now been changed by the addition of a novel foods standard.

[42]*Australia New Zealand Food Standards Code,* s 1.1.1-10.

characteristics, such as altered composition or nutritional profile, when compared with existing counterpart food that is not produced using gene technology. The generally applicable, non-safety, labelling requirements for GM food are discussed in Sect. 3.8 below.

3.3.1.2.1 Triggers for Food Regulatory Framework

The trigger for application of the food regulatory framework to food from GM crops is that the food was produced using gene technology. This term and gene technology are defined in the Code. Those definitions are in the box below. Explanatory notes in the Standard make it clear that food derived from animals fed feed produced using gene technology is not food produced using gene technology, unless the animals are themselves products of that technology. The term also does not include food from organisms administered GM agents, such as GM veterinary products, via non-genetic routes.

> Standard 1.1.2, Australia New Zealand Food Standards Code
> ***Food produced using gene technology*** means food which has been derived or developed from an organism which has been modified by gene technology.
> ***Gene technology*** means recombinant DNA techniques that alter the heritable genetic material of living cells or organisms.

The definition of food produced by gene technology is essentially the same in the Code as the definition of GMO in the GT Act. However, there is no separate reference to progeny in the Code's definition.

More importantly, the definition of the process used to trigger application of the food regulatory framework—gene technology—is different to the definition of the same term in the GT Act. The definition in the food Code is narrower than that in the GT Act, being limited to 'recombinant DNA techniques', a term that is not itself defined. FSANZ has advised that the definition of recombinant DNA techniques is generally taken to mean the recombining or joining of DNA from two or more sources and inserting it into an organism.[43] The core of the trigger is that the organism contains new pieces of DNA in its genome. That new piece of DNA can be derived from any source, including the same species.[44]

Further, unlike the GT Act, the food Code arguably divides methods to produce plants into only two groups—those that are gene technology and those that are not, which are called conventional breeding. This conclusion is suggested because of the definitions included in Schedule 26, including a definition for conventional breeding. Schedule 26 lists those foods produced using gene technology approved for sale.

---

[43] FSANZ (2018a), fn. 8.

[44] FSANZ (2018a), pp. 7 and 10.

The Schedule lists those foods by reference to the particular line tested and approved, which is described by reference to the particular genetic modification made to the plant. For the purposes of that list, the Schedule includes additional definitions that are instructive in understanding the meaning of gene technology.

Food produced by organisms created using conventional breeding methods are not covered by the standard although general principles of food safety continue to apply. Such food is presumed to be safe based on a history of safe use compared with that produced using gene technology, which is not presumed to be safe and needs a formal risk assessment before introduction into the market. That assessment is discussed in Sect. 3.4.

Schedule 26, Australia New Zealand Food Standards Code, section S26-2(2)
conventional breeding means all methods used to produce plants, excluding techniques that use gene technology.
line means:

(a) a plant, the genetic material of which includes a transformation event or events; or
(b) any plant, descended from the plant referred to in paragraph (a), that is the result of conventional breeding of that plant with:
    (i) any other plant that does not contain a transformation event or events; or
    (ii) any other plant that contains a transformation event or events, whether expressed as a line or event, that is listed in the table to section S26—3;
    (iii) but shall not be taken to mean any plant derived solely as a result of conventional breeding.

transformation event means a unique genetic modification arising from the use of gene technology.

Unlike the definition of food produced using gene technology, the definition of line clearly includes progeny. Further, progeny are included whether they carry the modification made to their parent(s) or not.

3.3.1.3 Agricultural and Veterinary Chemicals

The Australian Pesticides and Veterinary Medicines Authority (APVMA) is a federal agency and has responsibility for the evaluation, registration and review of agricultural and veterinary chemical products up until the point of retail sale. Plants form a subset of agricultural chemical products, known as biological agricultural products. A biological chemical product is an 'agricultural chemical product where the active constituent comprises or is derived from a living organism (plant, animal, micro-organism, etc.), with or without modification.'[45]

Genes, inserted in the genome of plants using gene technology, that code for the production of pesticidal substances are considered to be a pesticide and the plant with that particular trait must be registered by APVMA, as well as being regulated by the GTR.[46] Plants that have been conventionally bred to have, say, increased resistance to insect attack on the other hand do not.[47] Plants with naturally evolved resistance also do not require registration.[48]

GM plants other than those with traits for pest resistance do not require registration as agricultural chemical products. Nevertheless, some implications of their use may mean they still require registration by the APVMA. For example, herbicide tolerant plants do not fall under the definition of an agricultural chemical product and therefore need not be registered as such. However, where the use pattern of a chemical product changes in association with a GM crop plant, APVMA will assess the new use pattern of the chemical. For example, glyphosate-tolerant soybeans could result in glyphosate use in that crop and glyphosate products would require assessment of a major new use for the products.[49]

There is no definition of conventionally bred or genetic modification/manipulation/engineering technology, which are all terms used in the APVMA Guidelines for Biological Agricultural Products. No formal statement by APVMA on genome editing has been made. In any case, the meaning of gene technology is not crucial for the triggering of the agvet framework because the trigger is that the product be an agricultural chemical product.

3.3.2 Regulatory Classification of Genome Editing/Genome Edited Plants (Table 3.2)

3.3.2.1 Gene Technology

In light of the ambiguities in the GT regulatory framework, the GTR has adopted an interim approach to genome edited plants, while two ongoing reviews of the regulatory framework are completed. These reviews are discussed in Sect. 3.6 below. *General advice from the Regulator on coverage of new technologies*, was issued October 2016.[50] In it, the OGTR advises that it is limited in what it can do until the GT Act and/or Regulations are amended. Advice on a case-by-case basis is

45 APVMA (2018), section 2.1.
46 APVMA (2018), section 3.4.
47 APVMA (2018), section 2.2.
48 APVMA (2018), section 3.4.
49 APVMA (2018), section 2.2.4.
50 OGTR (2016a, b).

Table 3.2 Summary of interim approach to genome edited plants by Australian regulators

Technique	Regulated by Gene Technology Regulator	Regulated by Food Regulator
SDN-1	No	No
SDN-2	Yes	No
SDN-3	Yes	Yes
ODM	Yes	No

Table 3.3 Summary of interim approach to genome edited plants by Gene Technology Regulator

Technique	Regulated as a GMO
SDN-1	No
SDN-2	Yes
SDN-3	Yes
ODM	Yes

also offered by the GTR to those undertaking the techniques or using their products, with the OGTR advising that it will take a conservative approach consistent with promoting the Act's object and the broad scope of the legislation's definition of GMO.[51]

The following table summarises the GTR's current approach to genome edited plants in light of advice issued by the GTR so far and the papers produced by the OGTR during the ongoing reviews (Table 3.3).

The regulatory classification of plants created using new technologies is based on the particular genome editing technique used. The following general conclusions are suggested, based on advice issued by the GTR so far and the papers produced by the OGTR during the ongoing reviews.

3.3.2.1.1 Genome Editing Using SDN Is Gene Technology

The GTR has advised that the use of zinc finger nucleases (ZFN) is not 'chemical mutagenesis'.[52] This is significant because chemical-induced mutagenesis is expressly excluded from the definition of gene technology by the GT Regulations and no other exception to the meaning of gene technology is relevant to ZFN. For that reason, it is likely that all SDN techniques are gene technology because they modify an organism's genes or genetic material.

For the same reason, because during the development of the final plant, plant cells or whole plants would be created which contain the ZFN constructs, that development and use is a dealing with a GMO requiring authorisation. This means the undertaking of SDN must be authorised under the GT Act, even though the final crop

[51] OGTR (2016a, b).

[52] OGTR (2011).

may not be a GMO, depending on the type of SDN used, and commercial release of the final plant would not need authorisation.

3.3.2.1.2 Removal of DNA from a Plant Does Not Create a GMO (Provided the Construct and Vector Are No Longer Present)

Where the focus is on the final plant, the fact that during that plant's development, the plant contained foreign DNA is not sufficient to cause the *final* organism to be a GMO. The GTR has advised that crops developed using ZFN based EXZACT Delete technology developed by Dow AgroSciences Australia, where the ZFN construct and vector are no longer present in the final organism, are not GMOs. The GTR considers such crops fall within item 1 of Schedule 1 as being excluded from the definition of GMO. This is because the resulting crop is 'a mutant organism' as defined in the regulations, with no additional foreign nucleic acid having been added to its genome and is therefore not a GMO.[53]

Null segregants are also unlikely to be GMOs. It is suggested by the approach taken above that crops that undergo genome editing to induce early flowering, so that some breeding objective such as a new trait can be introduced without the use of a template, would not be GMOs. This is on the condition that the early flowering trait is later removed or progeny are selected that have not inherited the new DNA.

3.3.2.1.3 Addition of DNA to an Organism's Genome Is Not Sufficient of Itself to Create a GMO

SDN-1 can, but not always, result in a single or few base substitutions or localised insertions. That the inserted material is derived from the organism's own genome is not relevant of itself to determining whether the technique is gene technology and the resulting organism therefore a GMO. The determining factor is whether a repair template is provided. If the double-stranded DNA break is repaired by non-homologous end-joining (NHEJ), the technique is not gene technology and the resulting organism is not a GMO.

3.3.2.1.4 Use of a Template to Guide Modifications Results in GMOs That Are Subject to Regulation

Plants developed using Dow AgroScience Australia's EXZACT Insert or EXZACT Edit technology, where the mutational event involves the introduction of nucleic acid from template DNA with additional or altered nucleotides would be GMOs.[54]

---

[53]OGTR (2011).

[54]OGTR (2011).

Table 3.4 Summary of interim approach to genome edited plants by Food Regulator

Technique	Regulated as food produced using gene technology
SDN-1	No—provided not used to permanently introduce new gene
SDN-2	No—provided not used to permanently introduce new gene
SDN-3	Yes
ODM	No

The use of such techniques means the resulting organism does not come within the group of 'mutant organisms' excluded from the definition of GMO because foreign nucleic acid is introduced.

3.3.2.2 Food

FSANZ has held two scientific workshops (in 2012 and 2013) to provide advice on 'new plant breeding techniques' (NBTs). While FSANZ has not made any final determination on genome editing techniques, FSANZ has indicated that it will have regard to the conclusions of these workshops in considering any application made for foods produced using such techniques.[55] FSANZ also released a consultation paper in February 2018, seeking stakeholders' views on specific issues around its approach to genome editing and other NBTs which gives further indications of its proposed response to genome editing.[56] A summary of the workshop outcomes informed by the 2018 Consultation Paper, is outlined below (Table 3.4).

The scientific panel at the workshops and which is advising FSANZ during the current review is tasked with assessing the current science and potential food safety risks, if any, caused by the use of the techniques, rather than the classification of such techniques as gene technology for the purposes of the regulatory framework. The definitions are the subject of the broader review process, including the 2018 Consultation Paper.[57]

The Consultation Paper continues the FSANZ's use of its own classifications of plant breeding techniques. These classifications are (i) where the genome remains unchanged by gene technology, (ii) where the genome is changed but no new DNA is present in the organism from which food for sale is obtained, and (iii) where the genome contains new DNA.

This approach is taken because of the definition of gene technology used in the food Code. As FSANZ explains, the Code's definitions refer to gene technology techniques that result in the insertion of new pieces of DNA into a genome. FSANZ explains that new DNA in this context means a 'piece' of DNA that is new to the host organism in terms of its nucleotide sequence, genome location or orientation of

[55]Jones and Horticulture Innovation Australia (2016), p. 52. See also FSANZ (2012), p. 6; FSANZ (2013), p. 4.

[56]FSANZ (2018a).

[57]FSANZ (2018a), p. 4.

insertion.[58] Unfortunately, the line between a 'piece of DNA' and a 'change to DNA' is unclear. For example, it seems that a change to a single nucleotide is not a 'piece of DNA' because such changes are grouped into category (ii) above by the Consultation Paper. Category (ii) also includes those techniques that make defined changes to the genome without permanently introducing any new DNA although it may be present in the genome initially. The organism from which the food for sale is obtained may therefore contain genome changes but not new DNA.[59] Such food will therefore not be food produced using gene technology.

In regards to potential food safety risks, the scientific panel concluded that changes introduced using targeted mutagenesis, such as ODM or ZFN, would be typically small and definable and have predictable outcomes. Such changes would therefore be similar to those made by traditional mutagenic techniques used in conventional plant breeding, and do not present significantly greater food safety concerns than those from other forms of mutagenesis.[60]

The Consultation Paper observes that genome editing techniques (which the paper describes as being SDNs, ODM and base editing) change an organism's existing genome. However, the paper also observes that no new DNA is present in the organism from which food for sale is obtained even though changes can be made that include the insertion of one or a few nucleotides.[61] Therefore, food from such plants is not food produced using gene technology. Nevertheless, as part of the review, FSANZ is addressing whether the risks that may arise from these targeted changes or off-target changes mean that pre-market assessment and approval should still be required.[62]

Importantly in terms of understanding what the term 'recombinant DNA' includes, the scientific panel noted that because ODM relies directly on a chemically-synthesised oligonucleotide molecule, ODM is not a recombinant-DNA technique.[63] In regards to the possible amendment of the Code to include ODM, the scientific panel noted that there are no identified safety concerns associated with ODM's use both in terms of the nature and extent of the specific changes it can introduce to target plants as well as unintended effects. It was concluded that 'Foods derived from plants modified using ODM would be similar to that derived using traditional mutagenic techniques, or that occur naturally through spontaneous mutation.'[64] Therefore they should not be included in the scope of the term 'food was produced using gene technology' and therefore would not be regulated under the specific standard for such food. The recent Consultation Paper deals with ODM in a similar fashion.

---

[58]FSANZ (2018a), p. 4.
[59]FSANZ (2018a), p. 4.
[60]FSANZ (2013), p. 11.
[61]FSANZ (2018a), p. 12.
[62]FSANZ (2018a), p. 12.
[63]FSANZ (2012), p. 19.
[64]FSANZ (2012), p. 19.

In regards to null segregants, the scientific panel concluded that food derived from these plants should not be regarded as food produced using gene technology.[65] Such food was considered similar to food produced using traditional mutagenic techniques.[66] The Consultation Paper nevertheless raises for consideration whether food from null segregants organisms should be excluded from pre-assessment and approval.[67] On this point, the practice of FSANZ for many years has been to allow the use of null segregants as non-GM comparators for compositional analysis as part of a GM food safety assessment.[68]

FSANZ's final category for food derived using NBTs is where an organism's genome contains new pieces of DNA which remain in the organism from which food for sale is obtained. This includes transgenes, cisgenes and intragenes. While SDN-3 is not referred to in the Consultation Paper, SDN-3 is likely to fall into this group and food from organisms modified using that technology is therefore likely to be regulated as food produced using gene technology. This is because SDN-3 introduces a new gene into a specific site in the plant's genome and is therefore a recombinant DNA technique.[69]

3.4 Regulatory Prerequisites for Activities Relating to Genome Edited Plants

3.4.1 Gene Technology

For those genome edited plants which do not trigger the GT regulatory framework, generally applicable rules around import, export, and cultivation of agricultural plants and the sale of their products will apply. These include biosecurity controls that Australia and its states/territories impose.

For genome edited plants which do trigger the GT regulatory framework, the relevant use or dealing is important. If the proposed dealing will be contained experimental and development work, a licence for GMO work in contained facilities will be required. This is known as a licence for a dealing not involving intentional release of the GMO into the environment (DNIR licence). Some experimental genome editing work may even be classified as low risk. In that case, only notification to the GTR and compliance with certain regulatory requirements around containment rather than a licence, will be required. If the plant is to be released into the environment, whether in a field trial or for commercial release, a licence for a

65FSANZ (2012), p. 4.

66FSANZ (2013), p. 11.

67FSANZ (2018a), p. 11.

68FSANZ (2018a), p. 11.

69FSANZ (2012), p. 22; FSANZ (2013), p. 11.

dealing involving intentional release of GMOs into the environment (DIR licence) will be necessary.

Licence applications are made through the OGTR's Application Entry Point using the relevant application form. Prior to lodging the application, the application must have been reviewed by an Institutional Biosafety Committee. A risk assessment and risk management plan (RARMP) is prepared by the OGTR as part of the decision-making process for all licence applications.[70] For DIR applications, consultation on the RARMP and possible decision, with the public, the states/territories and other federal government authorities prescribed in the regulations, the Minister for the Environment and Energy and local governments is required.

The GT Act requires that licensing decisions be made within a prescribed time.[71] Licence conditions include a requirement that organisations conducting the dealings be accredited and maintain that accreditation.[72] DNIR licences usually require that the work be done in contained facilities that are certified physical containment facilities at level 2 or higher.[73] Additional conditions can be imposed by the GTR.

Licensing in all cases requires that the GTR be satisfied that any risks posed by the proposed dealing can be managed in a way that protects human health and safety and the environment. Assessed risks, namely risks to human health and safety and to the environment, are assessed in the context of the risks posed by the non-modified parental organisms. Accordingly, if the risks posed by a GMO are no greater than those posed by the conventional version of the organism, approval is likely. Socio-economic risks and potential benefits are not assessed under the national framework. DIR licence applications must provide information about proposed uses of the GMO or of products derived or produced from it because the GTR considers potential risks posed not only by the GMO but also by its products.[74] In light of those risks, limitations on how the products may be used may be imposed as licence conditions if thought necessary.[75] However, the GTR will usually not consider it necessary to impose a condition if another regulator of the same product has or will impose the same or similar condition.

[70]For DNIR, this is done by the Contained Dealings Evaluation Section of the OGTR; for DIR, by the Plant Evaluation Section of the OGTR.

[71]*Gene Technology Act 2000* (Cth) s 43(3). This is 90 business days for DNIR, 180 for a DIR licence for limited and controlled release (i.e. field trial) and 255 for a DIR licence for commercial release. A longer period is established where a significant risk is identified. Gene Technology Regulations reg 8. Note that the statutory timeframe clock can be stopped where the GTR seeks more information. Applications for accreditation and certification also have time-frames set by the regulations.

[72]Australian Department of Health (2017), p. 22. Accreditation is pursuant to *Gene Technology Act 2000* (Cth) s 91.

[73]Certification occurs pursuant to *Gene Technology Act 2000* (Cth) s 84. Classification relates to structural integrity of buildings and equipment, and to handling practices used by people working in the facility. Australian Department of Health (2017), p. 33.

[74]*Gene Technology Act 2000* (Cth) s 43(2) and Gene Technology Regulations 2001 (Cth) reg 7 and Sch 4.

[75]*Gene Technology Act 2000* (Cth) s 62(1).

Assessment of applications and decisions about licence conditions are based on current available science and a published Risk Analysis Framework to ensure consistent decision-making.[76] The Framework is based on the Australian/New Zealand Standard ISO 31000:2009 Risk Management—Principles and guidelines. Risk assessments include consideration of the following key questions.

- What could go wrong? Consideration is given to a range of circumstances where a GMO could harm people or the environment.
- How serious could the harm be? An assessment is made about the seriousness of potential harm using risk scenarios.
- How likely is the harm to occur? An assessment is made about the likelihood of potential harm using risk scenarios.
- What is the level of concern? The risk is assessed as negligible, low, moderate or high depending on the seriousness of the harm and the likelihood of it occurring.[77]

The data for the risk assessment is provided by the applicant but the GTR can also undertake independent research. Regulatory actions, such as decisions on licence applications, are not postponed due to a lack of scientific certainty, and are balanced with efficiently protecting human health and safety and the environment.[78]

3.4.2 Food

Approval for sale of a food produced using gene technology requires a variation to Schedule 26 in the *Australia New Zealand Food Standards Code* (the Code), to list the relevant food in the Schedule. FSANZ's functions include the development and variation of standards including Schedule 26.[79]

The primary objective of FSANZ in developing or varying a food regulatory measure is the protection of public health and safety.[80] Accordingly, the safety assessment is a central part of considering an application. The assessment does not address risks to the environment that may occur as the result of growing GM plants used in food production, or risks to animals that may consume feed derived from GM plants. However, the FSANZ does conduct a cost/benefit analysis of the approval

[76]Productivity Commission (2016), p. 266.

[77]Productivity Commission (2016), p. 266.

[78]*Gene Technology Act 2000* (Cth) s 4(aa).

[79]*Food Standards Australia New Zealand Act 1991* (Cth) s 13. Division 1 of Part 3 of the FSANZ Act specifies that the Authority may accept applications for the development or variation of food regulatory measures, including standards. This Division also stipulates the procedure for considering an application for the development or variation of food regulatory measures.

[80]*Food Standards Australia New Zealand Act 1991* (Cth) s 18.

(or not) of the food produced using gene technology. In that analysis, direct and indirect benefits that would arise from varying a food regulatory measure, as a result of an application, and the costs to the community, government or industry that would arise from the variation of that measure are considered.

FSANZ advises that the cost/benefit analysis of the regulatory options is not intended to be an exhaustive, quantitative financial analysis of the options as most of the impacts that are considered cannot be assigned a dollar value. Rather, the analysis seeks to highlight the qualitative impacts of criteria that are relevant to each option. These criteria are deliberately limited to those involving broad areas such as trade, consumer information and compliance. This is in part to address the other two of the three objectives of the food regulatory framework.[81] In addition to the protection of public health and safety, the legislation provides that the objectives of FSANZ (in descending priority order) in developing or reviewing food regulatory measures and variations of food regulatory measures are the provision of adequate information relating to food to enable consumers to make informed choices and the prevention of misleading or deceptive conduct.[82] Labelling requirements for some GM foods are intended to meet the second objective, while the provision of detection methodology by the applicant addresses the third. For this purpose, applicants must provide the full DNA sequence of the insert and adjacent genomic DNA. Using this information, any DNA analytical laboratory would have the capability to develop a PCR-based detection method.[83]

The applicant provides the data for the safety assessment but other available resource material including published scientific literature and general technical information is used in the safety assessment. Data requirements are set out in the FSANZ Application Handbook and, in turn reflect internationally-accepted GM food safety assessment guidelines. These safety assessments are characterised by:

- a case-by-case consideration of the food. This allows each food to be assessed according to its particular characteristics, including the type of genetic modification.
- consideration of both the intended and unintended effects of the genetic modification. For example, the intended effect of genetic modification of an organism may be a new trait such as insect protection, but unintended effects such as changed nutritional characteristics may also arise. Both of these effects are evaluated.

---

[81] *Food Standards Australia New Zealand Act 1991* (Cth) s 18(1).

[82] *Food Standards Australia New Zealand Act 1991* (Cth) s 18(1).

[83] The Food Regulation Standing Committee's Implementation Sub-Committee [now known as the Implementation Subcommittee for Food Regulation] has formed an Expert Advisory Group (EAG), involving laboratory personnel and representatives of the Australian and New Zealand jurisdictions, to identify and evaluate appropriate methods of analysis associated with all applications to FSANZ, including those applications for food produced using gene technology. FSANZ (2017), p. 6. 20 December 2017 [35-17] Call for Submissions – Application A1154. Food derived from insect-protected cotton line MON88702.

- comparisons with conventional foods with an acceptable standard of safety. This enables the identification of similarities and differences between the GM food and an appropriate comparator, and allows identified differences to be characterised to determine any potential safety or nutritional issues.[84]

3.4.3 Agricultural and Veterinary Chemicals

The APVMA Guidelines for Biological Agricultural Products state that products based on GMOs have extra data requirements for information concerning the genetic manipulation.[85] Applicants must provide details about:

- the host organism
- the donor organism
- genetic engineering techniques used in the genetic modification
- identity of the inserted or deleted gene segment (base sequence data or enzyme restriction map of the gene)
- information on the control region of the gene in question
- description of the new traits or characteristics that are intended to be expressed
- tests to evaluate genetic stability and exchange
- environmental expression and toxicology tests.

A comparison undertaken by CropLife Australia of the data requirements for assessment of GM products with incorporated pest and/or disease control by the APVMA, the GTR and FSANZ shows a high level of concordance. Product efficacy and resistance management considerations stand out as differentiators of the APVMA assessment.[86]

3.5 Status Quo of Genome Edited Plants and Products Derived from Them

No genome edited plants have been authorised by the GTR for contained use, field trial or cultivation in Australia and no applications for such authorisation are pending. No approvals have been granted by or sought from FSANZ for products of genome edited plants. Applications for foods derived using intragenesis have been approved by FSANZ. These concern potato lines developed by SPS International Inc to bruise less and have lower levels of the toxin acrylamide (line E12) as well as lines

[84] Productivity Commission (2016), p. 267.

[85] APVMA (2018), section 2.3.

[86] Croplife Australia (2017), p. 10.

with improved disease-resistance, reduced acrylamide potential and reduced browning (lines F10, J3, W8, X17 and Y9).

3.6 Reform Efforts

3.6.1 Timeline re Australian Regulatory Responses to Genome Edited Plants (Table 3.5)

Table 3.5 Timeline re Australian regulatory responses to genome edited plants

Date	Regulator	Action
May 2012	FSANZ	Report of workshop on New Plant Breeding Techniques (NBTs)
August 2013	FSANZ	Report of a 2nd workshop on NBTs
Oct 2016	GTR	Discussion Paper on Technical Review of GT Regulations: Options for regulating new technologies (Public comment 17 Oct 2016–16 Dec 2016) *General Advice from the Regulator on coverage of new technologies*
June 2017	FSANZ	Commenced review of NBTs including gene editing
July 2017	GTR	Background Paper on GT Regulatory Scheme Review (Public comment 25 Jul 2017–29 Sept 2017)
Nov 2017	GTR	Consultation Paper on GT Regulatory Scheme Review (Public comment 30 Nov 2017–21 Feb 2018) Re Technical Review of GT Regulations, the following were released for public consultation (30 Nov 2017–21 Feb 2018): Exposure draft of proposed amendments to the Regulations Consultation quick guide Regulation Impact Statement
Feb 2018	FSANZ	Consultation Paper. Food derived using new breeding techniques (public comment open until 12 Apr 2018)
March 2018	GTR	Preliminary Report on GT Regulatory Scheme Review (public comment open until 21 May 2018)
Mid 2018	FSANZ	Expected release of final report on review of new breeding techniques
Late 2018	GTR	Expected release of final report on GT Regulatory Scheme Review

3.6.2 Gene Technology

3.6.2.1 Introduction

The GTR periodically reviews the GT framework in order to advise the GT Legislative Forum (the policy setting body) about the framework's effectiveness.[87] Following reviews of the GT regulatory scheme in 2006 and 2011, a third review is now underway and is discussed in subsection 3 below. The 2006 review considered whether the technologies and organisms covered by the Act should be changed, but this predated the practical development of genome editing and no changes to the scope of the framework in this regard were recommended.[88] However, the 2006 review recommended that the issue be considered again in five years to ensure that the scheme continued to accommodate emerging trends.

The 2011 review did not however, examine the meaning of the relevant definitions. Instead, on the issue of new technologies, the 2011 review recommended that faster processes for amending the Regulations be developed as a way to address developing technology.[89] The existing process for amendment is cumbersome. The GTR seeks the formal agreement of the majority of states and territories through the Legislative Forum, after having finalised proposed amendments. The GTR then seeks the Minister's agreement that amendments to Regulations be proposed to the Executive Council. The Council must then recommend to the Governor-General that the amendment regulations be made, and those regulations must then be tabled in both Houses of Parliament for scrutiny and potential disallowance.

There is also a parallel, but separate, review of the GT Regulations currently being undertaken by the GTR. This review, the *Technical Review of the GT Regulations*, is intended to ensure the regulations fit developing technology and scientific knowledge.[90] The Background Paper advises that '[t]his is important to provide clarity about whether organisms developed using a range of new technologies are subject to regulation as genetically modified organisms, and to ensure that new technologies are regulated in a manner commensurate with the risks they pose.'[91] As to how this review of the GT Regulations fits within the broader review of the regulatory scheme, the GTR has said:

> In the ongoing Technical Review of the Regulations, the Regulator considers that an interim approach, to continue to regulate some new technologies based upon the process used, best supports the effectiveness of the legislative framework at this time. ... As an interim measure, the Technical Review would provide clarity while broader policy considerations are progressed in the Review of the Scheme.[92]

---

[87] Australian Department of Health (2017), p. 60. This is required pursuant to the Gene Technology Agreement between the national and state/territory governments, clause 44.

[88] Commonwealth of Australia (2006).

[89] Allen Consulting Group (2011), p. 27.

[90] The Legislative and Governance Forum on Gene Technology (2017), p. 7.

[91] The Legislative and Governance Forum on Gene Technology (2017), p. 4.

[92] OGTR (2017b), p. 19 citing OGTR (2016-17).

3.6.2.2 Technical Review of the Gene Technology Regulations

In October 2016, the GTR released a Discussion Paper on options for regulating new technologies, and the *General Advice on coverage of new technologies*, both of which are discussed in Sect. 3.3.2 above.

The Discussion Paper addresses the regulation of organisms produced using SDN techniques and ODM. RNAi was also investigated and is discussed in Sect. 3.12 below. The Review's focus is cases where the capture or exclusion of these techniques is not clear, whether those technologies should be regulated, and the scientific evidence relating to risks posed as a result of using the new technologies.[93] Four options for regulation of the new technologies were presented:

Option 1: no amendment to the Gene Technology Regulations.
Option 2: regulate certain new technologies (ODM, SDNs-1, -2 and -3).
Option 3: regulate some new technologies based on the process used (ODM, SDN-2 and SDN-3).
Option 4: exclude certain new technologies from regulation on the basis of the outcomes they produce (i.e. regulate only SDN-3).

Horticulture Innovation Australia (HIA) submission to the review was that only SDN-3 technologies should be regulated to 'enable the vegetable industry to use SDN-1 and SDN-2 technologies without the stigma or unnecessary costs triggered by Gene Technology regulations'.[94] HIA also noted that clarity in the Regulations is vital for the future implementation of the techniques by the Australian vegetable industry.[95]

After considering the issues raised by the submissions on the Discussion Paper,[96] as well as scientific understanding, potential risks, regulatory burden implications, whether regulatory burden would be commensurate with risk, and the policy intent of the GT Act, the OGTR has prepared draft proposals discussed below, adopting Option 3.[97]

Together with the exposure draft of proposed amendments, the OGTR released a Consultation Regulation Impact Statement (RIS)[98] and a Quick Guide to the Consultation.[99] These are intended to assist public submissions. A Recommendation Report and a Decision Regulatory Impact Statement, which analyses the public submissions and evidence on costs and benefits of proposed changes, will then be submitted by the GTR to the GT Legislative Forum for a decision on whether to adopt the final recommendations. It is important to remember that the Technical Review can only

[93]OGTR (2016a, b), p. 1.

[94]Jones and Horticulture Innovation Australia (2016), p. 57.

[95]Jones and Horticulture Innovation Australia (2016), p. 58.

[96]The OGTR received 741 submissions, 615 of which were received through a form on a website initiated by Friends of the Earth Australia. Australian Department of Health (2017), pp. 60–61. The submissions are available on the OGTR website.

[97]OGTR (2018e).

[98]OGTR (2017a).

[99]OGTR (2017c).

address possible amendments to the GT Regulations and not to the Act. The parallel Regulatory Scheme Review can recommend policy and legislative changes.

One major issue identified by the Technical Review was the ambiguity in the GT Regulations because 'new technologies for altering genetic sequence and gene expression are not specifically addressed in the legislation'. It was unclear whether (or not) organisms that have undergone several specific techniques are within, or excluded from, the scope of regulation under the GT Act.[100] Amongst the risks identified as possible if this ambiguity is not clarified, is that the use of the techniques will be inhibited because organisations are unsure of regulatory requirements or will not proceed because organisations mistakenly believe that there are prohibitive regulatory burdens. Possible consequences of these impacts are identified as including delaying the progress of basic research, and that products (such as food crops or human or animal therapeutics) may not be commercialised or will be delayed in benefiting the Australian public. 'In the longer term, if uptake of these technologies continues to be inhibited this could hamper industry development and affect the international competitiveness of Australian businesses.'[101]

Key proposals around clarifying the scope of regulation regarding the meaning of GMO are made. These are:

3.6.2.2.1 Organisms Modified Using SDN-1 Are Not GMOs

It is proposed that the exception to the definition of GMO, for mutant organisms (in item 1 of Schedule 1, GT Regulations) be deleted. A new exception would be added (as item 4). This new exception would exclude organisms that have undergone SDN-1 from being GMOs. In particular, it is proposed that the new exception would say:

> Proposed exception to the definition of GMO
> 'An organism modified by repair of single-strand or double-strand breaks of genomic DNA induced by a site-directed nuclease, if a nucleic acid template was not added to guide homology-directed repair'.[102]

The current exception excludes a broader group of organisms from the definition of GMO than the proposed replacement exception. However, the GTR considers that this will not cause problems because chemical and radiation-induced mutagenesis are excluded as techniques that are gene technology. Therefore organisms created

[100]OGTR (2017a), p. 4.
[101]OGTR (2017a), p. 5.
[102]OGTR (2017a), p. 9; OGTR (2017d).

using either of those techniques will continue to not be regulated as GMOs even if the existing mutagen exception to GMO is repealed.[103] The GTR has explained that SDN-1 is not treated in this way, that is, by exclusion of the technique from the definition of gene technology, because that would also exclude any intermediate GMOs produced in the course of SDN-1 which should continue to be GMOs. For example, organisms stably expressing a site directed nuclease should be GMOs.[104]

An important flaw in the proposed exception, though, is its reliance on language that is the name of a particular tool. This means the proposed exception may not be sufficiently future-proof. Product developers may replace SDNs with recombinases or other DNA modifying enzymes such as DNA methylases or deaminases (in base editing) with the intention of achieving the same effect as a SDN-1 mutation. Flexible language would accommodate these developments.

The OGTR's justifications for excluding SDN-1 from the definition of GMO are threefold.[105]

1. *Risk*—The extent of modifications to DNA and organisms' characteristics which are possible through SDN-1 are the same as that possible with currently excluded techniques of natural mutations, chemical mutagenesis and random mutagenesis. All of these rely on the same natural cellular repair mechanisms. The Consultation RIS notes that when the GT scheme commenced, the list of organisms excluded from being GMOs 'was intended to exclude techniques on the basis that they "give rise to organisms that can occur in nature and as such do not pose a particular biosafety risk to the environment or human health or safety." Excluding organisms modified using SDN-1 from regulation is consistent with this intention, and appropriate on the basis that these organisms do not pose different risks to natural mutants'.[106]
2. *Compliance enforceability*—Practical considerations for regulatory compliance arise because it is not possible to distinguish products of SDN-1 from naturally occurring mutation.
3. *Consideration of policy settings of the regulatory scheme*—Excluding SDN-1 maintains the policy settings of the GT scheme. Exclusion of SDN-1 is consistent with the approach of defining GMOs on the basis of the use of the process of gene technology. This means a process regulatory trigger continues to be used. The Consultation RIS says that the use of a template to direct sequence changes is the

[103]OGTR (2017a), p. 11. The GTR notes that there are two organisms excluded through item 1 that cannot take advantage of the proposed approach and will therefore be unintentionally reclassified as GMOs if item 1 is deleted. These are NoGall (*Agrobacterium radiobacter* strain K1026) and VaxSafe PM (*Pasteurella multicida* strain PMP1). It is proposed that they will be specifically listed in Sch 1 items 10 and 11 to ensure they continue to be excluded from regulation. OGTR (2017a), p. 11.

[104]OGTR (2017a), p. 11.

[105]OGTR (2017a), p. 9.

[106]OGTR (2017a), p. 10 citing Gene Technology Regulations Regulation Impact Statement Section 4 part (a), discussion of listing a limited class of organisms as not being GMOs, published as part of the 2001 Explanatory Statement.

hallmark of techniques considered gene technology since the inception of the regulatory scheme.[107] It also explains why organisms modified using SDN-2 or SDN-3 are intended to be regulated as GMOs. It is recognised that some outcomes of SDN-2 or 3 may be equivalent to sequence modifications possible through unregulated techniques. But the RIS explains that allowing such organisms to be unregulated means looking at the product as the trigger rather than the process and that is outside the ambit of the Technical Review's purview.[108] As a minor matter it is observed by this author that in light of that comment, the justification above around compliance enforceability should also be irrelevant with respect to SDN-1.

3.6.2.2.2 Organisms Modified Using SDN-2 and ODM Are GMOs

A new schedule, Schedule 1B, is proposed to be added which will list organisms that are GMOs for the purposes of paragraph (c) of the definition of GMO.[109] The organisms which will be expressly named as GMOs are set out in the box below.

> Proposed Schedule 1B, GT Regulations 2001 (Cth)
> Organisms that are genetically modified organisms
> Item 1—An organism that has had its genome modified by oligonucleotide-directed mutagenesis.
> Item 2—An organism modified by repair of single-strand or double-strand breaks of genomic DNA induced by a site-directed nuclease, if a nucleic acid template was added to guide homology-directed repair.

While the proposed reliance on the use of a template as a distinguishing factor of a GMO is consistent with the process trigger used by the current GT regulatory framework, it highlights the deficits of such a trigger. That the organisms produced using SDN-2 and ODM may in some cases pose equal or less risk than the outcomes of other forms of induced mutagenesis is not taken into account.

3.6.2.2.3 Organisms Derived from GMOs

To clarify that null segregants are not GMOs, it is proposed that two new items be added to the list in Schedule 1 of organisms that are not GMOs. These are set out in the box below.

---

[107]OGTR (2017a), p. 10.

[108]OGTR (2017a), p. 15.

[109]A new regulation, reg 4A, will also be added for this purpose.

Proposed new items 8 and 9, Schedule 1, GT Regulations 2001 (Cth)
Item 8—An organism that is descended from a genetically modified organism (the initial organism), but which has not inherited any traits that occurred in the initial organism because of gene technology.
Item 9—An organism that was modified by gene technology but in which the modification, and any traits that occurred because of the gene technology, are no longer present.

Unfortunately, these proposed amendments continue the ambiguity present in the existing definition of GMO. The term 'occurred because of gene technology' continues to be open to misinterpretation. The addition of site directed mutagenesis to Schedule 1A as a technique that is not gene technology would assist in clarifying this uncertainty but as noted above, the GTR does not want to pursue that approach because it will cause intermediate organisms produced in the course of the genome editing to be excluded.

3.6.2.3 GT Regulatory Scheme Review

The GT Regulatory Scheme Review is considering broader policy and contextual issues around the Scheme's response to emerging issues.[110] The Review's Consultation Paper, which gathered the key issues raised by public submissions on the earlier Background Paper, was released in November 2017. The March 2018 Preliminary Report presents 33 findings on those issues. A final report to the GT Legislative Forum (the body responsible for overseeing the GT scheme) is expected in the second half of 2018 following a further public consultation phase on the Preliminary Report's findings.

The Consultation Paper identifies key issues raised by the public: whether the GT regulatory scheme is sufficiently agile to adapt to or regulate new technologies and whether the scheme's trigger should be changed from a process to an end-product trigger.[111]

On the first issue, the Preliminary Report finds that existing definitions in the GT Act and Regulations 'may not appropriately classify a range of advances in technology'.[112] Further, that the agility of the Scheme could be increased through

[110]As the Consultation Paper explains 'While the Regulator can undertake reviews of the Regulations to improve the clarity of definitions and practices, any change in approach to what is to be regulated or not can only be done by the owners of the policy setting for the Scheme.' OGTR (2017b), p. 12.

[111]OGTR (2017b), p. 16. As in the 2011 review of the GT Regulations, the Consultation Paper notes that the complex and lengthy process for amending the regulatory framework challenges governance and needs to be overhauled.

[112]Australian Department of Health (2018), Finding 3.

mechanisms such as enabling the GTR and a standing committee to undertake certain activities including making determinations on the applicability of regulation to technological developments, resulting in faster responses by the regulatory framework.[113] However, the Preliminary Report states that 'additional work is required before a conclusion can be reached on how the relevant definitions in the Act and the Regulations should be amended, with this additional work needing to seek viewpoints from a wide range of stakeholders'.[114] The Report also notes that such work should also take into account work by the Technical Review and work internationally.[115]

The Review's willingness to consider the second key issue raised by the public submissions—a change from process to end-product trigger—is of great interest. Like the Technical Review, the Scheme Review is considering whether it is practical, efficient or appropriate to regulate the products of genome editing when, from an enforcement perspective, it may not be possible to distinguish those products from the products of conventional methods.[116] Industry has indicated that it supports a change of trigger. However, the Preliminary Report states that there are strong arguments in favour of maintaining the process-based trigger.[117] Nevertheless it found that there are opportunities for additional risk tiering within the Scheme as a way to reduce unnecessary regulatory burden and that additional work could be undertaken to determine such risk tiers and regulatory requirements assigned to each tier.[118] The Report suggests possible considerations for risk tiering, including whether foreign DNA is introduced or if edits only to the existing genome are made as well as the characteristics of the final organism.[119]

Other relevant findings by the Review include a finding that other factors, such as potential economic, environmental and health benefits, should not be considered in regulatory decision making regarding GMOs after some stakeholders had proposed that these factors should be included in governance decisions.[120] The Review also found that some aspects of gene technology may benefit from greater clarity around the Scheme's policy position. This provides industry and the community with certainty and greater transparency. These include the 'risk appetite for, and acceptance of multiple genetic modifications in an organism', 'acceptance of Low Level Presence (LLP) standards for GM products and their co-existence with non-GM crops' and 'linkages and interfaces with other regulatory schemes, domestically and

[113]Australian Department of Health (2018), Findings 13 and 14.

[114]Australian Department of Health (2018), p. 15.

[115]Australian Department of Health (2018), p. 16.

[116]OGTR (2017b), p. 19.

[117]Australian Department of Health (2018), Finding 8.

[118]Australian Department of Health (2018), Finding 9 and pp. 26–30.

[119]Australian Department of Health (2018), p. 29.

[120]Australian Department of Health (2018), Finding 20 and p. 52. See also OGTR (2017b), p. 31.

internationally (e.g. for harmonisation of definitions and regulatory requirements to reduce regulatory burden)'.[121]

3.6.3 Food

A formal review to consider how the Food Standards Code applies to new breeding techniques (NBTs), including genome editing, commenced in June 2017.[122] The review is expected to take about 12 months with a consultation paper released in February 2018, from which a final Review Report will be developed. After the review, FSANZ will consider whether to amend the Code. This will be a separate process to the review and involves further public consultation.

An Expert Advisory Group on New Breeding Techniques (EAG NBT) has been established to provide FSANZ with expert advice on issues relevant to the review, such as the current science and potential food safety issues associated with the use of NBTs.[123] FSANZ is also consulting with the states, as the entities that enforce the Code, and other interested stakeholders.

The FSANZ Review's objectives are to consider:

- what foods should be captured for pre-market assessment and approval under Standard 1.5.2
- whether the definitions for 'food produced using gene technology' and 'gene technology' in Schedule 1.1.2-2 need amendment.[124]

These objectives exclude consideration of the labelling provisions and the definitions relevant to GM food labelling obligations. However, it is possible that changes made in response to the review will be broader than the objectives indicate. During a workshop presentation by a FSANZ representative in late 2017, the speaker noted that if it is proposed to change the definitions in Standard 1.5.2, definitions in Standard 1.5.1 regarding novel foods would also be addressed.[125]

The same representative explained that the FSANZ approach to the review is to apply a risk-based approach to the question of whether foods should be captured, saying the 'focus is on characteristics of food itself' and that it 'no longer makes sense to make distinctions based on process, or use of a specific technique'. She observed that genome editing had put attention back on the GM standard—how and why we regulate them and whether it should be extended to new technologies?

[121] OGTR (2017b), p. 32. See Australian Department of Health (2018), pp. 4–6. See OGTR (2018) Findings 9, 15, and 24–26.

[122] Pursuant to *Food Standards Australia New Zealand Act 1991* (Cth) s 113.

[123] FSANZ (2018a), p. 4.

[124] FSANZ (2018b).

[125] Kelly (2017).

Answering these issues, she noted, required us to consider what had been learnt from 20 years of regulating GM foods.[126]

The review will consider the targeted changes that can be introduced and their impact on the food and characterise the potential food safety risk from off-target changes. In regards to targeted changes, it will be considered whether a presumption of safety can be applied to food from genome edited plants by assessing the products that can be produced now and in future and how these compare with foods already in the food supply. The potential to develop novel foods with potential food safety risks will also be evaluated. In regards to off-target changes, the frequency and potential food safety risks and how they compare to random changes that occur with chemical or radiation mutagenesis or occur spontaneously and whether they can be predicted and screened for will be addressed.[127] The FSANZ representative indicated that whether the general food law is sufficient to protect public health and safety or whether an additional process is needed to provide assurance such foods are safe will be answered.

The (in)compatibility of the definitions used by the food regulatory framework with those in other regulatory schemes and problems with enforceability and promoting an efficient and internationally competitive food industry were also referred to by the speaker. In this regard it should be noted that FSANZ is a bi-national agency, regulating food in both Australia and New Zealand. Any move to make the food regulatory definitions consistent with those of the Australian GT regulatory framework, could simultaneously raise the challenge of consistency (or not) with the New Zealand GMO regulatory framework, the *Hazardous Substances and New Organisms Act 1996* (HNSO Act).

The New Zealand Government acted in August 2016 to clarify that traditional chemical and radiation mutagenesis techniques are not GM for the purposes of its regulatory framework for GMOs but that at least some of the new techniques are not deregulated. This action followed a 2013 decision by the New Zealand regulator, the Environment Protection Authority of New Zealand (NZ EPA), that the use of custom ZFNs and TALENs, delivered as either mRNA or protein, did NOT result in organisms classified as GMOs. The New Zealand HNSO Act defines GMO as 'any organism in which any of the genes or other genetic material have been modified by in vitro techniques' unless expressly provided by regulations. 'In vitro techniques' are not defined. Regulation 3 at that time excluded from the definition of GMO '...organisms that are regenerated from organs, tissues, or cell culture, including those produced through selection and propagation of somaclonal variants, embryo rescue, and cell fusion (including protoplast fusion or chemical or radiation treatments that cause changes in chromosome number or cause chromosome rearrangements).' The NZ EPA's decision was appealed to the New Zealand High

[126]Kelly (2017).

[127]Kelly (2017).

Court which ruled that the NZ EPA could not itself expand the list of breeding techniques excluded from being GM.[128]

The New Zealand Government then reviewed the position, following public consultation. It was decided not to deregulate new techniques.[129] In any case, the New Zealand GMO regulatory framework does not use the term 'gene technology' and therefore changes to the definition of that term in the food regulations would not cause a direct inconsistency, as it does in the Australian frameworks.

3.7 Low Level Presence

Australia's GT and food regulatory frameworks have a zero tolerance to LLP in both seed for sowing and for food, feed and processing.

However, with respect to seed for sowing the GTR together with industry have developed an *Unintended Presence Strategy (Unapproved GMOs in seed for sowing).*[130] This strategy creates a 'risk-based national strategy to manage the unintended presence of unapproved GMOs in seeds imported for sowing in Australia'.[131] The OGTR has also worked with the Australian Seed Federation to develop a voluntary testing program of existing industry quality assurance measures and engages with other government departments regarding LLP of unapproved GMOs.[132]

With respect to LLP for food, feed and processing, the Australian Government actively participates in coordinated international discussions related to LLP and global trade efforts around LLP, including the Global LLP Initiative.[133] The Global LLP Initiative is developing an approach whereby a GM crop will undergo a shortened authorisation process if the amount of LLP is below a set tolerance and the crop has been approved in its exporting country. Genome edited plants have not been discussed as part of the Initiative. Such discussion would be premature until the regulatory status of such plants in Australia and other jurisdictions is clearer. If genome edited plants are not classified as GMOs by exporting countries, the Initiative's approach may need to be changed if sufficient importing nations classify genome edited plants as GMOs.

---

[128] *Sustainability Council of NZ v EPA* (heard 6 and 7 Nov 2013; judgment 20 May 2014).

[129] To clarify that these new techniques were not deregulated (and correct a drafting error identified by the High Court) the Regulations were amended by deleting from regulation 3(1)(b) 'or chemical or radiation treatments that cause changes in chromosome number or cause chromosome rearrangements' and inserting new regulation 3(1)(ba) 'organisms that result from mutagenesis that uses chemical or radiation treatments that were in use on or before 29 July 1998'.

[130] OGTR (n.d.).

[131] Australian Department of Health (2017), p. 53.

[132] Australian Department of Health (2017), p. 54.

[133] Australian Department of Agriculture and Water Resources.

The GT Act has also been amended to allow the GTR to grant Inadvertent Dealing Licences. These licences are one mechanism by which the GTR can deal retrospectively with incidents of LLP. Inadvertent dealing licenses allow importers to seek regulatory authorisation in a much shorter timeframe than the usual licensing process. However, such licences can only allow for the 'disposal' of a GMO. This term is not defined and the licensing process has so far not been used in relation to GM-derived crop materials (grain or seed). The extent of use that can be made under the term of 'disposal' is therefore unclear. The 2011 Review of the GT Act called for amendment to enable authorisation of other dealings during the disposal of inadvertently obtained GMOs, such as storing and testing. Industry groups have supported even broader amendments to clarify that disposal allows use in the course of manufacture, import and transport.

Industry has also requested that the Australian Government examine the impact of its current legislation in relation to LLP and develop specific policies to recognise its trading partners' systems for risk assessment and management, particularly in relation to import of GM-derived plant materials (grain or seed).[134] In particular, industry is seeking that the GTR be enabled to undertake proactive risk assessment. This would involve using a limited data set to make faster decisions and the introduction of LLP thresholds. If LLP presence is below a particular level and the plant concerned is approved in the exporting country, the GTR can then grant conditional approval to allow the shipment with LLP to enter trade and commerce.

3.8 Labelling

3.8.1 Gene Technology

No labelling obligations are directly imposed by the GT regulatory framework on GMOs or their products. However, the conditions of authorisation to release a GMO into the environment through a field trial or commercial release often effectively require such labelling. For example, the legislation requires licence holders to inform any other person covered by the licence of all relevant licence conditions.[135] Particular conditions imposed on the use of GM seed such as buffer zones for example, may need to be passed on to farmers who buy that seed from an authorised person and this could be required to be done through labelling.[136] The GTR can also impose licence conditions regulating how GM products derived from the relevant GMO may be used and this could also require labelling.[137]

---

[134]Croplife Australia (2017), p. 11.

[135]*Gene Technology Act 2000* (Cth) s 63(1).

[136]*Gene Technology Act 2000* (Cth) s 63(3).

[137]*Gene Technology Act 2000 (Cth)* s 62(1).

3.8.2 Food

As discussed in Sect. 3.3, certain foods produced using gene technology have mandatory labelling requirements.[138] However, whilst all foods produced using gene technology must be approved prior to sale, not all must be labelled.

Mandatory labelling is required for all 'genetically modified food' unless an exemption applies. Importantly, the class of food regulated by this section (using the term 'genetically modified food') is narrower than the class regulated by that part of the Standard relevant to safety assessment (which uses the term 'food produced using gene technology'). The relevant definitions are in the box below.

Section 1.5.2-5, Australia New Zealand Food Standards Code
Genetically modified food means food produced using gene technology that:
contains novel DNA or novel protein; or
is listed in section Schedule 26-3 as requiring such a label (i.e. during safety assessment of the food, a labelling condition was imposed on the sale of such food).
Novel DNA or novel protein means:
DNA or protein which, as a result of the use of gene technology, is different in chemical sequence or structure from DNA or protein present in counterpart food that has not been produced using gene technology, other than protein that:
is used as a processing aid/food additive; and
has an amino acid sequence found in nature.

There are also exceptions to the labelling requirements. These are set out in the box below.

Section 1.5.2-4(1), Australia New Zealand Foods Standards Code
Exemptions from the labelling requirements apply to:

- highly refined food where the refining process removes the novel DNA or novel protein and the food is not listed in Schedule 26-3(2) or (3) as requiring such a label;
- processing aids and food additives except where novel DNA or novel protein from them is present in the final food;
- flavours present in concentrations of no more than 1 g of flavour/kg of food; and

(continued)

[138]*Australia New Zealand Food Standards Code* Standard 1.2.1—Requirements to have labels or otherwise provide information. See also Standard 1.5.2—Food Produced Using Gene Technology.

- food intended for immediate consumption prepared and sold from premises and vehicles, including take-away outlets, restaurants or self-catering institutions.

Where a label is required because the food concerned is genetically modified food, the food's label must include the statement 'genetically modified' in conjunction with the name of that food (e.g. 'Soy flour – genetically modified').[139] For GM ingredients, food additives and processing aids, the required information may be in the product's table of ingredients. For GM foods not sold in packaging, such as fruit and vegetables, the required information must be displayed on or in connection with the display of the food.[140]

Non-GM foods do not require any statement as to genetic status.[141] However, should producers voluntarily choose to label such products, for example as being GM free, they must be able to verify the truth of that statement or risk penalties under consumer protection laws.[142] This may require the method used to verify that status be disclosed on the label. It is interesting to note here that the Australian consumer protection regulator, the Australian Competition and Consumer Commission, has indicated that claims that chicken was GM-free when the animals had been fed GM grain, could be misleading.[143] As discussed above, under the relevant labelling regulations, such chicken would not require labelling as GM. Nevertheless, the ACCC considered consumers may interpret the label as meaning GM feed had not been used and was therefore potentially misleading.

A tolerance is established for labelling purposes. Food can contain up to 10 g/kg (one percent) of *unintended* presence of an approved GM product without requiring labelling.[144] However, there is a zero tolerance for unapproved GM food, even where a GM food is exempt from labelling requirements or for intentional presence of approved GM product.

The Australian Productivity Commission, referred to above, recommended in 2016 that there no longer be mandatory labelling for GM food unless the labelling requirement was intended to address a safety concern. It said:

> The case for mandatory labelling of genetically modified (GM) foods is weak. Given that Food Standards Australia New Zealand assesses GM foods for their health and safety, GM labelling is a consumer value rather than a food safety issue. Where consumers prefer to purchase non-GM foods, the market is able to provide information through voluntary labelling.[145]

---

[139] *Australia New Zealand Food Standards Code* s 1.5.2-2.

[140] *Australia New Zealand Food Standards Code* Standard 1.2.1.

[141] *Australia New Zealand Food Standards Code* s 1.5.2-4(4).

[142] *Competition and Consumer Act 2010* (Cth), Australian Consumer Law (schedule to that Act) cl 18.

[143] Sylvan (2005).

[144] *Australia New Zealand Food Standards Code* s 1.5.2-4(1)(d).

[145] Productivity Commission (2016), p. 362.

The Government has not yet responded to this recommendation but has responded to a similar recommendation made in an earlier review. In 2011 a review of food labelling law and policy recommended that the exemption for food intended for immediate consumption be repealed.[146] The Government's response was to reject that recommendation on the basis that all GM food is safety assessed before it can be sold.

3.9 Identity Preservation System (Coexistence)

Coexistence measures have been introduced by industry and state and territory governments. The focus of these is largely on the responsibilities of those wanting to adopt GM crops although some guidance is given in regards to organic producers.[147] At the legislative level, government involvement differs between states and territories. For example, in Victoria segregation costs and measures are to be determined by market and consumer requirements whereas in New South Wales, segregation costs and measures are implemented under a system of ministerial approval by the relevant Government Minister and an Expert Committee.

3.10 Liability

Failure to comply with obligations imposed by the gene technology, food or agvet regulatory frameworks are offences under relevant legislation. For example, in regards to the GT regulatory framework it is a criminal offence to deal with a GMO in breach of licence conditions or otherwise as authorised in the legislation.[148] Such an offence requires the person concerned to know that what they are doing is not authorised or be reckless as to whether it is or not. There are also strict liability offences for technical breaches of the legislation where someone knowingly deals with a GMO whether or not they knew or were reckless as to whether they were unlicensed. Heavier penalties apply for 'aggravated offences', that is offences that cause or are likely to cause significant damage to health and safety of people or to the environment.[149]

---

[146] Blewett Report (2011).

[147] See, for example, Australian Department of Agriculture and Water Resources (2018).

[148] *Gene Technology Act 2000* (Cth) ss 32–38. The penalties for dealing without a licence and for breach of licence conditions include up to two years imprisonment, fines of up to $55,000 for an individual and from $275,000 for a corporation.

[149] The penalties for aggravated offences include imprisonment for up to five years, fines of up to $220,000 for an individual and up to $1.1 million for a corporation. Pursuant to s 188 in the case of corporate offenders, it is sufficient that conduct engaged in by a director, employee or agent of the corporation was within the scope of their actual or apparent authority and that the director,

The state moratoria legislation also create offences relevant in the particular jurisdiction concerned.[150] Some states/territories require intention or recklessness to commit an offence but others have no such requirement, creating strict liability offences. In all jurisdictions with moratoria legislation, GM crops not authorised under the state/territory legislation and crops commingled with such crops can be destroyed or ordered to be destroyed in some circumstances. Future use of land on which such crops were present can also be restricted in some jurisdictions.

Most importantly, neither the national GT regulatory framework nor state/territory moratoria legislation provides statutory immunity or ousts common law avenues for redress available to people who suffer as a result of actions taken by the GTR or those using gene technology. Further, licences under the GT Act and state moratoria legislation do not make legal what is illegal under other legislation. However, the national GT regulatory framework does not create liability or compensatory mechanisms in respect of harm caused by GMOs although some moratoria legislation provides for limited compensation to those who are inadvertently contaminated by a GMO and the Western Australian Parliament recently announced a new inquiry into mechanisms for compensation for economic loss to WA farmers 'caused by contamination by genetically modified material'.[151] It is clearly intended by the national framework that liability for harm arising from GMO releases be determined by common law principles. The courts will therefore be the forum that determines civil liability where a GMO spreads to others' property.

Common law liability in private nuisance and negligence is possible in Australian courts but unlikely particularly where only pure economic harm has been caused.[152] There has been one court judgment regarding GM crop liability in Australia. This is the case of *Marsh v Baxter*.[153] The GM crop considered in this case was legally released and it was not alleged that the GMO concerned was physically harmful or dangerous—the alleged damage was solely economic loss. The plaintiffs brought torts proceedings against their GM canola growing neighbour after Monsanto Roundup Ready™ canola plant was blown onto the plaintiff couple's property. The Marshes did not grow canola, there was no physical risk to their crops, livestock or property and negligible risk of GM material fertilising any of their crops.[154] The Marshes claimed that their certifier's private organic standards imposed a zero tolerance for GM crops by prohibiting both intentional and adventitious

employee or agent had the relevant state of mind. The corporation is then deemed to have engaged in such conduct unless the corporation establishes that it took reasonable precautions and exercised due diligence to avoid the conduct.

[150]In the Australian Capital Territory and New South Wales these can result in imprisonment for up to two years as well as a fine. The other jurisdictions only have fines as possible penalties.

[151]Western Australian Legislative Council (2017).

[152]Ludlow (2005b), p. 163; Ludlow (2005c).

[153]*Marsh v Baxter* [2014] WASC 187.

[154]*Marsh v Baxter* [2014] WASC 187 [216–218]. However, following the plaintiffs' failure to collect the material or allow others to do it for them until six months after its arrival, eight volunteer GM canola plants eventually grew on the plaintiffs' land. [138], [438] and [669].

'contamination' and that the spread of GM canola to their land caused them to breach those standards, resulting in economic loss to them.[155]

The judge at first instance found that the third party certifier had misunderstood its' own rules when it decertified the plaintiffs' land and that the defendant was not liable for any subsequent economic loss. One important reason why no duty of care was owed by the GM farmer to his organic farmer neighbours was that the plaintiffs had what was essentially a 'self-inflicted contractual vulnerability' that generated their claimed economic losses, particularly given the certifier's behaviour could be objectively assessed as unreasonable or even in breach of the contract between the certifier and plaintiffs.[156] This was confirmed on appeal[157] and the High Court of Australia refused the plaintiffs special leave to appeal to it.[158]

The peculiar facts of the above decision means that given appropriate facts, it remains possible that liability could be established in private nuisance and negligence following the inadvertent presence of GM crops on a third party's land.

3.11 Perception of Genome Editing

3.11.1 Position of Public Authorities

Australian regulators are responding differently to genome edited plants. This in part reflects the different frameworks they operate under. The GTR more quickly responded to calls to review the regulatory framework to take into account genome editing, than the food regulator. It is noted that the Australian Department of Agriculture and Water Resources preferred option 4 during the current review of the GT Regulations, which would regulate only SDN-3 as gene technology.

3.11.2 Public Opinion

The results of a 2017 national survey of community attitudes to gene technology, commissioned by the OGTR, were released in October 2017.[159] These show that while biotechnology is supported by the majority of Australians, the level of support for particular techniques was conditional, often based on regulation and safety being ensured and varied with the type of modification and its purpose.[160] For example, there are wide differences in support for GMOs in medical uses (63%), industrial

[155]*Marsh v Baxter* [2014] WASC 187 [739].

[156]*Marsh v Baxter* [2014] WASC 187 [321].

[157]*Marsh v Baxter* [2015] WASCA 169.

[158]*Marsh v Baxter* (P44/2015) Results of Special Leave Applications heard 12 February 2016.

[159]McCormack and Mercer (2017). Total sample size, n was 1255.

[160]McCormack and Mercer (2017), p. 185.

uses (55%), environmental uses (54%) and food and crops (38%).[161] About 13% of the Australian population was 'strongly opposed to GMOs, and these respondents stood out as having more extreme attitudes to food and agriculture than any other group, as well as low overall trust.'[162] In this regard, Friends of the Earth Australia (an environmental activist group) has been advocating for genome editing to be included within gene technology by the GTR and FSANZ.[163] This includes making and encouraging members of the public to make submissions to the two GTR reviews and the FSANZ review on this topic.

Most respondents to the national survey referred to above (71%) felt that biotechnology would improve Australian's future way of life, although only 46% felt that GMOs would do so.[164] In regards to modifying genes of plants to produce food, '[a]lmost a third indicated that it was acceptable, another third were less sure and were hedging their bets and a quarter clearly believed it was not acceptable. Only 7% indicated *don't know*.'[165] For GM crops in particular, key factors that could cause respondents to change their position on whether such crops should be grown in the respondent's home state were if the crops: provided positive benefits for human health (51% of those opposed would change their position), provided positive outcomes for the environment (47% would change their position), passed stringent health regulations (42% would change their position) and if it enhanced economic competitiveness (33% would change their position).[166]

While awareness of genome editing was lower amongst the respondents than GM[167] 'more than half of respondents (57%) indicated they thought gene editing would improve our way of life in the future'.[168] However, 17% thought genome editing might make things worse.[169] Some types of genome editing were more acceptable to the community than GM. In particular, the survey found genome editing 'received quite high acceptance (42%) relative to other techniques, when asked about *making a small change to an existing gene within a plant, as is done in gene editing*'.[170] On being asked to use a scale of 0–10, where 10 is completely acceptable and 0 is completely unacceptable, to indicate how acceptable modifying the genes of *plants* to produce food by a range of different techniques, the results showed that 'there is more support for modifications that are perceived to be less

---

[161] McCormack and Mercer (2017), p. 3.

[162] McCormack and Mercer R (2017), p. 3.

[163] Sales (n.d.).

[164] McCormack and Mercer (2017), p. 4.

[165] McCormack and Mercer (2017), p. 37.

[166] McCormack and Mercer (2017), p. 12.

[167] '17% of respondents stating that they could explain [genome editing] to a friend, 39% stating that they had heard of it but knew little or nothing about it, and another 39% stating that they had never heard of it.' McCormack and Mercer (2017), p. 21.

[168] McCormack and Mercer (2017), p. 11.

[169] McCormack and Mercer (2017), p. 4.

[170] McCormack and Mercer (2017), p. 4.

radical or extreme. So the highest levels of support were for *Introducing the genes of a plant of the same species (43%)* and *Making small changes to the existing genes within a plant, as is done in gene editing* (42%).'[171]

Australian business may not be as optimistic in relation to public opinion on genome editing. Horticulture Innovation Australia (HIA), which represents the Australian vegetable industry, interviewed a range of industry members (vegetable plant breeders and seed merchants) in preparing its response to the current gene technology regulation review. HIA concluded that 'across all groups consulted, there was consistent recognition that the critical hurdle for the implementation of NBTs in the Australian vegetable industry will be public/consumer acceptance.' HIA explains that its stakeholder concerns regarding the likely acceptance of NBTs by consumers relate to the current negative attitude of the public towards GM food and whether non-GM varieties developed using NBTs will be differentiated from GM varieties. The question of differentiation will have two components: (i) whether the non-integrative NBTs as a class of technologies will be perceived by society simply as an extension of existing GM technology and (ii) whether products of particular NBTs will be explicitly classified as GM by regulatory bodies. Differentiation from 'traditional' GM varieties was the most frequently expressed concern of researchers, plant breeders, growers and other industry operators according to HIA.[172]

3.12 Treatment of Other New Breeding Technologies

3.12.1 Gene Technology

The GTR's interim advice, the *General advice from the Regulator on coverage of new technologies*, identifies some RNA interference applications as problematic new technologies. Its' Consultation RIS following the Technical Review of the GT Regulations proposes to list the application of RNA molecules to induce RNAi as a technique that is *not* gene technology, provided the RNA cannot give rise to changes to genomic sequence and cannot be translated into proteins. This exclusion applies regardless of how the RNA is introduced—including spraying or soaking plant parts with RNA solutions, or exposing cultured cells to RNA solutions. RNAi techniques which involve inserting sequences into the genome or use of viral vectors would continue to result in GMOs which are subject to regulation.[173] Techniques involving infectious non-coding RNAs such as viroids, are also regulated (Table 3.6).

The GTR justifies the exclusion of RNA-delivered RNAi techniques from regulation as being consistent with the original intent of exclusions to regulation, because

[171]McCormack and Mercer (2017), p. 39.
[172]Jones and Horticulture Innovation Australia (2016), pp. 36–37.
[173]OGTR (2017e).

Table 3.6 Summary of approach to other new breeding technologies by Gene Technology Regulator

Technique	Regulated as GT
RNAi	No
Grafting	Yes

such organisms can occur in nature and therefore do not pose a particular biosafety risk to the environment or human health or safety.[174]

It seems that under current Australian regulations, the grafting of non-transgenic scions onto transgenic rootstock, causes the whole plant and its harvested products to be a GMO.[175]

3.12.2 Food

A summary of the FSANZ workshop outcomes and Consultation Paper discussed above, in regards to other new breeding technologies is outlined below (Table 3.7).

Cisgenesis/intragenesis FSANZ classifies cisgenes and intragenes with transgenes, as all introduce DNA into a novel site in the plant's genome.[176] Food from plants produced using such techniques is therefore regulated as food produced using gene technology. However, the source of the gene could potentially influence the type of safety assessment required. Genes from organisms commonly used as food and with a history of safe use may require less evidence to establish their safety.[177]

Grafted Plants Plants which have undergone GM rootstock grafting are regarded as a single organism. While food produced by the non GM scion of such an organism does not contain modified DNA, it (and the food it produces) may contain novel gene products (RNA or protein) and have altered characteristics because of the GM to the rootstock. Although FSANZ has acknowledged that such plants do not neatly fit into the same category as transgenes, it is likely to be classified as GMOs. Food from such plants is therefore food produced using gene technology and would need to undergo premarket safety assessment. However, as with cis/intragenic plants, the scientific panel concluded that a simplified safety assessment would be appropriate where there is no transmission of novel gene products to the food and no altered characteristics due to the genetic modification to the rootstock.[178] This is being explored further in the current review, which has also proposed that it may be appropriate to assess and approve a particular GM rootstock, which would allow

[174]OGTR (2017a, b, c, d, e), pp. 11–12.

[175]Jones and Horticulture Innovation Australia (2016), p. 10.

[176]FSANZ (2018a), p. 10.

[177]FSANZ (2012), pp. 4 and 15.

[178]FSANZ (2012), p. 4.

Table 3.7 Summary of approach to other new breeding technologies by Food Regulator

Technique	Regulated as GM food
Cisgenesis/intragenesis	Yes
Grafting	Yes
Agro-infiltration	Determined by whether plant is GM plant[a]
RNA-dependent DNA methylation (RdDM)	No conclusion
Reverse breeding	No conclusion

[a]FSANZ (2013), p. 5

grafting of any non-GM scion onto an approved GM rootstock without the need for individual assessment of the composite plant.[179]

Agro-infiltration FSANZ's scientific panel has noted that this technique most commonly uses somatic (non-germ line) cells and therefore the integrated DNA, if there is any, will not be inherited in the next generation. Such techniques will therefore not be gene technology pursuant to the definition in the food Code. FSANZ also notes that from a food perspective, the most likely substances to be produced using this technology will be food processing enzymes or food additives and potentially also protein supplements. FSANZ noted that food processing enzymes and food additives are not regarded as food, and are regulated under separate Standards in the Code, irrespective of whether or not they have been produced using GM techniques.[180] For substances that are food products, such as protein supplements, whether they are regarded as food produced using gene technology depends on whether the expression vector becomes stably integrated into the plant genome. Such integration events may occur at low frequency in the infiltrated area. However, the plants won't be used as food, only the purified proteins they produce and no food safety concerns are raised.[181] Whether such foods are regulated will depend on the outcomes of the current review.

Reverse Breeding It is unlikely the resulting organisms produced using reverse breeding and food derived from them will be food produced using gene technology. Whether such food should nevertheless undergo some form of pre-assessment and approval is nevertheless being considered in the current review.[182] The scientific panel recommended that whether food regulation is triggered should depend on criteria that it recommended to be developed, including whether there is a complete barrier/genetic separation between the early GM breeding lines and the non-GM food-producing lines.[183] FSANZ is currently seeking public input on these criteria, should it be decided that such food should be subject to pre-assessment and approval.[184]

[179]FSANZ (2012), p. 17.
[180]FSANZ (2013), p. 12.
[181]FSANZ (2013), p. 13.
[182]FSANZ (2018a), p. 11.
[183]FSANZ (2012), p. 13.
[184]FSANZ (2018a), p. 11.

3.13 Conclusion

Australia's regulatory response to genome edited plants is under close scrutiny as the two national regulators relevant to plant cultivation and the sale of their products, the GTR and FSANZ, complete reviews. Both regulators have indicated that the use of a process trigger for regulation to apply may no longer be suitable for their regulatory frameworks. Certainly a change of trigger seems to be an issue under consideration although it is not the first time such an idea has been recommended to government. In 2008, a paper prepared for the Australian Government recommended that the GT regulatory framework adopt an output rather than input trigger because, in part, of the next wave of 'agricultural biotechnology techniques'.[185] Interestingly it was observed in that paper that 'The policy approach underpinning the current regulatory system does not appear well placed to deal with an increasing proliferation of new biotechnology techniques and applications. If the principles upon which the current system is based continue to apply as new technology is introduced, replicas of the OGTR may well be established for each new technology. This outcome is unlikely to be a workable solution and could create significant distortions between the uses of new technologies in the future.'[186] Nevertheless, it was concluded that having a GT framework provided consumer confidence and so it was ultimately recommended that the framework remain. It is unlikely that a change of trigger, even if it is agreed upon, would occur in the next five years.

The approach of the food regulatory framework to GM foods has similarly been criticised in external reviews, with the Australian Productivity Commission recommending that GM labelling be abandoned. The Commission correctly observed that general GM labelling requirements are present in the Code only for consumer information purposes and not safety reasons. FSANZ does not seem to favour the expansion of such requirements. In respect of the GT regulatory framework, there is the added layer of regulation that could be imposed by the states and territories for trade or other socio-economic reasons, in the same way as has been done around GMOs. This may become a more significant risk if the public is not satisfied by the responses of the national regulators.

That plants produced using some genome editing techniques and their products may in some cases be indistinguishable from conventionally bred plants and their products is another factor that Australian regulators have taken express note of. Should such plants and products nevertheless be regulated, a labelling and certification or paper trail, such as that used for origin labelling, may have to be implemented.[187] The willingness of the regulators to take on such a methodology remains to be seen.

In any case, in the interim it is clear that the gene technology and food regulatory frameworks apply to at least some genome edited plants or their products.

[185]ACIL Tasman (2008), p. 59.

[186]ACIL Tasman (2008), p. 57.

[187]Fernandez Albújar and van der Meulen (2017), p. 11.

Unfortunately, which plants and products are or will be regulated differs between frameworks. This difference has been raised by both regulators but it seems unlikely that it will be sufficient to cause the regulators to ensure that they take the same approach, if that is the only driving factor for such a move. Consistency with international approaches may nevertheless drive the regulators in the same direction.

References

ACIL Tasman (2008) Biotechnology and Australian Agriculture. Towards the development of a vision and strategy for the application of biotechnology to Australian Agriculture (July). Report prepared for Biotechnology Australia and the Australian Department of Agriculture Fisheries and Forestry

Allen Consulting Group (2011) Review of the Gene Technology Act 2000, undertaken for the Australian Department of Health and Ageing, Canberra, August

APVMA (2018) Guideline for the regulation of biological agricultural products. In: Registrations and Permits. Australian Pesticides and Veterinary Medicines Authority. Available at https://apvma.gov.au/node/11196. Accessed 8 Jan 2018

Australian Department of Agriculture and Water Resources (2018) Guideline for responding to contamination by prohibited substances or materials in the organic supply chain (2018-01). www.agriculture.gov.au/export/controlled-goods/organic-bio-dynamic/organic-notices/2018/2018-01. Accessed 12 Apr 2018

Australian Department of Agriculture and Water Resources International Statement on Low Level Presence. http://www.agriculture.gov.au/ag-farm-food/biotechnology/international-statement-low-level-presence. Accessed 13 Apr 2018 and Joint statement on innovative agricultural production technologies, particularly plant biotechnologies, http://www.agriculture.gov.au/ag-farm-food/biotechnology/ag-production-technologies. Accessed 13 Apr 2018

Australian Department of Health (2017) Operations of the Gene Technology Regulator Annual Report 2016-17. ISSN 1833-6264

Australian Department of Health (2018) The Third Review of the National Gene Technology Scheme. Preliminary Report

Blewett Report (2011) Labelling logic. Review of food labelling law and policy

Brookes G (2016) Adoption and impact of GM Crops in Australia: 20 years' experience. Report prepared for CropLife Australia Ltd, Canberra

Commonwealth of Australia (2006) Statutory review of the Gene Technology Act 2000 and The Gene Technology Agreement, report prepared for the Gene Technology Ministerial Council, Canberra

Croplife Australia (2017) Submission to the 2017 Review of the National Gene Technology Regulatory Scheme, 29 September

Fernandez Albújar G, van der Meulen B (2017) The legal GMO concept. European Institute for Food Law Working Paper Series 2017/3

FSANZ (2012) New plant breeding techniques. Report of a Workshop hosted by Food Standards Australia New Zealand. http://www.foodstandards.gov.au/consumer/gmfood/Pages/New-plant-breeding-techniques-in-the-spotlight.aspx. Accessed 5 Jan 2018

FSANZ (2013) New plant breeding techniques. Report of a Workshop hosted by Food Standards Australia New Zealand. http://www.foodstandards.gov.au/consumer/gmfood/Pages/New-plant-breeding-techniques-in-the-spotlight.aspx. Accessed 5 Jan 2018

FSANZ (2017) 2017 Call for Submissions – Application A1154. Food derived from insect-protected cotton line MON88702, 20 December

FSANZ (2018a) Consultation Paper. Food derived using new breeding techniques (Feb)

FSANZ (2018b) Food derived using new breeding techniques - review. http://www.foodstandards.gov.au/consumer/gmfood/Pages/Review-of-new-breeding-technologies-.aspx. Accessed 10 Jan 2018

FSANZ Application Handbook. http://www.foodstandards.gov.au/code/changes/pages/applicationshandbook.aspx. Accessed 21 Feb 2018

Huttner E (1997) 1996: Transgenic crops debut on the world stage. In: McLean GD et al (eds) Commercialisation of transgenic crops: risks, benefit and trade considerations. Proceedings of a workshop, Canberra, March 1997. AGPS, Canberra, p 1

ISAAA (2016) Global Status of Commercialized Biotech/GM Crops: 2016. ISAAA Brief No. 52. ISAAA, Ithaca, NY

Jones M, Horticulture Innovation Australia (2016) Final Report. New breeding technologies and opportunities for the Australian vegetable industry ISBN 9780 7341 39887

Kelly L (2017) FSANZ, CSIRO Gene editing of crops Workshop, Kiama, NSW, 28–30 November

Ludlow K (2004a) Genetically modified organisms and pollution in Victoria. Environ Plann Law J 21:351–371

Ludlow K (2004b) Cultivating chaos: state responses to releases of genetically modified organisms. Deakin Law Rev 9:1–40

Ludlow K (2005a) Gene Technology Regulation and the Environment Protection and Biodiversity Conservation Act 1999 (Cth). Monash Univ Law Rev 30:165–198

Ludlow K (2005b) The economic impact of genetically modified organisms as actionable damage in torts. Torts Law J 13:159–187

Ludlow K (2005c) Genetically modified organisms and private nuisance liability. Tort Law Rev 13:92–121

McCormack C, Mercer R (2017) Community attitudes to gene technology. Consultancy Report for OGTR Ref no J 2537, October

OGTR (2011) Advice to Dow AgroSciences EXZACT technology. http://www.ogtr.gov.au/internet/ogtr/publishing.nsf/content/C2FFDD165F9AE9AECA257E56007F4B20/$File/Dow%20Exzact%20advice%20letter%20of%20May%202012.PDF

OGTR (2016-17) Technical Review of the Gene Technology Regulations 2001. Available at http://www.ogtr.gov.au/internet/ogtr/publishing.nsf/Content/reviewregulations-1

OGTR (2016a) Australian Department of Health, Technical Review of the Gene Technology Regulations 2001. Discussion Paper: Options for regulating new technologies (October)

OGTR (2016b) General advice from the Regulator on coverage of new technologies. http://www.ogtr.gov.au/internet/ogtr/publishing.nsf/Content/newtechnologies-htm. Accessed 16 Dec 2016

OGTR (2017a) Updating Gene Technology Regulation in Australia. Regulation Impact Statement for Consultation, Australian Government, Department of Health (released 30 Nov)

OGTR (2017b) Review of the National Gene Technology Scheme 2017. Consultation Paper: Overarching issues for consideration under the review (released 30 Nov)

OGTR (2017c) Updating Gene Technology Regulation in Australia. Consultation quick guide, 2017 Australian Government, Department of Health (released 30 Nov)

OGTR (2017d) Draft future law compilation of Gene Technology Regulations 2001 (released 30 Nov)

OGTR (2017e) Technical Review of the Gene Technology Regulations. In: About the OGTR. Office of the Gene Technology Regulator. http://www.ogtr.gov.au/internet/ogtr/publishing.nsf/Content/reviewregulations-1. Accessed 6 Dec 2017

OGTR (2018) Licence. In: Table of applications and authorisations for Dealings involving Intentional Release (DIR) into the environment. Available at http://www.ogtr.gov.au/internet/ogtr/publishing.nsf/Content/ir-1. Accessed 8 Jan 2018

OGTR (n.d.) Unintended Presence Strategy (Unapproved GMOs in seed or sowing). http://www.ogtr.gov.au/internet/ogtr/publishing.nsf/Content/mon-unintended-1. Accessed 13 Apr 2018

Productivity Commission (2016) Regulation of Australian Agriculture. Report no 79, Canberra, Australia

Sales L (n.d.) Friends of the Earth Australia, GM 2.0. Australian Regulators Engineering the Truth

Sylvan L (2005) Responsibly Informing Consumers – Experience from the Trade Practices Act, Highlands 2005 Forum Australian Food and Grocery Council Senior Executive Forum Gold Coast, May 6

The Legislative and Governance Forum on Gene Technology (2017) Review of the National Gene Technology Regulatory Scheme. Background Paper (July), Australia

Western Australian Legislative Council (2017) Media Release 7 December, Environment and Public Affairs Committee: New Inquiry into mechanisms for compensation for economic loss to farmers in Western Australia caused by contamination by genetically modified material

Chapter 4
Regulation of Genome Editing in Plant Biotechnology: Canada

Stuart J. Smyth

Abstract Gene editing technologies are the latest to emerge from the broader field of agricultural biotechnology. Canada's science-based regulatory system was adapted for genetically modified (GM) crops in the early 1990s and has proven sufficiently robust in responding to these new plant breeding techniques, having approved two varieties of gene edited canola. Canada does not have a mandatory labelling system and the products from these crops will have seamlessly entered the food supply system in Canada.

A 2017 industry organized workshop on the future of plant breeding regulations in Canada, acknowledged the excellence of Canada's regulatory system for GM crops. However, it was noted that Canada cannot afford to be complacent when it comes to regulatory competitiveness as multinational technology development firms are making investment decisions based on regulatory efficiency. To ensure that Canada's regulatory framework remains efficient, a two-tiered regulatory system has been suggested as a means of leveraging the 25 years of experience and knowledge gained through the safe regulation of GM crops. To date, no regulatory changes to move in this direction have been implemented by Canadian regulatory agencies.

4.1 Introduction

Canada is one of the leading countries in terms of the adoption of genetically modified (GM) crops, having first done so in 1995. The first GM crops approved for commercial production in Canada were two varieties of herbicide tolerant canola, followed quickly by varieties of corn and soybeans in 1996 and 1997, respectively. GM canola adoption was very rapid, with 12% in the initial year, 64% by 2002 and 93% by 2010 (Fig. 4.1). The remaining 7% of canola varieties are herbicide tolerant, but have been developed through mutagenic breeding techniques. Presently 100% of

S. J. Smyth (✉)
Department of Agricultural and Resource Economics, University of Saskatchewan, Saskatoon, SK, Canada
e-mail: stuart.smyth@usask.ca

H.-G. Dederer, D. Hamburger (eds.), *Regulation of Genome Editing in Plant Biotechnology*, https://doi.org/10.1007/978-3-030-17119-3_4

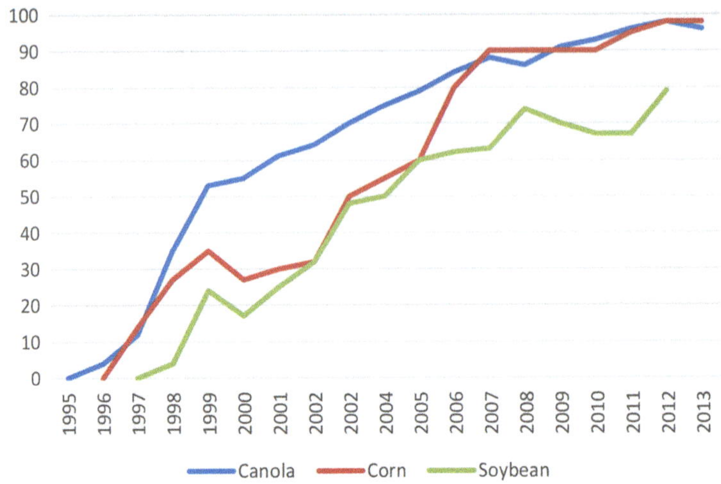

Fig. 4.1 Adoption percentage of GM crops in Canada (Smyth 2014, p. 196)

the canola produced in Canada is herbicide tolerant. Based on seed sales, virtually all corn varieties are either herbicide tolerant, insect resistant or stacked varieties with both traits. The adoption of GM soybeans has lagged the other two crops, largely due to the lack of specific soybean breeding programs dedicated within Canada. Once these programs were established, adoption rates increased substantially. GM canola and corn can be viewed to be at virtually full adoption while GM soybeans is between 75% and 80%. Acreage for the three dominant GM crops in Canada are 23 million acres for canola, 7.3 million acres of soy and 3.6 million acres of corn.

Varieties of GM apples and potatoes have also been approved in Canada for direct consumption and production, as well as GM salmon. There is also a small amount of GM sugar beet production. Canada has approved for import, a wide variety of GM crops commercialized in other countries that would not be able to be produced in Canada, such as GM papaya.

The majority of GM crops produced in Canada are exported. Just over half (56%) of the canola seed produced is exported, with 75% of the crushed oil and 90% of the canola meal being exported.[1] Approximately 70% of Canada's soy production is exported.[2] In the case of GM corn, the majority is mixed into livestock feed or enters the domestic ethanol industry.

In 2007, I led a survey of 580 canola producers to learn more about the producer level impacts of GM canola that were being observed one decade after commercialization. GM canola has been very profitable for Western Canadian producers, generating between $1.06 billion and $1.19 billion net direct and indirect benefits over the 2005–2007 period, partly attributed to lower input costs and partly to better

[1]Canola Council of Canada (n.d.).

[2]Soy Canada (2016).

Table 4.1 Gene editing variety approvals in Canada[a]

Crop	2013	2014	Total
Canola	1	1	2

[a]CFIA (2018)

weed control.[3] More than 94% of respondents reported that weed control was the same or had improved following the commercialization of GM canola, less than one quarter expressed any concern about herbicide resistance in weed populations, 62% reported no difference in controlling for volunteer GM canola than for regular canola and only 8% indicated that they viewed volunteer GM canola to be one of the top five weeds they needed to control. Control of volunteer canola was listed as an agronomic issue well in advance of the commercialization of GM canola. The adoption of GM canola varieties offered new options in weed control, allowing farmers to extend the number of years that they could go without having to till a field.

Conservation tillage practices changed substantially following the adoption of GM canola as in 1999, only 11% of canola acres practiced this form of land management, changing to a situation where 64% of producers were using zero or minimum tillage as their preferred form of weed control in 2007.[4] When comparing canola production in 1995 and 2006 the toxicity of herbicides applied to canola decreased by 53%, there was a decrease in producer exposure to chemicals of 55% and a decrease in chemical active ingredient application of 1.3 million kg.[5] The cumulative environmental impact per hectare (EI/ha) of the top five herbicides applied in 1995 was measured as a factor of 46,715, while the comparable factor for the top five herbicides applied in 2006 was 29,458. If GM canola had not been developed and Canadian canola producers continued to use previous production technologies, the amount of active ingredient applied to control weeds in 2007 would have been 38% above what was actually applied. The environmental benefits of GM canola were significant as 41% of farmers identified that they were now seeding canola onto land that they identified as erodible.[6]

Gene editing technologies have been utilized in the development of two canola varieties, both of which have received approval in Canada. Table 4.1 summarizes the varieties approved that were developed based on gene editing technologies.

These two canola varieties may not have been intended for release into commercial production, rather they may have been used as a technology test case. Private technology development firms would make the submission to learn about the process that the Canadian regulators would take in regard to the product and how it would fit with Canada's plants with novel traits classification. This test case concept would allow the firms to learn from regulators specifically about what information would be required that might have been different from previous submissions with transgenic variety submissions. Proceeding in this manner would help ensure that when gene

[3]Gusta et al. (2011), p. 11.
[4]Smyth et al. (2011a), p. 407.
[5]Smyth et al. (2011b), p. 499.
[6]Smyth et al. (2011a), p. 407.

Table 4.2 Legislation governing biotechnology in Canada[a]

Agency	Product	Act
Canadian Food Inspection Agency (CFIA)	Plants with novel traits Novel fertilizers and supplements Novel livestock feeds Veterinary biologics	Seeds Act Fertilizers Act Feeds Act Health of Animals Act
Health Canada	Novel foods Pest control products	Food and Drug Act Pest Control Products Act
Environment and Climate Change Canada	All animate products of biotechnology for uses not covered under other federal legislation	Canadian Environmental Protection Act (1999)

[a]CFIA (2005)

editing varieties intended for commercialization did come forward, that the process would be as efficient as possible.

4.2 The Regulatory Framework for Genetically Modified Organisms: An Overview

Canadian regulators established a new classification of plants to deal with the potential risks that had a probability of developing following the application of new genetic technologies to the science of plant breeding. The development of Canadian regulations for the initial innovative crops were based on science and subsequent regulatory changes have continued to be science-based. In accordance with recommendations from various international scientific societies, the regulations focus on the end product, not the process used to create the product. To this end, Canada developed regulations for plants with novel traits (PNTs). Plants that are classified as PNTs are plants that do not have a history of production and safe consumption in Canada. They may have been introduced from elsewhere, or be genetically modified using genetic engineering, mutagenesis, or any other breeding method.[7]

The regulation of products created via biotechnology is the responsibility of several federal government agencies: the Canadian Food Inspection Agency (CFIA), Health Canada and Environment and Climate Change Canada (Table 4.2). Using legislation from four different Acts, the CFIA is responsible for plants, animal feeds, fertilizers and veterinary biologics. The Office of Plant Biosafety was established within the CFIA to co-ordinate the safety evaluation of novel foods. Through the Food and Drugs Act and the Pest Control Act, Health Canada oversees the regulation of foods, drugs, cosmetics, medical devises and pest control products. All safety assessments are conducted based upon scientific principles developed

[7]CFIA (2004).

through expert international consultations with the World Health Organization (WHO), Food and Agriculture Organization (FAO) and the Organisation for Economic Co-operation and Development (OECD).[8] Environment and Climate Change Canada acts as a regulatory safety net for products of biotechnology, where they have the regulatory mandate for all animate products of biotechnology for uses not covered under other federal legislation. Environment and Climate Change Canada regulates biotechnology within the scope of the Canadian Environmental Protection Act (1999).

All novel trait products, prior to receiving registration approval, are thoroughly tested by the CFIA and Health Canada officials using scientific approaches. Officials from both departments work together on new variety applications. Officials do not conduct or redo the scientific experiments and research information that is submitted by the applicant (usually a private company or public university); rather, they analyze the data submitted and may redo portions of the experimentation to corroborate results. Frequently, government officials will ask the applicant to provide them with additional information regarding specific segments of the application, which may result in the applicant conducting additional scientific experiments. Upon the review of all information, the variety is accepted if all conditions are fully met and rejected if any condition is not deemed to be acceptable.

While Environment and Climate Change Canada (ECCC) is listed as a government department that has regulatory authority over GM crops and biotechnology, in practice, they have no role. Scientists with the CFIA and Health Canada review all of the documentation submitted by the variety developer as part of the risk assessment process. At no time in the risk assessment process are officials, experts or scientists from ECCC involved. To the best of my knowledge, in 25 years of regulating GM crops and PNTs in Canada, ECCC and previously Environment Canada, have never been involved in the risk assessment of any crop variety risk assessment process.

4.3 Regulatory Status of Genome Edited Plants

As the technologies of GM crops progressed their way from laboratory proof of concept in the early to mid-1980s, to greenhouse trials in the mid-1980s, open field trials in the mid to late 1980s, regulatory assessment in the early 1990s and finally commercial production, the regulatory systems in many jurisdictions were in development, becoming standardized when the initial GM varieties were approved in the US in 1994 and in Canada in 1995. Canada has remained committed to the scientific principles laid down in its domestic regulatory framework for plants with novel traits dating back to the late 1980s and early 1990s. All commercialized GM plants to date have been considered to contain novel traits and, therefore, have been assessed for safety. However, the approach used by the CFIA does not mean that all PNTs are

[8] Harrison (2001).

developed through genetic modification. Novel traits can be developed through various techniques (other than genetic modification) such as mutagenesis, somaclonal variation and other forms of what in other countries are considered 'traditional' breeding.

4.3.1 Applicability of the Regulatory Framework for GMOs

The governance system for crop agriculture is based on an extensive horizontally-based public/private regulatory system.[9] Risks are managed by various stakeholders depending on the stage of the variety development. Private and public breeders are responsible for managing any risks in their research programs as long as the materials remain in isolated conditions (e.g. in laboratories or under glass), once the breeder has developed a cultivar that is genetically stable and unique, it is ready to be examined for registration and the formal system takes over. In the production system, the public sector has tended to establish the general environment for private actors to effect transactions. The Food and Drugs Act (1985) sets rules for human consumption, the Feeds Act (1983) sets maximum tolerances of nutrients for live-stock feed and the Seeds Act (1985) specifies the performance standards for new germplasm.

These three Acts are designed to establish standards for risks related to plant agriculture. The main quality attributes of the Seeds Act are uniformity, stability and uniqueness. However, this Act also establishes thresholds for environmental safety risk aspects such as: the potential of the plant to become a weed or to be invasive of natural habitats; the potential for gene flow to wild relatives; the potential for a plant to become a plant pest; the potential impact of a plant or its gene products on non-target species; and the potential impact on biodiversity.[10] The Feeds Act defines the thresholds for the potential risks due to allergenicity, toxicity, digestibility and dietary exposure relating to animal feeding. The Food and Drugs Act establishes risk thresholds for allergenicity, toxicity, metabolization, nutrition and dietary exposure relating to human consumption. The integration of these three Acts into the regulatory framework for new plant varieties is designed to identify all potential risk categories and ensure that any new plant variety is benchmarked to existing varieties already determined to be safe for human and animal consumption. This is known as 'substantial equivalence', whereby as long as any new variety equals the physical properties of existing varieties, they are deemed to be substantially equivalent to existing varieties and approved for commercial production.

Due to the above definition and the subsequent assessment categories, every herbicide tolerant (HT) variety application that the CFIA has received, has been treated as a PNT, regardless of the technology used to create the HT variety, due to

[9]Smyth et al. (2004), pp. 70–73.

[10]CFIA (2004).

the novelty of herbicide tolerance. Herbicide tolerant varieties and subsequent applications with other, and stacked traits, have been assessed for variety approval under the following CFIA directives:

- Directive 94-08: Assessment Criteria for Determining Environmental Safety of Plants with Novel Traits;
- Directive 95-03: Guidelines for the Assessment of Novel Feeds: Plant Sources;
- Directive D-96-13: Import Permit Requirements for Plants with Novel Traits, and
- their Products; and
- Directive 2000-07: Guidelines for the Environmental Release of Plants with Novel Traits within Confined Field Trials in Canada.

Using these directives, the CFIA assesses all PNT variety applications for environmental release and use as animal feed.

Unlike the CFIA, which uses a product trigger, Health Canada defines novel foods as foods resulting from a process not previously used for food, products that do not have a history of safe use as a food or foods that have been modified by genetic manipulation, genetically engineered foods or biotechnology-derived foods.[11] Health Canada assesses the safety of all GM and other novel foods proposed for sale in Canada. Companies are required to submit detailed scientific data for review and approval by Health Canada, before such foods can be sold or as animal feed if the modified feed has the potential to introduce harmful components into the portion of the animal being consumed as food.

Health Canada does not review all foods new to the Canadian market but only those that are deemed novel. Therefore, the concept of prior safe use as a food was introduced to exclude foods new to the Canadian market which have a history of safe food use in other countries, from being the target of a novel food notification. Secondly, the concept of 'major change' was introduced into the novel food definition in order to avoid the potential of a minor processing change to trigger a novel food notification. This approach intended to restrict novel food notifications due to introduction of new processes only to those that are truly new and cause substantial changes in the composition of the food.

While no formal standard or definition for novel exists, Canadian plant breeders use a rule of thumb that if the specific trait they are selecting for expresses at 20–30% higher or lower than conventional varieties, the plant breeder initiates discussions with regulators regarding the applicability of PNT regulations in the specific instance. The PNT regulations apply to all plant varieties having a novel trait, regardless of how they were developed, meaning that the variety could be developed by gene editing, genetic modification, mutagenesis or even old fashioned pollen transfer breeding. It is expected that some of the gene editing technologies may create products that are PNTs, while some of them may create products that are not PNTs.

[11]Health Canada (2006).

4.3.2 Regulatory Classifications of Genome Editing/Genome Edited Plants

Canada's approach to gene editing technologies is no different from the technologies that have preceded it, in that if the technology creates a novel product, then Canada's PNT regulations are triggered, resulting in additional regulatory oversight on allergenicity, toxicity and impacts on non-target organisms.

New breeding techniques have been employed to develop new varieties of crops that have been submitted to Canadian regulators for risk assessment over the past few years. While it is likely still too early for CRISPR developed crops to have reached the stage where they have completed field trials and the requisite data has been gather to compile a regulatory submission package, they will likely be reaching this stage quickly. Further information on this is nonexistent as this information is treated as confidential business information by the technology development firms. Once greenhouse variety lines have been assessed, selected lines are then put forth into field trials to gather agronomic data required to inform regulators. It typically takes 3 years of field trials to gather the pertinent information. Based on this, the first CRISPR-based varieties might have entered field trials at the very earliest 2016, however, 2017 or 2018 are more likely. This would result in CRISPR-based varieties being at the stage of variety approval submission around 2020.

While the regulatory system has not been flooded with NBT submission, it has received, assessed and approved two varieties of canola (Table 4.1). Canada's PNT system does not differentiate between any of the gene editing technologies. Given that Canadian regulators treat PNT submissions on a case-by-case basis, they believe that the PNT system is sufficiently flexible to deal with increased submission of gene edited varieties. This position is similar to that of South Africa's Academy of Sciences that conducted a review of South Africa's regulatory framework for compatibility with gene editing variety submissions.[12]

4.4 Regulatory Prerequisites for Activities Relating to Genome Edited Plants

Canada does not differentiate between any plant breeding method or technology that results in a PNT, including the various gene editing technologies. All of the variety development procedures and protocols for gene editing varieties are the same as they are for any other crop variety that is classified as a PNT.

All of the regulatory requirements for GM crops discussed in Sect. 4.3.1, would be equally applicable for gene edited developed crop varieties. Identical information documents would be required to be submitted by the variety developer, addressing

[12] ASSAf (2017).

Table 4.3 Gene editing techniques and traits

Crop	Approvals	Technique	Trait	Comments
Canola	2	Oligonucleotide-directed mutagenesis; site-directed mutagenesis	Herbicide tolerance	In the CFIA decision document for the ODM variety, the trait introduction method is listed as mutagenesis and conventional breeding[a]

[a]CFIA (2013)

the requirements of the three Acts governing variety approval. Gene edited varieties would have to conform to the Seeds Act for uniformity, stability and uniqueness, as well as environmental risk aspects including, the potential to become a weed or to be invasive, the potential for gene flow, the potential to become a plant pest, the potential impact on non-target species and the potential impact on biodiversity. Compliance with the Feeds Act would include addressing the risk potentials for allergenicity, toxicity, digestibility and dietary exposure relating to animal feeding. The Food and Drugs Act information would need to address allergenicity, toxicity, metabolization, nutrition and dietary exposure relating to human consumption. This data would be assessed by scientists within the CFIA and Health Canada. If the review of the data from a gene edited variety was deemed to be substantially equivalent to that of existing varieties, gene edited varieties would be found to be substantially equivalent and approved for commercial production.

Given that Canada has approved two gene edited varieties to date (Table 4.1), the decision documents publicly available do not identify that any additional risk assessment information was required, or provided, that deviated from the data required to be provided for any other PNT submission. Risks are quantified and scientifically assessed, with the process being the same for any plant breeding method resulting in a PNT variety. Under the Canadian regulatory framework for PNTs, if the variety being assessed has risks that are substantially equivalent to crop varieties previously approved and presently in commercial production, it is approved. A gene edited variety under assessment would not be subject to additional data requirements or risk assessment protocols and upon review of the scientific risk assessment documentation, if it was found to possess risks similar to existing varieties, it would be approved for commercial production, as have occurred on two previous instances.

4.5 Status Quo of Genome Edited Plants and Products Derived from Them

As of the end of 2017, Canada has approved two canola varieties developed by gene editing technologies (Table 4.3). Both of the variety traits approved have been for herbicide tolerance. Interestingly, the oligonucleotide-directed mutagenesis (ODM) canola variety is listed as having been developed via mutagenesis and conventional breeding, yet it is treated as a PNT, showing that it is the product that is regulated, not the process.

Given that these varieties were approved in 2013 and 2014, it should be expected that they would be available for commercial production. The one caveat to this is that the canola industry association has agreement that for a variety to be commercially released, it would have to be approved to be imported into the key canola export markets. This means that each of Canada's key canola export countries would have to have approved these varieties of canola for import, prior to it being commercially available in Canada. Given the uncertainties regarding regulation of gene editing technologies in many countries, it is doubtful that import approval has been received in all of the key Canadian canola export markets and that these varieties have yet to be made commercially available.

The CFIA does not provide information regarding variety submissions, treating all submission data as confidential business information. In talking to the biotech firms, they have gene editing varieties presently under regulatory review, but no information is provided by the CFIA. Only when decision documents are posted online by the CFIA is information about the variety publicly available.

4.6 Reform Efforts

An industry workshop on the future of plant breeding in Canada, was organized by CropLife Canada, the Canada Grains Council and the Canadian Seed Trade Association and held in Ottawa on 30 May, 2017. The workshop brought together experts from industry, academia and federal government policy and regulatory departments and agencies. The objectives of this multi-stakeholder event were: to clarify the present regulatory framework; assess the system's strengths and weaknesses; identify the science advancements that are driving the need for change; identify options for improving the regulatory framework; and establish some next steps.[13]

One of the strengths of the Canadian regulatory system is that it has safely regulated and approved GM crops beginning in 1995, establishing a 23 year period of correct risk assessment decisions. In addition to this, Canada is recognized as a leading regulatory agency given its long history of GM crop regulation. The consensus of the plant breeding workshop was that Canada is well positioned to adopt gene edited plants given this history of science-based regulation of focusing on the product, not the process.

While Canada does have an effective and efficient regulatory system for GM crops, this should not prevent the ongoing assessment for ways to increase regulatory efficiency. Increasingly, companies are making strategic investment decisions to invest new resources into countries with the most efficient regulatory system.[14] The

[13]CropLife (2017), p. 2.

[14]Smyth et al. (2014), p. 1.

workshop identified several opportunities for improving the Canadian regulatory system. It was felt that given the depth of knowledge and experience in regulating PNTs, that the CFIA and Health Canada could leverage this experience and familiarity to make the regulatory system more transparent and accessible. There was a view that with an expected increase in the number of gene editing varieties being submitted for approval in the coming years, that the previous experiences and knowledge could result in the development of a tiered regulatory system. A tiered regulatory system would have a reduced regulatory requirement for crops and traits that have lengthy production histories. Given that one gene edited canola variety has been listed as mutagenesis and conventional breeding in the CFIA's decision document, if the trait did not express outside of a range deemed to be significant by regulators, then these varieties could receive variety approval and be commercialized without regulatory oversight, similar to how conventional breed non-PNT varieties are approved and commercialized in Canada. In Canada, some non-PNT, new crop varieties are not assessed or evaluated by the CFIA or Health Canada, rather they are approved by specific commodity recommending committees.

This will be particularly important for public breeding institutions as there is some concern with public breeders that if new varieties of typically non-GM crops that are bred using gene editing techniques could somehow be viewed as being genetically modified organisms (GMOs), thus restricting export potential for the variety and also adoption. For example, if gene editing was used to create a new flax or wheat variety and it was viewed as a GMO, this would be viewed negatively by typical flax and wheat export markets. If new gene edited varieties were viewed as conventional plant breeding, this would reassure many public sector plant breeders. Presently, some public sector plant breeders are using gene editing techniques in their research laboratories to establish proof of concept and then using older mutagenic technologies to try and reproduce the gene edited trait as closely as possible, as the breeders know regulatory approval and market acceptance of these older technologies is not an issue of concern for foreign commodity importers.

The industry workshop identified that complacency is one of the biggest threats to the regulation of gene editing. Cost and transparency are also concerns that could be addressed by the consideration of a tiered regulatory system that leverages existing knowledge, thus lowering the data requirements, as well as the time and cost of regulatory approval. Regulatory competitiveness is one of the leading drivers of agriculture investment and countries that exhibit signs of an unpredictable and lengthy regulatory review and assessment process, will suffer from reduced innovation investments.

While there are no official federal government plans to change the regulatory system in Canada regarding gene editing technologies, there is an indication from plant breeders, both public and private for increase transparency and for recognition of knowledge gained through 30 years of safe GM crop regulation. Complacency is not an option for regulatory agencies as multinational technology development firms will invest in new variety development programs where the regulatory system provides timely and repeatable decisions. Europe's share of global agriculture research and development (R&D) investment has dropped from one-third at the start of GM crop commercialization in 1995, to less than 10% 20 years later.[15] To ensure that Canada does not experience a loss of R&D investment by international technology development firms, periodic assessments and if needed, revisions to Canada's regulatory framework, will be essential to maintaining Canada's regulatory competitiveness.

4.7 Low Level Presence

In 2012, Canada was the lead country for the establishment of the Global Low Level Presence Initiative, which resulted in the International Statement on Low Level Presence.[16] This agreement included 15 different importing and exporting countries that recognized the importance and the need to ensure that science-based principles were developed to avoid trade disputes. The objective of this Statement was to highlight the need to work collaboratively at an international level to maintain the existing system of commodity trade and that the low level presence (LLP) of some GM events could provide considerable disruption.

In order to stimulate international discussion, Canada developed a policy model that can be used to manage LLP in commodity imports.[17] This model is designed to mitigate the situation where a GM crop or trait has not been approved by the importing country, yet small amounts of it are comingled in an import shipment. It is designed to ensure that borders do not get slammed shut to commodity imports for a period of time, as happened between Canada and the European Union (EU) over flax imports and the LLP detection of GM flax.[18]

Two LLP thresholds are proposed in the policy model, 0.2% and 3% (Fig. 4.2). The first threshold of 0.2% is designed to deal with minute amounts of an unapproved event in a commodity shipment. The higher level of 3%, is designed to manage situations where regulatory approval documents have been submitted for the GM crop or trait, but approval has yet to be granted. This higher level is to act as a stopgap measure to allow international commodity trade to continue between the exporting and importing country in the commodity in question until the GM crop or trait is ultimately approved for import.

Gene edited varieties would meet the requirements of this LLP strategy if they are treated as equivalent to GM crops in the importing country. In countries where gene edited crops are treated as conventional plant breeding, there would be no need to have to followed this model. Canada would not treat the LLP of gene edited crop varieties any differently than it would the LLP detection of any other variety not approved for import.

[15]Little (2015).
[16]AAFC (2012).
[17]AAFC (2017).
[18]Ryan and Smyth (2012), p. 21.

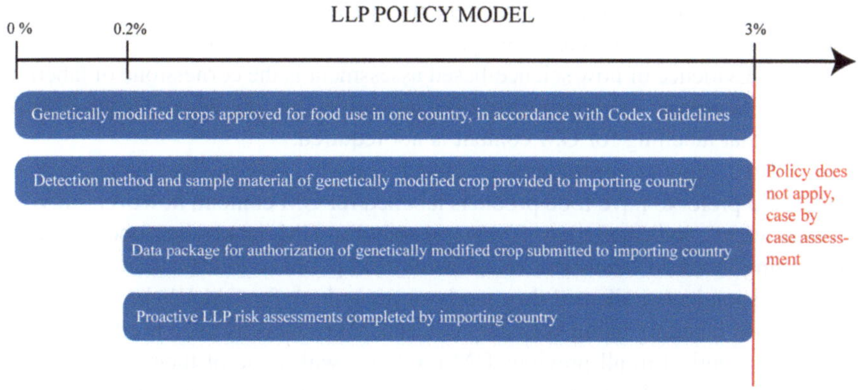

Fig. 4.2 Managing LLP (The figure is an adaption of the corresponding figure in AAFC (2017) which is licensed under the Open Government Licence—Canada. The Open Government Licence—Canada can be found at https://open.canada.ca/en/open-government-licence-canada)

The most significant issue about commodity trade of gene edited crops and import markets that have rigorous testing protocols and that have not, or do not, approve GM crops for import, is what happens when gene edited crops cannot be identified. Due to the absence of marker genes, there will be no detection tests available to distinguish gene edited crop from other crop types. Nor will there be any detection method available to determine if the change is a naturally occurring mutation or one that has been specifically, or randomly, generated. If the exporting country treats a gene edited crop variety as conventional plant breeding, requiring no regulatory review, importing countries that regulate based on process will not be able to know for certain what process was used. I asked a European regulator exactly this question in 2015. I asked what the default position would be for their country should import testing reveal a variety whose breeding process could not be determined and they admitted that the default in instances like this would be to reject the shipment. This represents a significant concern for international commodity trade as the potential to have shipments rejected in instances such as this, will raise the transaction cost of international commodity trade, resulting in higher food prices.

4.8 Labelling

In Canada, it has been estimated that GM foods and food ingredients are detectable in 11% of foods consumed and might be present (but often not detectable) in up to 75–80% of the processed foods sold in stores. Examples range from GM papaya and GM sweet corn that are directly consumed, to sucrose and fructose from GM corn and GM sugar beets that are used as sweeteners in numerous products, GM enzymes that are used in cheese production and GM yeast used in the baking industry.

Labelling for GM content or ingredients in Canada is voluntary and has been since labelling standards were developed in 1999.[19] These standards were reaffirmed in 2016.[20] Evidence of how science-based assessment is the cornerstone of labelling in Canada is seen in the approval of GM salmon in Canada, where Health Canada indicated that labelling for GM content is not required.[21]

In a 2016 study commissioned by Health Canada, 78% of Canadians indicated they would prefer to have food products labelled for GM content, however, only 7% of respondents indicated they always search for GM labels on food products. Confirming that GM labels are not an issue of importance to Canadian consumers, 45% of respondents indicated they rarely or never look for GM labels.

With GM labelling being voluntary in Canada, products from gene editing would be treated identical to all previous GM products, with none of them requiring any special or specific label identification. While no GM labels are mandated, there are voluntary non-GM labels on food products in Canada. This labelling scheme is promoted by the Verified Non-GMO Project, that frequent label products as non-GMO when no comparable GM product exists, such as wheat, tomatoes or carrots. The Verified Non-GMO Project is not a science-based process and does not improve the safety of food products, instead it attempts to create confusion and fear among an uninformed public. Complaints about this form of fraudulent labelling have obviously increased as in 2016, the CFIA established a web portal to lodge complaints regarding the fraudulent labelling of food products.[22]

4.9 Identity Preservation System (Coexistence)

Canada was the first country to have identity preserved a GM crop, doing so with the first 2 years of GM canola production in 1995 and 1996. About 300,000 acres of GM canola were grown over these 2 years, which was contained within Canada and the USA.[23] Once Japan approved GM canola for import, the system was discontinued upon agreement of the canola industry. It is estimated this system cost C$33–41 per metric tonne, a cost increase of 12–15% above the market price at this time. Following this initial identity preservation system, one other identity preservation system has been established. This is one that exports non-GM soybeans, largely produced in Ontario and Quebec.

Crop varieties developed through gene editing technologies may be treated differently upon receiving variety approval, depending on the regulatory treatment by other nations that have been deemed to be key export markets for the commodity

[19]Government of Canada (2004).

[20]CFIA (2016).

[21]Hui (2016).

[22]CFIA (2017).

[23]Smyth and Phillips (2001), p. 51.

receiving approval. Currently, some of the GM commodity industries and organizations have voluntarily agreed that variety approval for a new GM trait is required in a described number of key export markets prior to being commercialized. For example, a new variety of GM canola would need to be approved for import by the EU prior to it being commercially available in Canada due to commodity exports from Canada to the EU. Detection of an unapproved event is costly to international commodity trade. The list of key countries would be different for each GM commodity.

Where this would affect gene edited crop is whether they will be treated as equivalent to GM crops in other countries, thus requiring additional, and costly, regulatory oversight. If gene edited crops are viewed as conventional plant breeding in key export markets, this would reduce the cost of commercializing new crop varieties. It would also reduce the likelihood that a commodity shipment is rejected by an importing country if regulators are unable to determine which process was used to develop the crop variety.

4.10 Liability

A legal dispute arose following the commercial production of GM canola whereby claims of lost organic markets due to the comingling of GM canola with organic canola, resulted in demands for financial compensation. Two Saskatchewan organic farmers, Hoffman and Beaudoin, on behalf of all registered organic farmers in the province of Saskatchewan filed a class action lawsuit against the developers of GM canola, Bayer CropScience and Monsanto Canada. The two organic farmers sought damages for all of the members of the Saskatchewan Organic Directorate (SOD) under claims of negligence, nuisance, trespass and strict liability.[24] In particular, they argued that the comingling of GM canola had destroyed the export market for organic canola due to the inability to certify that canola exports (particularly to Europe) were free of GM canola.

The plaintiffs sought damages on behalf of all 1250 member of the SOD, regardless of whether they had ever grown organic canola or not. Indeed, it was revealed in the trial that many organic farmers had never grown organic canola. During the trial, the court heard from an organic farmer that still produced and exported organic canola. Based on this evidence, the judge opined that the evidence did not demonstrate that a majority or even a significant minority of the proposed class of organic farmers had suffered loss because of the inability to produce canola sufficiently free from GM comingling to be marketed as organic, as 10 years after the introduction of GM canola, some organic farmers were still growing organic canola and finding markets for it.[25]

---

[24]Hoffman v Monsanto (2005).

[25]Khoury and Smyth (2007), p. 222.

In Canada, each province has an Agricultural Operations Act. Essentially, these Acts describe and define current farming practices, making it illegal to sue a farmer for conventional farming practices. While comprehensive in nature, these Acts do not prevent lawsuits against farmers, allowing lawsuits in exceptional or unusual circumstances. Lawsuits regarding negligence, strict liability nuisance, trespass, pollution, reasonable foreseeability and duty of care have been heard by Canadian courts.[26] With each GM crop varieties having successfully undergone a risk assessment by the CFIA and Health Canada prior to approval, the potential of many of these aspects are removed as viable legal actions. To date, there has not been a successful lawsuit within the realm of liability against the commercializer of a GM technology, nor the producer of a GM crop variety.

4.11 Perception of Genome Editing

Perceptions of gene editing will be a contributing factor to the regulation of this technology in many markets. Decisions by domestic and regional governments as to the regulatory status of gene editing will be critical in the overall potential success. The stakes for the environmental non-governmental organizations (eNGOs) are high and in realizing this, they have been intensely lobbying for several years now to have gene editing banned. In an open letter to the European Commission, leading eNGO groups across Europe have attacked new breeding techniques, calling for every technology to be classified as a GMO technology and therefore rejected for use within the EU.[27] Should this occur, the prospects for gene editing would be greatly diminished in many developing countries that rely on exporting crops and food products to European Union countries.

4.11.1 Position of Public Authorities

Canada's two regulatory agencies responsible for the day-to-day regulation of gene editing, continue to stand behind the robustness and adaptability of the science-based system that has been the core of regulatory decision making for GM crops in Canada for 25 years. The importance of Canada's regulatory system is that it balances the need for science-based regulation with the knowledge that regulatory approval should not be burdensome to industry.[28]

As identified above in Sect. 4.6, representatives from the federal regulatory departments were invited to, and participated in, the industry led workshop in May

[26]Smyth (2005), pp. 53–68.

[27]Panella et al. (2015).

[28]Macdonald (2014), p. 63.

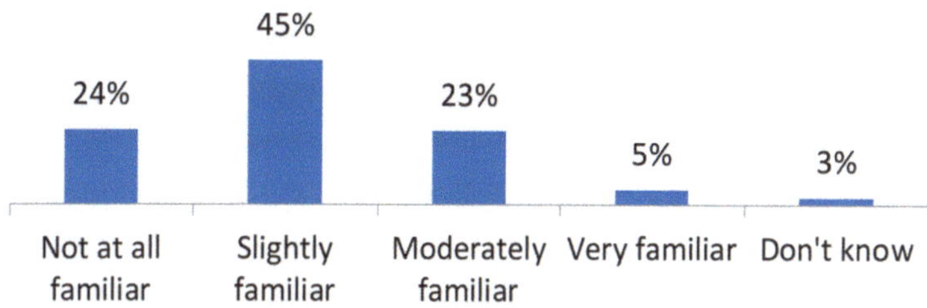

Fig. 4.3 Familiarity with plant breeding

2017 to discuss the future of plant breeding technologies in Canada. Having already approved two gene edited varieties (Table 4.1), Canada clearly supports the commercial application of the technology. Given existing approvals and willingness to participate in an industry workshop on the future of gene editing regulations, signals that federal regulators appreciate the concerns about regulatory burden and are willing to engage in discussion about reform that leverages the knowledge and experience gained to assess the nature of gene editing. At the very least, regulators must be of the view that the existing regulatory framework functions for gene editing varieties and that shifting to increase the regulatory oversight is not a topic that has shown any signs of being considered, expressing an opinion for the status quo and possibly for reassessment.

4.11.2 Public Opinion

In the summer of 2017, under a project I lead, two online surveys were administrated. Both surveys were completed by 500 English speaking Canadians and the results are representative of English speaking Canada. The first survey was completed in July and the second in August. The first survey focused on questions about food security and plant breeding. The second survey focused on benefits of modern plant breeding. The term 'modern plant breeding' was used to include GM and gene edited crops. Given the potential for preconceived perceptions about GM crops or gene editing, the more neutral term was chosen.

The good news is that close to half of participants (45%) claim to be slightly familiar with the concept of plant breeding, a further 23% indicating they are moderately familiar and 5% say they are very familiar (Fig. 4.3). The concern is that one-quarter of respondents are not at all familiar with plant breeding. A reasonable interpretation of this is that 69% of the Canadian public probably does not have an understanding of plant breeding that could be viewed at the level of being able to accurately explain what plant breeding is to someone. When we asked about familiarity with the farming methods of organic, conventional and genetically modified, between 87% and 94% responded that they had heard of these. It would appear that Canadians are generally familiar with broad farming practices but this

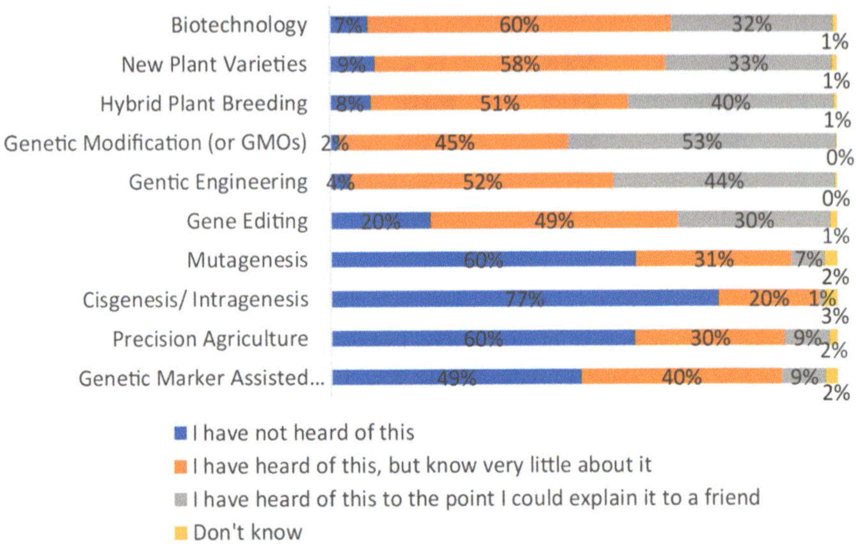

Fig. 4.4 Familiarity with various methods of plant breeding

declines rapidly as one moves away from this towards more specific and focused questions about the plant breeding.

When asked about familiarity with the different modern plant breeding techniques, the responses varied (Fig. 4.4). It should not come as a surprise to anyone in the agriculture or agriculture regulation sectors, that virtually all of participants had not heard of, or knew very little about, cisgenesis and intragenesis (97%). More troubling is that very similar results are observed for awareness of mutagenesis, a plant breeding technology that has been used for decades, with 91% report little or no awareness. This may be more troubling for the organic industry than any other sectors of the agriculture industry as virtually all of the seed used by the organic industry come from conventional, or mutagenic, plant breeding programs. By comparison, gene editing looks quite favourable with a mere 69% reporting little to no awareness. The only option to receive a 'passing grade' was GMOs, with 53% of respondents indicating they were familiar enough with this term to be able to explain it to someone. Two responses contribute to illustrating the broad, general lack of public knowledge about plant breeding are evident with 67% reporting little to no understanding of new plant varieties and 59% reporting the same for hybrid plant breeding. Basic agricultural concepts such as seed sterility, the need to purchase new seed each planting season and that new plant varieties are commercially released every year have virtually no public recognition or understanding. This is troubling as it allows the eNGOs to manipulate the issue and launch misinformation campaigns designed to convince the public that buying new seed every season is a bad thing and that new gene editing hybrid varieties should be rejected, in spite of hybrids being a common part of agriculture dating to the 1950s.

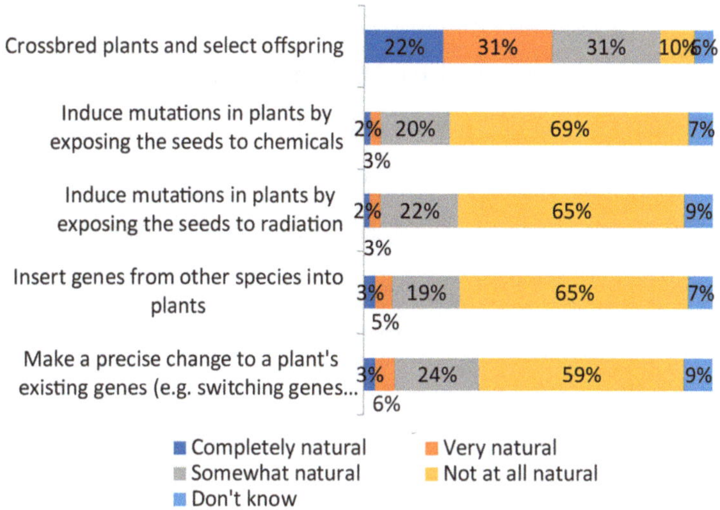

Fig. 4.5 Plant breeding perceptions of natural

The annual use of certified seed is on the rise among farmers, at least this is the case in Saskatchewan. Although not all farmers are happy purchasing new seed each year, the majority of farmers view this as an investment in producing a higher yielding crop. Based on the percentage of seed sales sold that are protected by patents, virtually 100% of canola and corn farmers in Canada buy new seed each year. The thousands of farmers in Canada that grow canola and/or corn accept the tradeoff of saving seed and are willing to spend the extra money to buy new seed if it means higher yields.

Farmers investing in new seed each growing season see this as a means of improving the agronomic quality of their crop. A new variety may have a better disease or insect resistance, higher yield potential and other traits. Whereas reusing seed from one year to the next may result in lower yields. As a seed moves from one generation to the next, it slowly begins to express a lower level of the trait that made it a unique and valued variety.

Participants were asked about the natural status of different types of plant breeding techniques (Fig. 4.5). While having 84% report that crossbred plants and select offspring are viewed as completely, very or somewhat natural, what is troubling is that one in ten view this process that agriculture has relied on for millennia as unnatural. This exemplifies the true level of public ignorance about agricultural practices, plant breeding and simply, where food comes from. Chemical and radiation mutation were first applied to plant breeding in the 1930s,[29] yet over 80 years of use does not resonate with the public. Two-thirds of respondents found both of these technologies to be not at all natural. Transgenic modification is viewed

[29]Qaim (2016), p. 27.

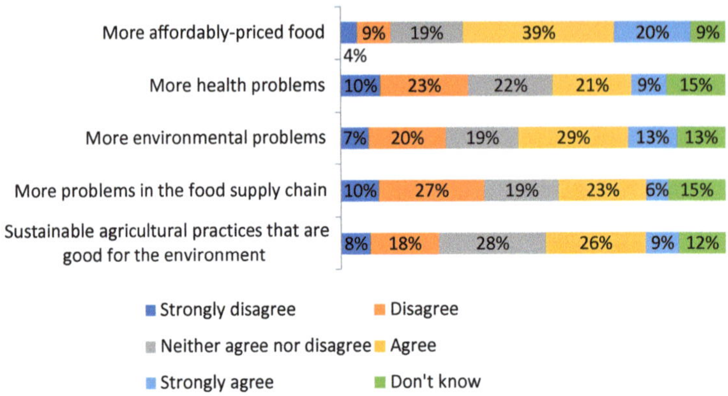

Fig. 4.6 Potential consequences of modern plant breeding

slightly more favourably, with 65% reporting this is not natural. Interestingly, gene editing scored the most favourable of the breeding techniques, with 59% reporting it is not natural, while one-third reported that it is natural.

When asked about potential consequences of modern plant breeding, high levels of uncertainty were exhibited, with 9–15% not knowing and 19–28% having no opinion (Fig. 4.6). The most positive aspect across all five questions was that 59% agreed that modern plant breeding could provide more affordably priced food. Potential for health consequences was more evenly split, with 33% disagreeing that modern plant breeding would result in more health problems, while 30% agreed with the option. When asked if modern plant breeding would create more problems in the food supply chain, 37% of respondents rejected this, with 29% agreeing. Somewhat contradictory are the results regarding the environment. When asked if modern plant breeding would create more environmental problems, 42% agreed, yet when we asked if modern plant breeding would create sustainable agricultural practices that benefit the environment, only 26% disagreed. Framing the environmental option negatively, in that more problems would arise, induced a higher positive response. When sustainability was framed positively to benefit the environment, we would have expected 42% to have disagreed with this option, however we see those responding accordingly to the previous environmental option, dropped by 16%. Evidently, the framing used for questions and responses, either positive or negative, does influence respondents. Awareness of this needs to be given consideration in the framing of questions for public opinion as while the framing effect cannot be said to be significant, it will influence responses.

When it comes to the potential benefits, uncertainty increased from the consequences. Responses of no opinion ranged from 13% to 31%, with do not know ranging from 7% to 27% (Fig. 4.7). In one response, the non-response rate reached 53%. This reinforces the earlier acknowledgement that the public knows very little about plant breeding and are choosing response options that do not require them to make a choice. Increased productivity and reduced weeding are leading benefits.

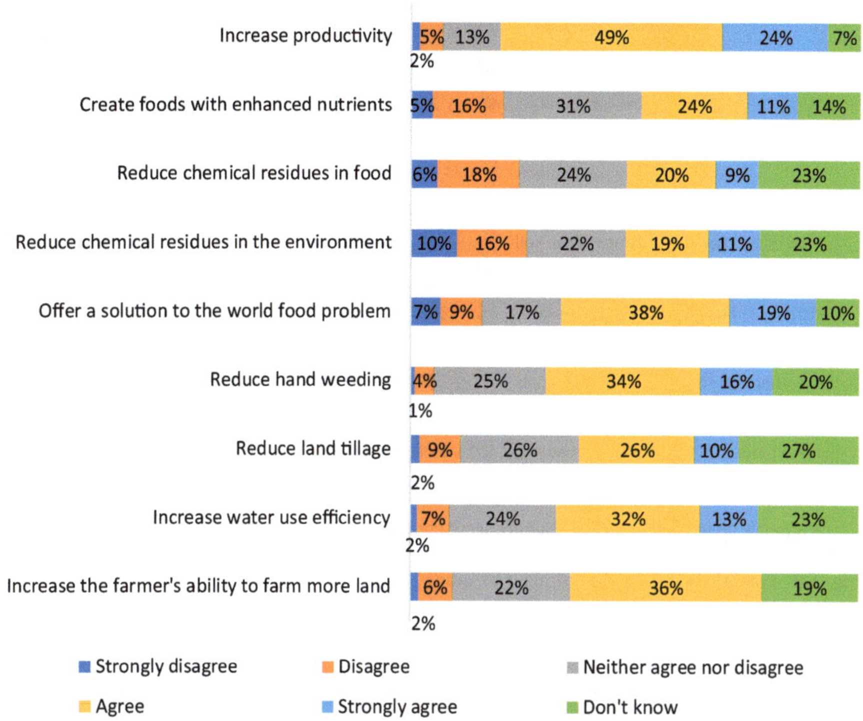

Fig. 4.7 Benefits of modern plant breeding

When asked about the potential environmental risks of modern plant breeding techniques, the rates in which participants did not know was higher than that of the previous questions, rising to 19–34% (Fig. 4.8). Those with no opinion rose marginally (19–23%), while narrowing from the previous question. Combined, these options account for 45–55% of respondents, indicating that roughly half of Canadians lack either the knowledge to make a decision or the confidence to express one. Opinion across all seven examples were evenly spread across neutral and agree, with loss of biodiversity agreed/strongly agreed with the most (51%), and increased cases of farmer pesticide poisoning agreed/strongly agreed with the least (22%).

When asked about the equity issues of modern plant breeding techniques, respondents provided views that differ substantially from much of the recent literature on these options. Identical positive responses (48%) were given indicating that modern plant breeding only benefits large, multinational corporations and does not benefit small-scale farmers (Fig. 4.9). When it comes to farmer freedom of choice, 42% believe that farmers in developing countries have no option and that new crop varieties are forced upon farmers. Respondents were evenly split when it came to whether or not consumers benefit from new plant breeding, with 32% agreeing with the statement that consumers do not benefit and 32% disagreeing with the statement.

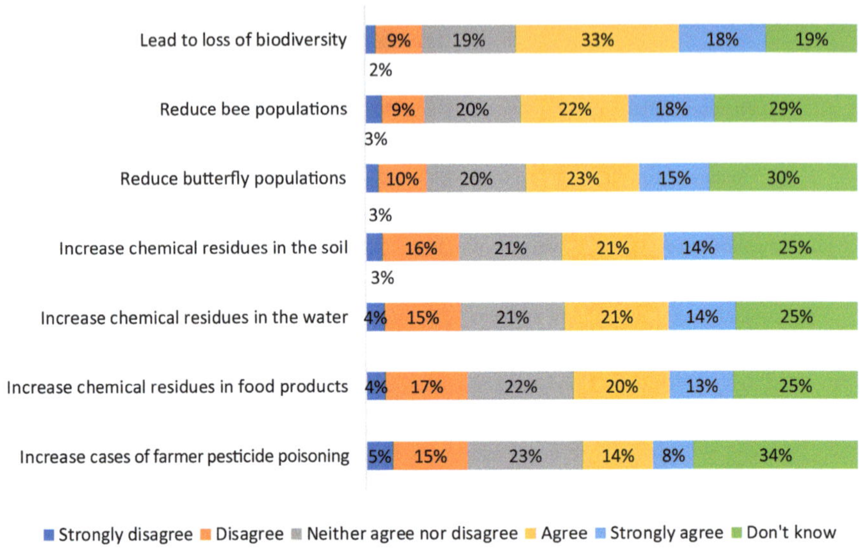

Fig. 4.8 Potential risks of modern plant breeding

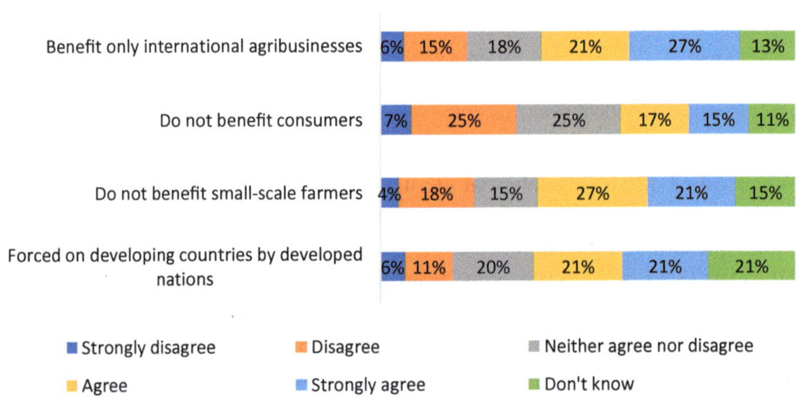

Fig. 4.9 Equity of modern plant breeding

While the Canadian public sees some benefits of new plant breeding techniques, this is over shadowed by environmental risks, the perceived unnaturalness of plant breeding and the unequal distribution of benefits from the technologies. The key observation based on the results from the two surveys conducted is that the public lacks a great deal of basic information regarding how plant breeding has been done over the past 50 years, let alone any comprehension of new gene editing techniques. There is a glimmer of hope in that some responses indicated the potential for increased benefits, particularly within the environmental realm.

4.12 Treatment of Other New Breeding Technologies

With Canada's regulatory system focusing on the novelty of the trait, no breeding technique is defined as requiring additional regulatory oversight. Plant breeders are encouraged to reach out to the CFIA prior to submitting a variety registration package, to discuss the breeding technique and the expression changes of the trait or traits. Based on these informal conversations, regulators are able to suggest appropriate courses of action. While the CFIA has regulated all GM crop variety submissions as PNTs, this has been scientifically rationalized due to the novelty of the trait. The regulation of any plant breeding method in Canada would be based on the novelty of the trait, rather than the breeding method that was used to create the plant variety being assessed.

4.13 Conclusions

One of the first meetings held in Canada regarding the regulation of agricultural biotechnology and GM crops was held in 1988, 30 years ago. It was decided at this meeting that the regulation of this new, innovative technology would be based on science and that the product, not the process, is what would be regulated. Having commercialized dozens of GM canola, corn and soybean varieties, in addition to numerous other crop types for import as food products (i.e. papaya), Canada's regulation of these technologies has provided timely and repeatable decisions.

Canada did not develop a new regulatory framework from scratch, but rather adopted existing governing protocols for assessing varieties, determining that the novelty of the trait being expressed would be the trigger mechanism, regardless of the plant breeding method used to develop the variety. This has resulted in mutagenesis bred varieties and genetically modified varieties being classified as PNTs. The Canadian Food Inspection Agency regulates novel products based on the novelty of the trait (or traits as the case may be in stacked varieties), however all GM varieties have been regulated as PNTs, novel feeds and novel foods. The two gene editing varieties that have been approved to date, have also been approved as PNT varieties.

While the plant breeding sector believes that the CFIA and Health Canada have done an exceptional job of regulating the technology thus far, there are concerns beginning to be expressed that perhaps it is time to reassess what has been learned about the regulation of GM crops and to re-evaluate the existing regulatory framework. Considerable knowledge and experience could be leveraged that could lead to the structuring of a two-tiered regulatory system, whereby new products and new traits would undergo the compete regulatory risk assessment process, while new plant breeding technologies involving existing crops and traits would be subject to a reduced risk assessment. It is unclear at the point of writing as to how, or even whether, Canada's regulatory bodies are receptive to this concept suggested by industry.

Regardless of how, and what, is regulated, it is abundantly evident that the Canadian public possesses little to no knowledge about plant breeding. The results from two public surveys demonstrate troublingly high levels of the lack of awareness of technologies that in some cases, have been applied to plant breeding for decades. Also evident, is a general lack of awareness about the benefits of new plant varieties, particularly regarding environmental impacts and sustainability.

Having approved two gene edited canola varieties, Canada is well positioned to continue applying the PNT regulatory framework to novel varieties developed via these breeding techniques. This process will be science-based and focus on the product, not the process. Time will tell whether the regulatory framework is revised to account for 25 years of safe regulation, to ensure that Canada continues to have a proven and safe regulatory system, but also one that is competitive with the regulatory systems of competing agriculture producing countries.

References

Academy of Science of South Africa (2017) Regulatory implications of new breeding techniques. http://research.assaf.org.za/bitstream/handle/20.500.11911/29/2017_%20assaf_newbreedingtechniques.pdf. Accessed 18 Jan 2018

Agriculture and Agri-Food Canada (2012) International statement on low level presence. http://www.agr.gc.ca/eng/industry-markets-and-trade/agri-food-trade-issues/technical-trade-issues-in-agriculture/international-statement-on-low-level-presence/?id=1422037306229. Accessed 22 Jan 2018

Agriculture and Agri-Food Canada (2017) Policy model – managing low level presence of genetically modified crops in imported grain, food and feed. http://www.agr.gc.ca/eng/policy-model-managing-low-level-presence-of-genetically-modified-crops-in-imported-grain-food-and-feed/?id=1472836695032. Accessed 22 Jan 2018

Canadian Food Inspection Agency (2004) Directive Dir2000-07: conducting confined research field trials of plants with novel traits in Canada. http://www.inspection.gc.ca/english/plaveg/bio/dir/dir0007e.shtml. Accessed 18 Jan 2018

Canadian Food Inspection Agency (2005) Regulation of agricultural biotechnology in Canada. Queen's Printer, Ottawa

Canadian Food Inspection Agency (2013) DD 2013-100: determination of the safety of Cibus Canada Inc.'s canola (Brassica napus L.) event 5715. http://www.inspection.gc.ca/plants/plants-with-novel-traits/approved-under-review/decision-documents/dd-2013-100/eng/1427383332253/1427383674669. Accessed 18 Jan 2018

Canadian Food Inspection Agency (2016) Labelling of genetically engineered foods in Canada factsheet. http://www.inspection.gc.ca/food/labelling/food-labelling-for-industry/method-of-production-claims/genetically-engineered-foods/eng/1333373177199/1333373638071. Accessed 26 Jan 2018

Canadian Food Inspection Agency (2017) Report a food safety or labelling concern. http://www.inspection.gc.ca/food/information-for-consumers/report-a-concern/eng/1364500149016/1364500195684. Accessed 22 Jan 2018

Canadian Food Inspection Agency (2018) Decision documents – determination of environmental and livestock feed safety. http://www.inspection.gc.ca/plants/plants-with-novel-traits/approved-under-review/decision-documents/eng/1303704378026/1303704484236. Accessed 30 July 2018

Canola Council of Canada (n.d.) Statistics. https://www.canolacouncil.org/markets-stats/statistics/. Accessed 17 Jan 2018

CropLife Canada (2017) Future of plant breeding oversight in Canada. http://cdnseed.org/wp-content/uploads/2017/09/Proceedings-Future-of-Plant-Breeding-Oversight-May-2017.pdf. Accessed 26 Jan 2018

Government of Canada (2004) Voluntary labelling and advertising of foods that are and are not products of genetic engineering. http://www.tpsgc-pwgsc.gc.ca/ongc-cgsb/programme-program/normes-standards/internet/032-0315/documents/commite-commmittee-eng.pdf. Accessed 17 Jan 2018

Gusta M, Smyth SJ, Belcher K, Phillips PWB, Castle D (2011) Economic benefits of genetically-modified herbicide-tolerant canola for producers. AgBioForum 14(1):1–13

Harrison B (2001) Regulating novel foods in Canada. In: Presentation to the conference The Convergence of Global Regulatory Affairs: its potential impact on international trade and public perception. Saskatoon, Canada, March 19–20

Health Canada (2006) Genetically modified (GM) foods and other novel foods. https://www.canada.ca/en/health-canada/services/food-nutrition/genetically-modified-foods-other-novel-foods.html. Accessed 5 Feb 2018

Hoffman v. Monsanto Canada Inc. (2005) S.K.Q.B. 225

Hui A (2016) Genetically modified salmon approved for consumption in Canada. https://www.theglobeandmail.com/news/national/health-canada-approves-genetically-modified-salmon-as-safe-for-consumption/article30094235/. Accessed 26 Jan 2018

Khoury L, Smyth SJ (2007) Reasonable foreseeability and liability in relation to genetically modified organisms. Bull Sci Technol Soc 27(3):215–232

Little J (2015) Smart regulation and innovation for EU agriculture. Paper presentation to the 7th European Innovation Summit. Brussels, Belgium, 7–10 December

Macdonald P (2014) Genetically modified organisms regulatory challenges and science: a Canadian perspective. J Consum Prot Food Saf 9(Suppl 1):59–64

Panella F, Holland N, Steinbrecher R, Ferrante A, Schimpf M, Wallace H, Richartz S, Then C (2015) Open letter to the Commission on new genetic engineering methods. http://www.genewatch.org/uploads/f03c6d66a9b354535738483c1c3d49e4/New_Breeding_Techniques___Open_Letter_27_Jan_2015.pdf. Accessed 31 Jan 2018

Qaim M (2016) Genetically modified crops and agricultural development. Palgrave Macmillan, London

Ryan CD, Smyth SJ (2012) Economic implications of low-level presence in a zero-tolerance European import market: the case of Canadian triffid flax. AgBioForum 15(1):21–30

Smyth SJ (2005) A decade of regulating agricultural biotechnology liability in Canada: a case study from 1994–2004. Dissertation, University of Saskatchewan

Smyth SJ (2014) The state of genetically modified crop regulation in Canada. GM Crops Food 5 (3):195–203

Smyth SJ, Phillips PWB (2001) Competitors co-operating: establishing a supply chain to manage genetically modified canola. Int Food Agribusiness Manag Rev 4(1):51–66

Smyth SJ, Phillips PWB, Kerr WA, Khachatourians GG (2004) Regulating the liabilities of agricultural biotechnology. CABI Publishing, Wallingford

Smyth SJ, Gusta M, Belcher K, Phillips PWB, Castle D (2011a) Environmental impacts from herbicide tolerant canola production in Western Canada. Agric Syst 104(5):403–410

Smyth SJ, Gusta M, Belcher K, Phillips PWB, Castle D (2011b) Changes in herbicide use following the adoption of HR canola in Western Canada. Weed Technol 25(3):492–500

Smyth SJ, McDonald J, Falck-Zepeda J (2014) Investment, regulation and uncertainty: managing new plant breeding techniques. GM Crops Food 5(1):1–14

Soy Canada (2016) Canadian soybean industry research and innovation strategy workshop. http://soycanada.ca/wp-content/uploads/2016/10/Soy-Canada-Research-Strategy-Workshop-Report-June-2016.pdf. Accessed 17 Jan 2018

Chapter 5
Regulation of Genome Editing in Plant Biotechnology: European Union

Brigitte Voigt and Ansgar Münichsdorfer

Abstract Plants, animals and microorganisms obtained by any type of genome editing technology (SDN-1, SDN-2, SDN-3, ODM) are regulated through the EU's GMO regime. A judgment of the European Court of Justice in July 2018 provided regulatory certainty about their GMO status. It ended more than a decade of legal debates.

The EU's GMO regime is harmonised at the EU level and encompasses authorisation requirements regarding contained use, field trials and the placing on the market of GMOs as well as post-market monitoring, labelling, traceability and identity preservation obligations. Thus, GMOs are governed in a comprehensive, detailed and rigorous manner. In addition, GMOs are subject to widespread political and societal rejection.

The impossibility to distinguish certain genetic alterations induced by genome editing from those that are induced naturally or by traditional breeding techniques leads to problems as yet unresolved. It might hamper GMO authorisation, the EU's zero tolerance policy for unauthorised GMOs and GM labelling.

It cannot be excluded that amendments to the GMO framework will be introduced in the aftermath of the European Court of Justice's judgment. Without any, it will take several years until the first genome edited plants are commercially cultivated or imported.

Ansgar Münichsdorfer wrote Sect. 5.10 ("Liability"). Brigitte Voigt wrote the remaining sections. The final version of this Chapter was submitted on 01 February 2019 with minor updates in June 2019.
Consolidated versions of the EU legislation cited in this report can be accessed in all official EU languages at the website for European Union law EUR-Lex, https://eur-lex.europa.eu.

B. Voigt (✉) · A. Münichsdorfer
Faculty of Law, University of Passau, Passau, Germany
e-mail: Brigitte.Voigt@uni-passau.de

H.-G. Dederer, D. Hamburger (eds.), *Regulation of Genome Editing in Plant Biotechnology*, https://doi.org/10.1007/978-3-030-17119-3_5

5.1 Introduction

The European Union is both the world's leading exporter and importer of agri-food products.[1] It is therefore an important agri-food trading partner.

The present situation in EU agriculture regarding biotechnology is characterised by an **ambivalent relationship**: Due to strong reservations hardly any GM crops are cultivated (Fig. 5.1). Regarding GM imports, in contrast, the EU animal feed sector is largely dependent on importation of GM feed, making the EU a major importer of GM soybean and corn.[2]

Only a single GM event is currently authorised for commercial **cultivation** in EU agriculture, which is the Bt-maize line MON810.[3] It is cultivated in Spain (95% of cultivation area) and Portugal at an area of about 120,000 ha (as of 2018).[4] This means the EU's total GM cultivation area amounts to less than 0.1% of the global GM cultivation area.[5] In two-thirds of the member states, GM crop cultivation is not allowed at all, following an "opt-out" possibility provided by the EU's GMO framework.[6] Spain stands out, as it has been successfully adopting GM crops for cultivation for 20 years now.[7] The major developers BASF, BayerCropScience, KWS, Limagrain and Syngenta have become reluctant to develop crops for the EU market and continue to relocate GMO research facilities to non-EU countries.[8]

---

[1]European Commission (2018a), pp. 1, 4. The main export destinations of agri-food products are the US (by far), China, Switzerland, Russia and Japan. Main importers into the EU regarding agri-food products are Brazil, the US, Argentina, Ukraine and China, European Commission (2018a), pp. 5–6; Eurostat (2018).

[2]The group of FAS Biotechnology Specialists in the European Union (2018), pp. 9–10.

[3]Until today, only six GM crops have obtained an approval for commercial cultivation in the EU (one only for seed production), the first in 1997. The MON810 maize line is the only one remaining. As for the other five, the approval expired, was withdrawn by the European Commission or annulled by the EU General Court. The most famous case is the GM potato "Amflora" developed by BASF. After an approval procedure which lasted almost 14 years, it was cultivated in the EU in 2010 and 2011. It was withdrawn from the EU market in the beginning of 2012. In 2013, the EU General Court overturned the permission. Cf. International Service for the Acquisition of Agri-biotech Applications (2017), p. 92; McEldowney (2015), p. 1; Schauzu (2011), pp. 60–61; Hunt (2011), pp. 140–141; Davison and Ammann (2017), pp. 16–17.

[4]Corresponding to about 300,000 acres. The group of FAS Biotechnology Specialists in the European Union (2018), p. 6.

[5]International Service for the Acquisition of Agri-biotech Applications (2017), pp. 3, 92; Canadian Biotechnology Action Network (2015), p. 9 (as of 2015).

[6]See Fig. 5.1 and text to n. 108.

[7]International Service for the Acquisition of Agri-biotech Applications (2016), p. 74; International Service for the Acquisition of Agri-biotech Applications (2017), p. 94.

[8]Davison and Ammann (2017), pp. 16–17; The group of FAS Biotechnology Specialists in the European Union (2018), p. 5; Baulcombe et al. (2014), p. 3; European Academies Science Advisory Council and German National Academy of Sciences Leopoldina (2013), pp. 11–12.

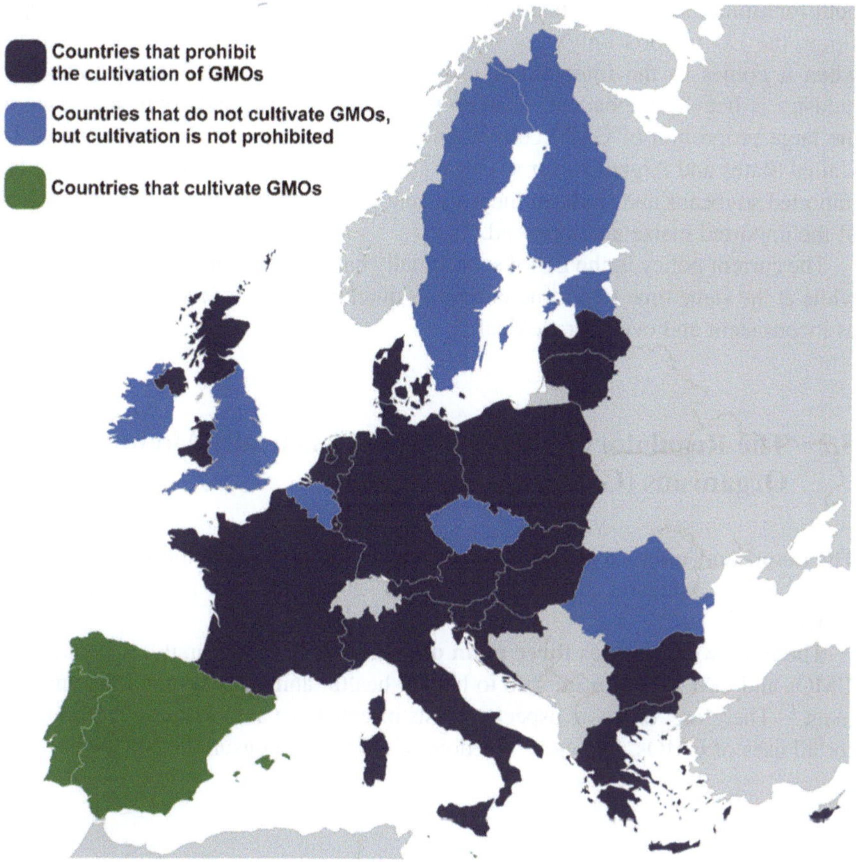

Fig. 5.1 GM crop cultivation in the EU

This fact that the EU cultivates hardly any GM crops is exceptional in regard to the worldwide trend towards increasing GMO cultivation. It can be explained both by the harsh EU regime on GMOs and by the widespread anti biotech attitude in the EU.[9]

The particularly cautious and critical attitude towards GM cultivation is contrasted by the role of the EU as a major **importer** of GM products. 111 GM plant lines are currently approved for import as food and feed into Europe (as of 31 August 2018).[10] Although the authorisation covers the importation of GM crops

[9]Hartung and Schiemann (2014), pp. 744–745.

[10]EU register of GM food and feed, European Commission (n.d.-g). GM lines containing stacked events are counted separately. In terms of single transformation events, 44 events are currently approved for import. Only 62 approval decisions have been made for the currently authorised lines as some of these approvals authorise both a GM line containing stacked events and lines containing subcombinations of these events.

both for animal feed and human consumption, due to the strong consumers' opposition, the EU imports only very little GM food.[11] However, the picture changes when it comes to the importation for the EU's animal feed industry. The feed industry is heavily dependent on imports, mainly on soya and soymeal.[12] Due to the large proportion of GMO cultivation in the supplier countries (e.g. Brazil, the United States and Argentina for soybean and soybean meal), more than 90% of the imported soybeans and soybean meal are genetically modified as well as about 20% of the imported maize and rapeseed.[13]

The current policy in the EU of substantially banning the cultivation of GM crops while at the same time importing enormous quantities of GM feed can be described as inconsistent and even paradoxical.[14]

5.2 The Regulatory Framework for Genetically Modified Organisms (GMOs): An Overview

The European Union's regulatory framework for GMOs was established in the early 1990s[15] and amended between 2000 and 2003, driving it to even greater stringency.[16]

The framework pursues **three main objectives** (Fig. 5.2). Firstly, it ensures that GMOs and GM products are safe to human health, animal health and the environment.[17] Therefore, one core aspect consists of extensive authorisation requirements for all uses of GMOs. Secondly, freedom of choice for consumers and producers is

[11]European Commission (2015a).

[12]European Commission (2016), pp. 2, 5 ("The EU is 70% dependent on imports of protein-rich crops"); The group of FAS Biotechnology Specialists in the European Union (2018), pp. 9ff; Nábrádi and Popp (2011), pp. 10ff, 17ff; Masip et al. (2013), p. 319; Davison and Ammann (2017), p. 21. Soybean and soybean meal represent more than 60% of the EU's total protein-rich feed materials and are to a very large extent derived from imports, European Commission (2016), p. 3; European Commission (2015a); de Visser et al. (2014), p. 2.

[13]The group of FAS Biotechnology Specialists in the European Union (2018), pp. 10, 12, 14. Slightly differing figures in European Commission (2016), p. 4.

[14]Tagliabue (2015), p. 57; Masip et al. (2013), p. 319; European Academies Science Advisory Council and German National Academy of Sciences Leopoldina (2013), p. 35.

[15]Cf. Council Directive 90/220/EEC of 23 April 1990 on the deliberate release into the environment of genetically modified organisms [1990] OJ L117/15 (not in force any more).

[16]Plan and van den Eede (2010), p. 3; Schauzu (2011), p. 58; Devos et al. (2006), pp. 133–143.

[17]Directive 2001/18/EC, art. 1; Regulation (EC) No 1829/2003, art. 1.

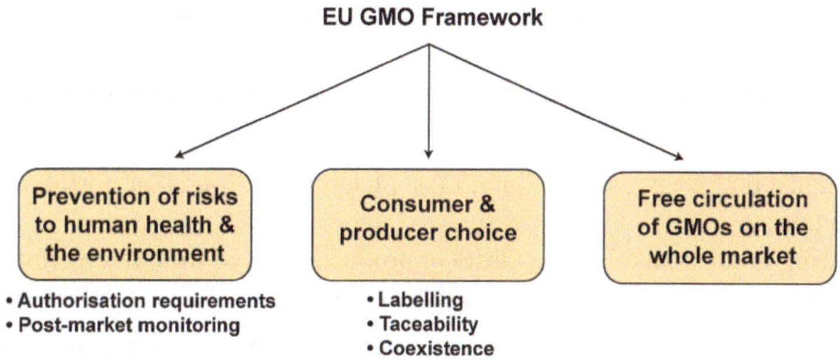

Fig. 5.2 Objectives of the EU's GMO framework

provided by coexistence, labelling and traceability rules.[18] Thirdly, free circulation of GMOs on the whole EU market is secured.[19]

The regulatory framework intends to govern GMOs and all uses of GMOs comprehensively. As compared to other countries, the regulatory requirements are **strict** and sometimes viewed as **complicated, cumbersome** and even **hostile** towards GMOs.[20] In particular, the approval procedure for placing GM food and feed on the European market is lengthy and costly: It usually takes between 4 and 7 years[21] and entails costs of 7–15 million euros (corresponding to

[18]Regulation (EC) No 1829/2003, recitals 17, 20, 21, art. 1; Regulation (EC) No 1830/2003, recitals 4, 11, art. 1.

[19]Directive 2001/18/EC, art. 1; Regulation (EC) No 1829/2003, art. 1.

[20]Devos et al. (2012), p. 10771: "most stringent and wide-ranging regulations on GM products and commodities in the world"; Masip et al. (2013).

[21]Cf. the (somewhat differing) figures in Smart et al. (2017), pp. 182, 190–192; Hartung and Schiemann (2014), p. 744; The group of FAS Biotechnology Specialists in the European Union (2018), pp. 23; Lappin (2018b), p. 2; The Netherlands Commission on Genetic Modification (2009b), p. 21; McDougall (2011); Madre and Agostino (2017). Regarding the approval times for the most recent approvals cf. Jany (2018b) (in German). The procedure takes considerably longer than e.g. the petitions for determination of nonregulated status in the USA, cf. Smart et al. (2017), p. 192; Kalaitzandonakes et al. (2016), p. 226; Lucht (2015), p. 4257 ("on average, it takes at least 15 to 20 months longer than [. . .] in the U.S., Brazil, and Canada"). With respect to the disruption of imports through asynchronous approvals cf. n. 229.

The legal timescales for GM food/feed applications (Regulation (EC) No 1829/2003, arts. 6(1), 7 (1); 18(1), 19(1), 35(2); cf. Roiz (2014), pp. 2–3) and for commercial cultivation (Directive 2001/ 18/EC, arts. 13(1), 14(2), (4), 15(1), (3), 18) allow for a much shorter duration of the authorisation procedure, Jany (2018a) (in German). However, there are possibilities for delay, e.g. provisions to "stop the clock" whilst additional information is required from the applicant, cf. Regulation (EC) No 1829/2003, arts. 6(1), 18(1); Directive 2001/18/EC, arts. 14(4), 15(1), 18(1). Furthermore, the length of the community procedure (see n. 100) is not prescribed. Therefore, the actual timescale is much longer and unpredictable, cf. Lusser et al. (2011), p. 49; Jones (2015), p. 3.

8.7–18.6 million US dollar).[22] Approvals for cultivation have stagnated entirely.[23]

The European GMO framework follows a **step-by-step approach** (Fig. 5.3 and Table 5.1).[24] It distinguishes between three steps:

(1) Contained use, i.e. use in laboratories, plant growth rooms or greenhouses
(2) Field trials[25]
(3) Placing on the market of the GMO or products derived from it; e.g. importation, cultivation for commercial purposes, marketing of food/feed, marketing of drugs

It is not compulsory to pass through all steps.[26] GMOs intended for importation into the EU often only apply for a step 3 authorisation whilst contained use and field trials have been carried out in other countries.

Different **sources of law** apply to the different steps. At all steps, GMOs are regulated at the European level. Contained use is governed by Directive 2009/41/EC[27] (Contained Use Directive). In fact, the Contained Use Directive only covers genetically modified microorganisms. Other GMOs are within the legislative competence of the member states. However, most member states have implemented the Directive with respect to other organisms on a voluntary basis.[28] Field trials are governed by Directive 2001/18/EC[29] (Deliberate Release Directive). The placing on the market, in contrast, is not governed by a single framework. Instead, each intended use is governed by its own framework: Cultivation is governed by Directive 2001/18/EC.[30] GM food and feed are governed by Regulation (EC) No

[22]There is strong variation in the estimation of regulatory costs, OECD (2018b), p. 62, see e.g. the estimates in Hartung and Schiemann (2014), p. 744; Kalaitzandonakes et al. (2007); Tait and Barker (2011), p. 766; The Netherlands Commission on Genetic Modification (2009b), pp. 10, 20–21; Madre and Agostino (2017); Lusser et al. (2011), p. 49. Of course, costs vary with the crop, the introduced trait, the intended use and the studies performed by the variety developers, Kalaitzandonakes et al. (2007), pp. 509, 510; McDougall (2011), p. 23. The costs are estimated to be 25% higher than in the US, The Netherlands Commission on Genetic Modification (2009b), p. 21.

[23]Several GM maize varieties are in the pipeline for approval for commercial cultivation (some of them for more than a decade), see The group of FAS Biotechnology Specialists in the European Union (2017), p. 11.

[24]Regarding the step-by-step concept OECD (1986); Directive 2001/18/EC, recital 24; Dederer (2016b), pp. 143–147.

[25]In the EU Directives and Regulations, field trials are referred to as "Deliberate Release of GMOs for any other purpose than for placing on the market", cf. Directive 2001/18/EC, part B (art. 5ff).

[26]Cf. Directive 2001/18/EC, recitals 23, 24; von Kries and Winter (2011), p. 33; Gross (2006), pp. 90–93.

[27]Directive 2009/41/EC of the European Parliament and of the Council of 6 May 2009 on the contained use of genetically modified micro-organisms [2009] OJ L125/75.

[28]Winter (2016b), p. 182; Dederer (2016b), p. 144; Friant-Perrot (2010), p. 82.

[29]Directive 2001/18/EC of the European Parliament and of the Council of 12 March 2001 on the deliberate release into the environment of genetically modified organisms and repealing Council Directive 90/220/EEC [2001] OJ L106/1.

[30]As a result, any release into the environment is governed by Directive 2001/18/EC.

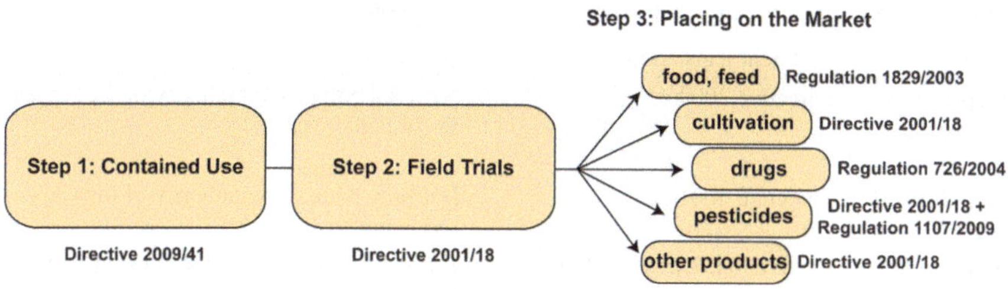

Fig. 5.3 Basic regulatory concept: the step-by-step approach

1829/2003[31] (GM Food and Feed Regulation), GM drugs by Regulation (EC) No 726/2004[32] (Drug Regulation) and so on. If there is no specific regime, Directive 2001/18/EC applies.

Contained use, field trials and the placing on the market each have their own **authorisation regimes**. Any authorisation procedure includes a scientific assessment of the risks to human health and the environment. In general, any form of placing GMOs on the market needs its own authorisation. However, a single authorisation procedure can be followed for GMOs intended for both cultivation and use as food/feed ("one door, one key" principle).[33] This option is not often used, though.[34]

As regards the **competent authorities**, authorisations/notifications for contained uses and authorisations for field trials are granted by the member states and accordingly only valid in the respective state. All forms of placing on the market, by contrast, are authorised at the EU level in a procedure involving both the EU institutions (in particular the European Commission in the decision making and the European Food Safety Authority in the risk assessment) and all member states. The authorisation is valid throughout the EU.

GMOs that have been placed on the market have to be **traceable** and **labelled** (Table 5.2).[35]

[31]Regulation (EC) No 1829/2003 of the European Parliament and of the Council of 22 September 2003 on genetically modified food and feed [2003] OJ L268/1.

[32]Regulation (EC) No 726/2004 of the European Parliament and of the Council of 31 March 2004 laying down Community procedures for the authorisation and supervision of medicinal products for human and veterinary use and establishing a European Medicines Agency [2004] OJ L136/1.

[33]Regulation (EC) No 1829/2003, art. 5(5). Still, the criteria for both regimes have to be met. Plan and van den Eede (2010), p. 8; Dederer (2016b), p. 147.

[34]Yusuf (2014), pp. 23, 36.

[35]Labelling: Regulation (EC) No 1830/2003 of the European Parliament and of the Council of 22 September 2003 concerning the traceability and labelling of genetically modified organisms and the traceability of food and feed products produced from genetically modified organisms and amending Directive 2001/18/EC [2003] OJ L268/24, art. 4(6); Directive 2001/18/EC, art. 21; specific provisions for GM food/feed labelling in Regulation (EC) No 1829/2003, arts. 12ff and arts. 24ff; traceability: Regulation (EC) No 1830/2003 arts. 4(1)-(5) and 5.

Table 5.1 Summary: Pertinent legislation and responsible public authorities

Step	Pertinent legislation[a]	Authorities
Step 1: Contained use	Directive 2009/41	National competent authority of the respective member state
Step 2: Field trials	Directive 2001/18	National competent authority of the respective member state
Step 3: Placing on the market	Cultivation: Directive 2001/18 + e.g., Directives 2002/53 + 2002/55[b]	European competent authorities, in particular European Commission, European Food Safety Authority; involvement of national competent authorities of all member states
	Food, feed: Regulation 1829/2003	
	Drugs: Regulation 726/2004	
	Pesticides: Directive 2001/18 + Regulation 1107/2009[c]	
	Other use: Directive 2001/18	

Notice: European Regulations are directly applicable in the member states [Consolidated version of the Treaty on the Functioning of the European Union [2012] OJ C326/47 (TFEU), art. 288(2).] European Directives are in general not directly applicable in the member states, for exceptions see ECJ, Case 41/74 *Van Duyn/Home Office* [1974] ECLI:EU:C:1974:133, para. 12. Member states implement them through national law, TFEU, art. 288(3).

[a]Supplemented by implementation rules, recommendations and guidance documents, cf. for details Sect. 5.4 ("Regulatory Prerequisites for Activities Relating to Genome Edited Plants")

[b]Directive 2002/53/EC of 13 June 2002 on the common catalogue of varieties of agricultural plant species [2002] OJ L193/1; Directive 2002/55/EC of 13 June 2002 on the marketing of vegetable seed [2002] OJ L193/33; both forming part of the EU's seed legislation

[c]Regulation (EC) No 1107/2009 of the European Parliament and of the Council of 21 October 2009 concerning the placing of plant protection products on the market and repealing Council Directives 79/117/EEC and 91/414/EEC [2009] OJ L209/1

Post-market **monitoring** is mandatory.[36] It consists of both a general surveillance and a GM variety-specific surveillance.[37]

Cultivators have to adhere to the rules regarding **coexistence**[38] the member states have introduced for GM plants.

Member states have two options to **restrict authorised GMOs**: **Safeguard clauses** allow them to provisionally restrict or even ban cultivation of a GMO or marketing of a GM product if there is new information on potential risks.[39]

---

[36]Directive 2001/18/EC, arts. 13(2)(e) and 20, annex VII; Council Decision 2002/811/EC of 3 October 2002 establishing guidance notes supplementing Annex VII to Directive 2001/18/EC of the European Parliament and of the Council on the deliberate release into the environment of genetically modified organisms and repealing Council Directive 90/220/EEC [2002] OJ L280/27.

[37]Cf. Directive 2001/18/EC, annex VII.

[38]Cf. Directive 2001/18/EC, art. 26a; further: European Commission (2010b).

[39]Cultivation: Directive 2001/18/EC, art. 23; cf. also the general safeguard clause for seed, Directive 2002/53/EC, art. 18; GM food: Regulation (EC) No 1829/2003, art. 34; cf. also the general safeguard clause for food, Regulation (EC) No 178/2002 of the European Parliament and of

Table 5.2 Requirements for GMOs placed on the market

Requirement	Pertinent legislation
Labelling	Regulation 1830/2003; Regulation 1829/2003 (food/feed)
Traceability	Regulation 1830/2003
Monitoring	Directive 2001/18
Coexistence	Member states' legislation; European Commission 'Recommendation on guidelines for co-existence measures'[a]

[a]European Commission (2010b)

Since 2015, member states can permanently restrict or prohibit the cultivation of authorised GMOs on their territory ("**opt-out**") with the consent of the applicant or if certain material conditions are met.[40]

At the level of **public international law**, the EU as well as all EU member states are members of the World Trade Organization (WTO). Consequently, the GMO regulatory framework also complies with international food safety standards, e.g. those set by the FAO/WHO Codex Alimentarius Commission.[41] Furthermore, the EU signed the Cartagena Protocol on Biosafety[42] in 2000 and ratified it in 2002. Transboundary movements of GM products are regulated by Regulation (EC) No 1946/2003.[43] It signed the Nagoya – Kuala Lumpur Supplementary Protocol[44] in 2011 and ratified it in 2013.

the Council of 28 January 2002 laying down the general principles and requirements of food law, establishing the European Food Safety Authority and laying down procedures in matters of food safety [2002] OJ L31/1, arts. 53, 54. Further Norer and Preisig (2016), pp. 34–36.

[40]Directive 2001/18/EC, art. 26b; in detail Sect. 5.4 ("Regulatory Prerequisites for Activities Relating to Genome Edited Plants"), text to n. 108.

[41]Cf. the WTO Agreement on the Application of Sanitary and Phytosanitary Measures (SPS Agreement), Introduction, art. 12.3, annex A(3)(a), obliging members to base their sanitary or phytosanitary measures on international standards, guidelines or recommendations. These international standards lead to a de facto harmonisation of GMO health risk assessments of all WTO members, cf. text to n. 127.

[42]Cartagena Protocol on Biosafety to the Convention on Biological Diversity, Montreal, 29 January 2000, in force 11 September 2003, 2226 U.N.T.S. 208; 39 ILM 1027 (2000); UN Doc. UNEP/CBD/ExCOP/1/3. http://bch.cbd.int/protocol/text/. Accessed 22 August 2018 [hereinafter Cartagena Protocol].

[43]Regulation (EC) No 1946/2003 of the European Parliament and of the Council of 15 July 2003 on transboundary movements of genetically modified organisms [2003] OJ L287/ 1.

[44]Nagoya – Kuala Lumpur Supplementary Protocol on Liability and Redress to the Cartagena Protocol on Biosafety, Nagoya, 15 October 2010, in force 5 March 2018, UN Doc. UNEP/CBD/BS/COP-MOP/5/17. https://bch.cbd.int/protocol/NKL_text.shtml. Accessed 22 August 2018.

5.3 Regulatory Status of Genome Edited Plants

All types of genome edited plants (i.e. plants obtained by SDN-1, SDN-2, SDN-3 and ODM)[45] are regulated through the EU's regulatory framework for GMOs.

5.3.1 Applicability of the Regulatory Framework for GMOs[46]

The scope of applicability of the EU's GMO framework is determined by its GMO definition (Fig. 5.4 and Table 5.3). What constitutes a GMO is defined in the Deliberate Release Directive (Directive 2001/18/EC), Articles 2, 3, Annex IA, Annex IB. The GMO definition encompasses all organisms except human beings. All other pieces of legislation refer to that GMO definition.[47] The only exception is the Contained Use Directive. It defines GM microorganisms distinctly, but in a similar way.[48]

The legal definition of a GMO starts with an abstract definition of what constitutes a GMO: A GMO is "an organism, with the exception of human beings, in which the genetic material has been altered in a way that does not occur naturally by mating and/or natural recombination". That definition is supplemented by three lists:

- Examples of techniques resulting in GMOs ("black list")
- Techniques *not* resulting in GMOs ("white list")
- Techniques resulting in GMOs, but these are *exempted* from the scope of the GMO regulatory framework ("grey list")

The EU's regulatory framework for GMOs is "**process-based**": The method of an organism's production triggers the application of the regulatory regime.[49]

---

[45]The different types of site-directed nuclease techniques (SDN-techniques) are understood as defined in the standardised outline to this report:

SDN-1: generation of site-specific random point mutations
SDN-2: generation of site-specific desired point mutations using template DNA
SDN-3: site-specific introduction of long stretches of donor DNA.

[46]Cf. further Devos et al. (2012), pp. 10770–10771; European Commission New Techniques Working Group (2011), para. 2.0.

[47]Cf. e.g. Regulation (EC) No 1829/2003, art. 2(5).

[48]Cf. Directive 2009/41/EC, arts. 2(b), 3, annex I, annex II part A. The only exception to that similarity is that Directive 2009/41/EC additionally excludes self-cloning from the scope of legislation, Directive 2009/41/EC, annex II part A(4); European Commission New Techniques Working Group (2011), para. 2.0.

[49]Cf. The Netherlands Commission on Genetic Modification (2009b), p. 3; Marchant and Stevens (2015), p. 234. All states party to the Cartagena Protocol have a process-based approach, Devos et al. (2012), p. 10770, but many approaches are not *purely* process-based, i.e. not only the method of production matters but also the genetic result.

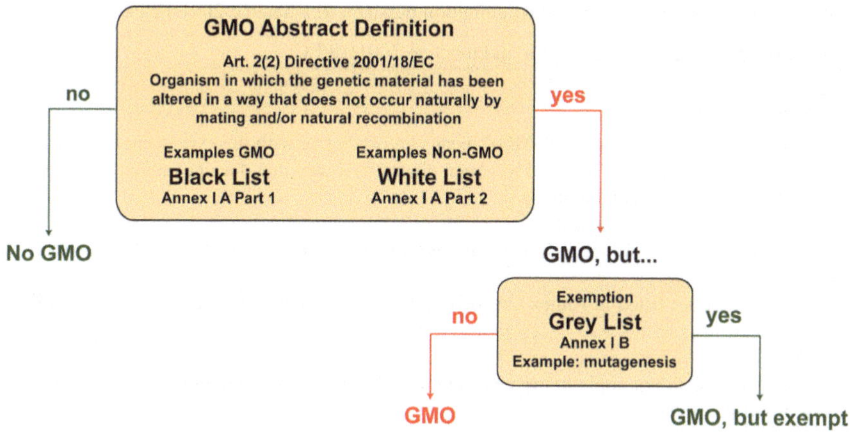

Fig. 5.4 The EU's definition of a GMO

5.3.2 Regulatory Classification of Genome Edited Plants

5.3.2.1 SDN-1, SDN-2, ODM: Yield GMOs

In July 2018, the European Court of Justice decided as part of Case C-528/16[50] that the EU's GMO regulatory regime is applicable to organisms obtained by directed mutagenesis techniques.[51] As examples of directed mutagenesis techniques, SDN-1 and ODM were mentioned in the proceedings.[52] It is debatable whether SDN-2, which is similar to ODM, could also be included in the term "directed mutagenesis techniques".[53] In any case, SDN-2 yields GMOs due to the general interpretations the Court made regarding the GMO definition.[54]

[50]ECJ, Case C-528/16 *Confédération paysanne and Others* [2018] ECLI:EU:C:2018:583; cf. also the press release, Court of Justice of the European Union (2018).

[51]ECJ, Case C-528/16 *Confédération paysanne and Others* [2018] ECLI:EU:C:2018:583, paras. 38, 51, 54.

[52]Cf. ECJ, Case C-528/16 Confédération paysanne and Others [2018] ECLI:EU:C:2018:583, para. 28 ("techniques/methods of mutagenesis such as those at issue in the main proceedings"); para. 23; Bobek (2018), para. 46 ("As [...] explained by the referring court [...] targeted mutagenesis methods applying new genetic engineering techniques have been devised, such as oligonucleotide-directed mutagenesis (ODM) or directed nuclease mutagenesis (SDN1)"; Conseil d'État, 3e et 8e ch., 3 oct. 2016, n°388649, ECLI:FR:CECHR:2016:388649.20161003, Confédération paysanne et autres, para. 23 (referral decision by the French Conseil d'État).

[53]Cf. also the description of directed mutagenesis used by the applicants in the proceedings before the French Conseil d'État, Conseil d'État, 3e et 8e ch., 3 oct. 2016, n°388649, ECLI:FR:CECHR:2016:388649.20161003, *Confédération paysanne et autres*, para. 23: "De nouvelles techniques, dites de mutagénèse dirigée ou d'édition du génome, consistent aujourd'hui, grâce au génie génétique, à provoquer une mutation précise dans un gène cible sans introduction de gène étranger." (directed mutagenesis is understood as the induction of a precise mutation in a target gene without introduction of a foreign gene).

[54]Cf. text to n. 61ff.

Table 5.3 Legal wording of the GMO definition pursuant to Directive 2001/18/EC

Legal wording of the GMO definition in Directive 2001/18/EC:
Abstract Definition
Art. 2(2) "'Genetically modified organism (GMO)' means an organism, with the exception of human beings, in which the genetic material has been altered in a way that does not occur naturally by mating and/or natural recombination; Within the terms of this definition: (a) genetic modification occurs at least through the use of the techniques listed in Annex I A, part 1; (b) the techniques listed in Annex I A, part 2, are not considered to result in genetic modification;" **Art. 3(1)** "This Directive shall not apply to organisms obtained through the techniques of genetic modification listed in Annex I B."
Black List
Annex I A Part 1 "Techniques of genetic modification referred to in Article 2(2)(a) are inter alia: (1) recombinant nucleic acid techniques involving the formation of new combinations of genetic material by the insertion of nucleic acid molecules produced by whatever means outside an organism, into any virus, bacterial plasmid or other vector system and their incorporation into a host organism in which they do not naturally occur but in which they are capable of continued propagation; (2) techniques involving the direct introduction into an organism of heritable material prepared outside the organism including micro-injection, macro-injection and micro-encapsulation; (3) cell fusion (including protoplast fusion) or hybridisation techniques where live cells with new combinations of heritable genetic material are formed through the fusion of two or more cells by means of methods that do not occur naturally."
White List
Annex I A Part 2 "Techniques referred to in Article 2(2)(b) which are not considered to result in genetic modification, on condition that they do not involve the use of recombinant nucleic acid molecules or genetically modified organisms made by techniques/methods other than those excluded by Annex I B: (1) in vitro fertilisation, (2) natural processes such as: conjugation, transduction, transformation, (3) polyploidy induction."
Grey List
Annex I B "Techniques/methods of genetic modification yielding organisms to be excluded from the Directive, on the condition that they do not involve the use of recombinant nucleic acid molecules or genetically modified organisms other than those produced by one or more of the techniques/methods listed below are: (1) mutagenesis, (2) cell fusion (including protoplast fusion) of plant cells of organisms which can exchange genetic material through traditional breeding methods."

The European Court of Justice's ruling came as a surprise.[55] Both the Advocate General,[56] whose task is to propose a legal solution to the Court,[57] and many EU entities as well as member states' authorities and advisory committees[58] were of the opinion that organisms developed through SDN-1, SDN-2 and ODM are *not* subject to the regulatory regime for GMOs. The judgment is (at least de facto[59]) binding on all national courts.

In Detail The European Court of Justice decided on two questions.[60] First, do organisms obtained by mutagenesis constitute GMOs within the meaning of Art. 2 (2) Directive 2001/18/EC? And second, assuming they do, which organisms does the "mutagenesis exemption" of the Directive (exemption of organisms obtained by mutagenesis) encompass—all organisms or only those obtained by mutagenesis techniques that existed before the adoption of the Directive? Regarding these two questions, the Court decided the following:

First, all organisms obtained by mutagenesis are GMOs within the meaning of Art. 2 (2) Directive 2001/18/EC.[61]

Background The Court decided on the general structure of the GMO definition. Furthermore, it implicitly decided an interpretational dispute relevant for GEOs: The legal GMO definition (Art. 2(2) Directive 2001/18/EC) refers to organisms in which the genetic material 'has been altered in a way that does not occur naturally'. This can be read in two ways[62]:

- the process of altering the genome is unnatural (**process-based interpretation**) or
- additionally, the result of the altering of the genome is unnatural (**product-based interpretation**)

Organisms obtained by the new directed mutagenesis techniques, i.e. SDN-1, SDN-2 and ODM, have a genetic constitution that can occur naturally as well.[63]

[55] Cf. Lappin (2018b), p. 2.

[56] The Advocate General was of the opinion that the new directed mutagenesis techniques are "mutagenesis techniques" in the legal sense of the term and thus exempt from the EU's GMO regulatory framework, cf. Bobek (2018), paras. 107, 86ff.

[57] The Court often follows this proposal, Craig and de Búrca (2015), p. 61.

[58] Cf. the overview in Eriksson (2018), p. 5.

[59] The legally binding effect on third parties is debated, cf. Craig and de Búrca (2015), pp. 465, 478; European Parliamentary Research Service (2017), p. 11.

[60] ECJ, Case C-528/16 *Confédération paysanne and Others* [2018] ECLI:EU:C:2018:583, para. 25 no. 1.

[61] ECJ, Case C-528/16 *Confédération paysanne and Others* [2018] ECLI:EU:C:2018:583, paras. 27–38, 54.

[62] Cf. e.g. European Commission New Techniques Working Group (2011), para. 4.1; Sprink et al. (2016), pp. 1497–1498; Custers (2016), p. 2; Callebaut (2015), pp. 20–22; 42–56.

[63] German Federal Office of Consumer Protection and Food Safety (2017), p. 6.

Accordingly, they are GMOs if a process-based interpretation is applied but they are not GMOs if a product-based interpretation is applied.

The Court did not mention this dispute. However, it implicitly followed a process-based interpretation. It argued that organisms obtained by mutagenesis are GMOs because the process of altering the genome is unnatural.[64] Therefore, it is now clear that any unnatural process of altering the genome yields a GMO no matter what the genetic result is.

Second, the mutagenesis exemption only applies to organisms obtained by mutagenesis techniques that have conventionally been used in a number of applications and have a long safety record.[65] The factual conditions in 2001 are decisive ("frozen" interpretation).[66] This means the mutagenesis exemption only applies to organisms obtained by mutagenesis techniques that were routinely used in 2001. Consequently, organisms obtained by the new directed mutagenesis techniques are *not* exempted from the scope of the GMO regime.[67]

Background The issue in question was whether all mutagenesis techniques within the scientific meaning are covered by the mutagenesis exemption or only some techniques, e.g. techniques that were in use at the time when the EU's GMO regulatory regime was established or techniques that are "safe".[68]

5.3.2.2 SDN-3: Yields GMOs

Before the judgment of the European Court of Justice (Case C-528/16), there had already been consensus that the stable integration of a transgene into the plant genome via genome editing (transgenesis[69]) gives rise to GMOs.[70] It is now also

[64]ECJ, Case C-528/16 *Confédération paysanne and Others* [2018] ECLI:EU:C:2018:583, para. 29.

[65]ECJ, Case C-528/16 *Confédération paysanne and Others* [2018] ECLI:EU:C:2018:583, paras. 39ff, 54.

The Court gives two main reasons: First, it refers to a recital of the Directive (Directive 2001/18/EC, recital 17). Second, according to the Court, "the risks linked to the use of those new techniques/methods of mutagenesis might prove to be similar to those which result from the production and release of a GMO through transgenesis" as "the direct modification of the genetic material of an organism through mutagenesis makes it possible to obtain the same effects as the introduction of a foreign gene into that organism and [. . .] the development of those new techniques/methods makes it possible to produce genetically modified varieties at a rate and in quantities quite unlike those resulting from the application of conventional methods of random mutagenesis."

[66]ECJ, Case C-528/16 *Confédération paysanne and Others* [2018] ECLI:EU:C:2018:583, para. 51.

[67]ECJ, Case C-528/16 *Confédération paysanne and Others* [2018] ECLI:EU:C:2018:583, para. 51.

[68]Bobek (2018), paras. 68–78.

[69]Transgenesis refers to the introduction of genes derived from organisms which are sexually incompatible with the engineered plant, Voigt and Klima (2017), p. 320; Schaart et al. (2016), p. 439.

[70]Sprink et al. (2016), p. 1497.

Table 5.4 Regulatory classification of genome edited plants

	Regulatory classification of resulting organism
SDN-1	GMO
SDN-2	GMO
ODM	GMO
SDN-3	
Cisgenesis	GMO
Intragenesis	GMO
Transgenesis	GMO

certain that other uses of SDN-3 (cisgenesis,[71] intragenesis[72]) yield GMOs because the European Court of Justice decided to interpret the GMO definition in a process-based manner.[73] As solely the application of an unnatural technique is decisive, it does not matter that cisgenesis introduces genes derived from sexually compatible organisms and therefore induces genetic results that could occur naturally or through traditional breeding methods.

5.3.2.3 Summary

See Table 5.4.

5.4 Regulatory Prerequisites for Activities Relating to Genome Edited Plants

As illustrated in Sect. 5.3.2 ("Regulatory Classification of Genome Edited Plants"), organisms developed through all types of genome editing are GMOs. Thus, the regulatory prerequisites for uses of GMOs apply to them.

Contained use, field trials and the placing on the market of GMOs have separate authorisation regimes, which are described in the following. Subsequently, general principles of all authorisation regimes will be elucidated. Finally, problems inherent to the authorisation of GEOs will be illustrated.

---

[71]Cisgenesis refers to the introduction of genes derived from organisms which are sexually compatible with the engineered plant, Voigt and Klima (2017), p. 320; Schaart et al. (2016), p. 439.

[72]Intragenesis refers to the introduction of newly arranged genes derived from organisms that are sexually compatible with the engineered plant, Voigt and Klima (2017), p. 320; Schaart et al. (2016), p. 439.

[73]Apart from that, they might fall within the black list-catalogue, Directive 2001/18/EC, annex IA part 1, European Commission New Techniques Working Group (2011), para. 5.3.5 A; UK Advisory Committee on Releases to the Environment (2013b), p. 13.

Table 5.5 Contained use: regulatory prerequisites

Class	Notification requirement	Information required for notification	Degree of containment	Responsible agencies
1 (no or negligible risk)	Notification the first time GMOs are used in the premises	Less strict		National competent authority of the respective member state
2 (low risk)	Notification every time GMOs are used in the premises			
3 (moderate risk)	Notification and prior consent of the competent authority			
4 (high risk)	Notification and prior consent of the competent authority	strict		

5.4.1 *Step 1: Contained Use (Directive 2009/41/EC)*

5.4.1.1 Authorisation Procedure; Responsible Agencies

Contained uses of GMOs are classified in four classes according to their hazardousness (Table 5.5).[74] Each class corresponds to a containment level[75] and a notification level.[76] While notification suffices for classes 1 and 2,[77] prior consent (authorisation) is needed for classes 3 and 4.[78]

5.4.1.2 Risk Assessment to Classify Contained Uses[79]

The contained use is classified in advance by the person responsible for the contained use of GMOs following the risk assessment that is laid down in the

[74]Directive 2009/41/EC, art. 4.

[75]Directive 2009/41/EC, art. 5, annex IV.

[76]Directive 2009/41/EC, arts. 6–9.

[77]Directive 2009/41/EC, arts. 7, 8.

[78]Directive 2009/41/EC, art. 9.

[79]Directive 2009/41/EC, art. 4(2),(3), annex III; Commission Decision 2000/608/EC of 27 September 2000 concerning the guidance notes for risk assessment outlined in Annex III of Directive 90/219/EEC on the contained use of genetically modified micro-organisms [2000] OJ L258/43.

Table 5.6 Contained use: information required for class 3 & 4 notification

Example: Information required for class 3 & 4 notification (see Directive 2009/41/EC, art. 9, annex V part C)
Date of submission of the notification before first use of premises for contained uses; names of the persons responsible for supervision and safety and information on their training and qualification
The recipient or parental organism(s) to be used; the host-vector system(s) to be used (where applicable); the source(s) and intended function(s) of the genetic material(s) involved in the modification(s); the identity and characteristics of the GMO; the culture volumes to be used
A description of the containment and other protective measures to be applied; the purpose of the contained use, including the expected results; a description of the parts of the installation
Information about accident prevention and emergency response plans
Copy of the risk assessment

Contained Use Directive.[80] In case of doubt as to which class is appropriate for the contained use, the more stringent protective measures are applied.[81]

5.4.1.3 Information Required for the Notification

The amount of information required for the notification increases with the classes (regarding the highest amount of information, i.e. class 3 and 4 notification, see Table 5.6).[82]

5.4.1.4 Requirements for Granting an Approval

The competent national authorities control the correctness of the risk classification of the contained use as well as the suitability of the containment and other protective measures, waste management and emergency response measures. If required, i.e. for classes 3 and 4, they then grant an approval.[83]

[80]This risk assessment thus differs fundamentally in function and approach from the risk assessments for GMO field trials and placing on the market. It serves to determine which level of regulatory oversight is needed and is carried out by means of theoretical considerations based on the available knowledge. Risk assessments for field trials and any placing on the market, in contrast, aim at proving an organism's safety with the highest possible certainty by generating data, mostly by performing studies.

[81]Directive 2009/41/EC, art. 4(4).

[82]Cf. Directive 2001/18/EC, arts. 6, 8, 9, annex V.

[83]Directive 2009/41/EC, art. 10(2), (3).

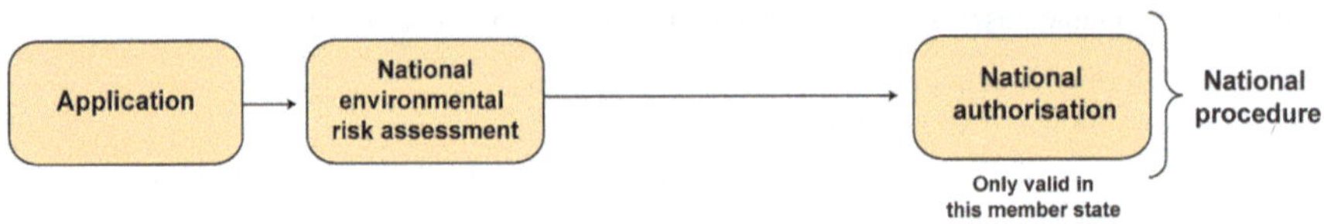

Fig. 5.5 Field trials: authorisation procedure

5.4.2 Step 2: Field Trials (Directive 2001/18/EC)

5.4.2.1 Authorisation Procedure; Responsible Agencies

Authorisations for deliberate release of GMOs into the environment for experimental purposes, i.e. field trails, are granted by the member states following an EU-harmonised procedure (Fig. 5.5). The individual authorisation is only valid within the respective member state.[84]

5.4.2.2 Information Required for the Application[85]

See Table 5.7.

5.4.2.3 Environmental Risk Assessment

For field trials, an environmental risk assessment is required (Table 5.8).[86] The environmental risk assessment is defined as the "evaluation of risks to human health and the environment, whether direct or indirect, immediate or delayed, which the deliberate release or the placing on the market of GMOs may pose".[87]

Both the data required for the environmental risk assessment[88] and the methodology[89] are harmonised at the EU level. As the EU environmental risk assessment approach follows internationally agreed guidelines,[90] they are similar to those in other countries.[91]

---

[84]Directive 2001/18/EC, art. 6(5); cf. Friant-Perrot (2010), p. 84; Plan and van den Eede (2010), p. 5.

[85]The application is called "notification" in the terminology of Directive 2001/18/EC, cf. art. 2(5).

[86]Directive 2001/18/EC, art. 6(2)(b).

[87]Directive 2001/18/EC, art. 2(8).

[88]Cf. Directive 2001/18/EC, art. 6(2)(a), annex III.

[89]Cf. Directive 2001/18/EC, art. 6(2)(b), annex II(C).

[90]E.g. by the United Nations Environment Programme (UNEP) or the Organisation for Economic Co-operation and Development (OECD), e.g. United Nations Environment Programme (1995), OECD (1986), pp. 24ff; OECD (2000b).

[91]Cf. Craig et al. (2008), p. 855.

Table 5.7 Field trials: application dossier

Application dossier: Field trials (see Directive 2001/18/EC, art. 6(2))
Technical dossier (Details: see Table 5.9)
Environmental risk assessment (Details: see Table 5.8)

Table 5.8 Environmental risk assessment

Environmental risk assessment (see Directive 2001/18/EC, annex II; EFSA Guidance on the environmental risk assessment of genetically modified plants[a])
Areas of risk to be addressed in the environmental risk assessment[b]:
• Persistence and invasiveness including plant to plant gene transfer • Plant to micro-organisms gene transfer • Interactions of the GM plant with target organisms • Interactions of the GM plant with non-target organisms • Impacts of the specific cultivation, management and harvesting techniques • Effects on biogeochemical processes • Effects on human and animal health

[a]European Food Safety Authority (2010a). In fact, this guidance of the European Food Safety Authority is not applicable to the environmental risk assessment of field trials (because the European Food Safety Authority does not evaluate field trial environmental risk assessments). It is only applicable to the environmental risk assessment for the placing on the market of GM plants. However, it is based upon the requirements in Directive 2001/18/EC, which cover both commercial cultivation and field trials and therefore correct as regards content for any environmental risk assessment of GM plants

[b]Directive 2001/18/EC, annex II(D.2); European Food Safety Authority (2010a)

The risk assessment is carried out by the applicant, building on the information required for the technical dossier (Table 5.9), and then reviewed by the national competent authority as part of the authorisation process.

The environmental risk assessment follows the six steps problem formulation including hazard identification, hazard characterisation, exposure characterisation, risk characterisation, risk management strategies and overall risk evaluation.[92] It is conducted in a tiered approach.[93] The required information covers all levels (molecular, cellular, organism, population, ecosystem).[94]

5.4.2.4 Requirements for Granting an Approval

The field trial must not induce adverse effects on human health and the environment.[95] "Adverse effects" mean effects exceeding the acceptable risk level.[96] The acceptable

[92]Directive 2001/18/EC, annex II(C.3).

[93]Cf. European Food Safety Authority (2010a), Devos et al. (2012), pp. 10789–10791.

[94]von Kries and Winter (2012), p. 573.

[95]Cf. Directive 2001/18/EC, art. 4.

[96]von Kries and Winter (2012), pp. 579–580; Craig et al. (2008), p. 854.

Table 5.9 Field trials: technical dossier

Technical dossier (see Directive 2001/18/EC, art. 6(2)(a), annex III B(I))
General information
Information relating to the recipient or parental plants
Molecular characterisation • Information relating to the genetic modification • Information relating to the GM plant
Information on specific areas of risk (see areas of risk in Table 5.8)
Information on control, monitoring, post-release and waste treatment plans
Description of detection and identification techniques for the GM plant
Information about previous releases of the GM plant, if applicable

risk level is a political question and influenced by social, economic and political factors such as public concerns or costs and benefits of GMOs.[97] It is therefore determined by the risk managers, i.e. the authorities granting or rejecting the approval.

5.4.3 *Step 3: Placing on the Market*

Both national authorities and EU institutions are competent for authorising the placing on the market of a GMO. Authorisations granted are valid throughout the EU.

Any placing on the market (cultivation, food/feed etc.) requires an environmental risk assessment.[98] Specific uses may require additional assessments. GM food and feed e.g. additionally require a human and animal health risk assessment.

5.4.3.1 Cultivation (Directive 2001/18/EC)

5.4.3.1.1 Authorisation Procedure; Responsible Agencies[99]

The company asking for a market authorisation for cultivation submits the application dossier to a national competent authority (Fig. 5.6). The authority evaluates the environmental risk assessment carried out by the applicant and issues an opinion (risk assessment report). If there are objections by other member states that remain (and in practice there always are), an EU safety assessment is carried out by the European Food Safety Authority (EFSA), taking into account observations by

[97]Cf. European Commission (2000), summary no. 5.

[98]Cf. Directive 2001/18/EC, art. 12(1); Regulation (EC) No 1829/2003, arts. 5(5), 6(4)/arts. 17(5), 18(4) (food/feed); von Kries and Winter (2012), p. 572.

[99]Directive 2001/18/EC, arts. 13–15, 18–19, 28(1); Hartung and Schiemann (2014), p. 744; Devos et al. (2012), pp. 10775–10776.

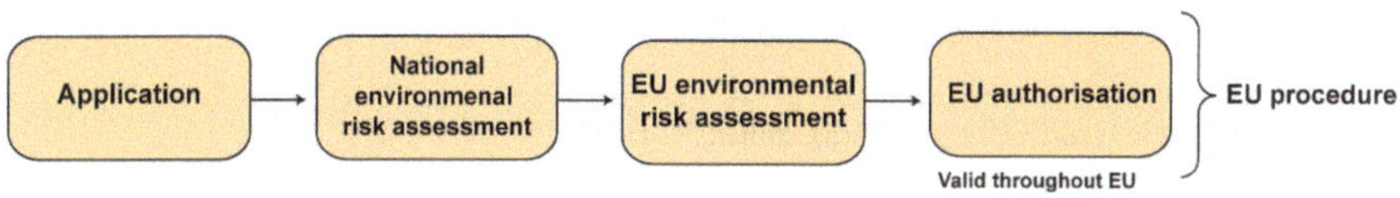

Fig. 5.6 Placing on the market—cultivation: authorisation procedure

Table 5.10 Placing on the market—cultivation: application dossier

Application dossier: Cultivation (see Directive 2001/18/EC, art. 13(2))
Technical dossier (Details: see Table 5.11)
Environmental risk assessment (Details: see Table 5.8)
Additional information, e.g. on the use of the product (see Directive 2001/18/EC, annex IV)
Conditions for the placing on the market of the product, including specific conditions of use and handling
Proposed period of consent (max 10 years)
Monitoring plan (see Directive 2001/18/EC, annex VII)
Proposal for labelling
Proposal for packaging
Summary

member states. The approval decision is then taken at the EU level, but involves all member states.[100] The approval is granted for 10 years and can be renewed following a procedure similar to the initial authorisation procedure.[101]

5.4.3.1.2 Information Required for the Application[102]

See Table 5.10.

[100]Cf. Directive 2001/18/EC, arts. 18(1), 30(2) (comitology procedure). In detail: The comitology procedure is the standard procedure for technical decisions. The European Commission issues a draft authorisation decision. A Committee composed of representatives of the member states adopts or rejects the decision by qualified majority. If the necessary majority cannot be reached ("no opinion"), the decision is referred to an appeal committee. In the event the necessary qualified majority cannot be reached in the appeal committee, either, the Commission adopts the final decision; cf. the infographic European Commission (2015b); cf. further Dederer (2016b), p. 150. In practice, a qualified majority is never reached, so the Commission takes all the final authorisation decisions, cf. European Commission (2015c), p. 3; Dederer (2016b), p. 150; Mühlböck and Tosun (2018), p. 387.

[101]Directive 2001/18/EC, arts. 13(2)(d), 15(4), 17; the data requirements are less burdensome than those for the initial authorisation, cf. Directive 2001/18/EC, art. 17(2).

[102]The application is called "notification" in the terminology of Directive 2001/18/EC, cf. art. 2(5).

Table 5.11 Placing on the market—cultivation: technical dossier

Technical dossier (see Directive 2001/18/EC, art. 13(2)(a), annex III B(II))
General information
Information relating to the recipient or parental plants
Molecular characterisation • Information relating to the genetic modification • Information relating to the GM plant
Comparative analysis of agronomic and phenotypic characteristics and of composition
Specific information for each area of risk (see areas of risk in Table 5.8)
Description of detection and identification techniques for the GM plant
Information about previous releases of the GM plant, if applicable

5.4.3.1.3 Environmental Risk Assessment

The legal provisions for the technical dossier as well as for the environmental risk assessment are similar for any placing on the market as for field trials.[103] However, the technical dossier for the placing on the market covers more aspects (notably a comparative analysis) and requires more extensive data (especially on the GM plant[104] and on each area of risk[105]) as a basis for the environmental risk assessment (Table 5.11). The applicant submits and builds upon data and results obtained from step 1 and 2 (research and field trials).[106]

5.4.3.1.4 Requirements for Granting an Approval

The cultivation of the GMO must not induce "adverse effects on human health and the environment".[107]

5.4.3.1.5 Member States' Possibility to "Opt Out" from GMO Cultivation[108]

Since 2015, member states can **opt out** from the commercial cultivation of an authorised GMO, i.e. prohibit or restrict the cultivation of this GMO on their

---

[103] See text to n. 86ff as well as Tables 5.8 and 5.9.

[104] Directive 2001/18/EC, annex III B(II)(B)(2)(b).

[105] Directive 2001/18/EC, annex III B(II)(B)(4).

[106] Directive 2001/18/EC, art. 13(2)(a), annex III B(II)(B)(6); cf. further von Kries and Winter (2011), p. 34.

[107] Cf. Directive 2001/18/EC, art. 4; regarding the notion "adverse effects" cf. text to n. 96.

[108] Cf. Dederer (2016b), pp. 151–159; European Commission (n.d.-f).

territory.[109] This means that even though the GMO is authorised for cultivation in the EU, it is not allowed to be cultivated in this member state. Member states that want to opt out have two options[110]:

- Option 1: Opt-out during the authorisation process. This requires the applicant's consent.
- Option 2: Opt-out after the granting of an authorisation. The member state has to invoke compelling grounds. Such compelling grounds are e.g. environmental policy objectives; town and country planning; land use; socio-economic impacts; avoidance of GMO presence in other products; agricultural policy objectives or public policy. The compelling grounds are thus not risk related.[111]

At present, 17 member states and parts of Belgium and the United Kingdom (hereinafter UK)[112] have opted out from cultivation of the GM maize MON810, which is authorised for cultivation in the EU, and of some or all GM varieties that are currently in the pipeline,[113] all with the applicant's/authorisation holder's[114] consent.[115] Accordingly, two-thirds of the EU member states are banning GMO cultivation on their territory.

However, there is a chance that less member states will seek to opt out from the cultivation of genome edited GM varieties, especially if people accept them better than traditional GM crops.[116]

5.4.3.2 Food and Feed (Regulation (EC) No 1829/2003)

Regulation (EC) No 1829/2003 regulates GM food and feed in an identical manner. The authorisation covers both food and feed use.[117]

---

[109]Cf. Directive 2001/18/EC, art. 26b. The aim of the opt-out possibility is to overcome member state obstructions of GMOs (deadlocks in the authorisation procedures of GMOs, invocation of safeguard clauses, overly restrictive coexistence measures etc.) by re-nationalising responsibilities, cf. Dederer (2016b), pp. 151–153. However, looking at recent authorisation decisions, it appears that this aim has not been reached, cf. Eriksson et al. (2018b). The conformity of the opt-out mechanism with EU law and WTO law is debated, cf. Dederer (2016b), pp. 159ff; Winter (2016a), pp. 132ff; Ferer (2016).

[110]Dederer (2016b), pp. 152ff.

[111]Cf. The group of FAS Biotechnology Specialists in the European Union (2017), p. 12.

[112]Wales, Scotland and Northern Ireland have opted out.

[113]European Commission (n.d.-j); The group of FAS Biotechnology Specialists in the European Union (2018), pp. 8–9; cf. also Fig. 5.1. Slovakia intends to opt out soon and is therefore already indicated as opted out in Fig. 5.1.

[114]Cf. the transitional measures laid down in Directive 2001/18/EC, art. 26c.

[115]McEldowney (2015).

[116]Cf. text to n. 269.

[117]Regulation (EC) No 1829/2003, art. 27.

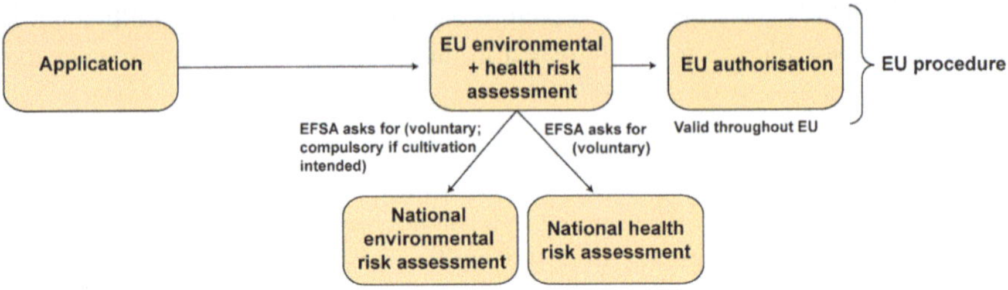

Fig. 5.7 Placing on the market—food/feed: authorisation procedure

The authorisation requirement applies to food and feed containing, consisting of or produced from[118] a GMO.[119]

5.4.3.2.1 Authorisation Procedure; Responsible Agencies[120]

The application dossier is submitted to a national competent authority and passed on to EFSA, which carries out an environmental risk assessment and a health risk assessment taking into account observations by the member states (Fig. 5.7). To fulfil this task, EFSA may ask the competent authority of a member state to carry out a food safety assessment or environmental risk assessment.[121] In case the GMO is also intended for cultivation, an environmental risk assessment of a national competent authority is compulsory.[122] The approval decision is then taken at the EU level, but includes all member states.[123] The approval is granted for 10 years and can be renewed following a procedure similar to the initial authorisation procedure.[124]

[118]"Produced from GMOs "means" derived, in whole or in part, from GMOs, but not containing or consisting of GMOs", Regulation (EC) No 1829/2003, art. 2(10). Food/feed is e.g. produced from GMOs if the GMOs have been destroyed during the production process (like highly refined oil). By contrast, products obtained from animals fed with GM feed are not produced "from", but only produced "with" GMOs and therefore do not have to be authorised, Regulation (EC) No 1829/2003, recital 16. "The determining criterion is whether or not material derived from the genetically modified source material is present in the food or feed", Regulation (EC) No 1829/2003, recital 16.

[119]Regulation (EC) No 1829/2003, art. 3(1).

[120]Regulation (EC) No 1829/2003, arts. 5ff/arts. 17ff (food/feed); cf. also the detailed description in European Food Safety Authority (2013), pp. 9ff.

[121]Directive 2001/18/EC, art. 6(3)(b), (c).

[122]Directive 2001/18/EC, art. 6(3)(c).

[123]Cf. Regulation (EC) No 1829/2003, arts. 7(3), 35(2) (comitology procedure), in detail see n. 100.

[124]Regulation (EC) No 1829/2003, arts. 7(5), 11/arts. 19(5), 23 (food/feed). The data requirements are less burdensome than those for the initial authorisation, cf. Regulation (EC) No 1829/2003, art. 11(2)/art. 23(2) (food/feed); Dederer (2016a), p. 108.

Table 5.12 Placing on the market—food/feed: application dossier

Application dossier: GM food and feed (see Regulation (EC) No 1829/2003, arts. 5(3),(5)/17(3),(5); Regulation (EU) No 503/2013,[a] annex I; EFSA guidance on the submission of applications for authorisation of genetically modified plants[b])
General information
Scientific information (Details: see Table 5.13)
Cartagena Protocol[c] (see Regulation (EC) No 1946/2003, annex II)
Labelling proposal
Methods of detection, sampling and identification and reference material (see Regulation (EU) No 503/2013, annex III)[d]
Environmental Assessment: Technical dossier; environmental risk assessment; Additional information[e] (see Directive 2001/18/EC, annexes II, III, IV) OR Authorisation under Directive 2001/18/EC
Environmental Monitoring Plan (see Directive 2001/18/EC, annex VII)
Summary

[a]Commission Implementing Regulation (EU) No 503/2013 of 3 April 2013 on applications for authorisation of genetically modified food and feed in accordance with Regulation (EC) No 1829/2003 of the European Parliament and of the Council and amending Commission Regulations (EC) No 641/2004 and (EC) No 1981/2006 [2013] OJ L157/1; only applicable to GM plants and food/feed derived from them; with respect to GM animals and micro-organisms cf. Commission Regulation (EC) No 641/2004 of 6 April 2004 on detailed rules for the implementation of Regulation (EC) No 1829/2003 of the European Parliament and of the Council as regards the application for the authorisation of new genetically modified food and feed, the notification of existing products and adventitious or technically unavoidable presence of genetically modified material which has benefited from a favourable risk evaluation [2004] OJ L102/14
[b]European Food Safety Authority (2013)
[c]Information required by the Cartagena Protocol; relevant for transboundary movements of GMOs
[d]This requirement constitutes a problem for organisms developed through SDN-1, SDN-2 and ODM, cf. text to n. 163ff
[e]See text to n. 98, 103–106 as well as Tables 5.9 and 5.11

5.4.3.2.2 Information Required for the Application

See Table 5.12.

Table 5.13 Placing on the market—food/feed: human and animal health risk assessment

Scientific information (see Regulation (EC) No 1829/2003, arts. 5(3)(e)/17(3)(e); Regulation (EU) No 503/2013, annex I part II, annex II; EFSA Guidance for risk assessment of food and feed from genetically modified plants[a] as well as other EFSA guidance documents on specific topics[b])
Hazard identification and characterisation • Information relating to the recipient and parental plants • Molecular characterisation • Information relating to the genetic modification • Information relating to the GM plant
Additional information relating to the genetically modified plant required for environmental safety aspects • Comparative analysis (compositional, agronomic and phenotypic characteristics of GM plant compared to its conventional counterpart) • Toxicological,[c] allergenicity and nutritional assessment
Exposure assessment
Risk characterisation
Post Market Monitoring
Additional information related to the safety of the GM food or feed

[a]European Food Safety Authority (2011a)
[b]Cf. the overview in European Food Safety Authority (n.d.)
[c]Including whole food/feed testing via a 90-day feeding study in rodents, cf. Regulation (EU) No 503/2013 recitals 10, 11, annex II part II 1.4.4.1

5.4.3.2.3 Human and Animal Health Risk Assessment[125]

The harmonised EU health risk assessment (Table 5.13) takes international standards into account.[126] Therefore, both the studies required and their methodologies are very similar to the safety assessment of GM foods in other countries.[127]

The health risk assessment follows the four steps hazard identification, hazard characterisation, exposure assessment and risk characterisation.

5.4.3.2.4 Requirements for Granting an Approval

The authorisation requires that the food/feed does not have adverse effects on human health, animal health or the environment, does not mislead the consumer and is not nutritionally disadvantageous compared to a food/feed it is intended to replace.[128]

[125]Cf. Schauzu (2011), pp. 73ff; European Food Safety Authority (2011a), pp. 5ff.

[126]In particular the Codex Alimentarius principles and guidelines on foods derived from biotechnology (which are binding on the EU via World Trade Law, see text to n. 41), e.g. Codex Alimentarius Commission (2003b); Codex Alimentarius Commission (2003a), as well as the pertinent OECD guidelines, e.g. OECD (2000a); cf. also Regulation (EU) No 503/2013, recitals 9, 21.

[127]Cf. Schauzu (2011), p. 73; Paoletti et al. (2008), p. 77.

[128]Food: Regulation (EC) No 1829/2003, arts. 7(1), 4(1); slightly different regarding feed, Regulation (EC) No 1829/2003, arts. 19(1), 16(1); cf. further Dederer (2016a), pp. 101–102.

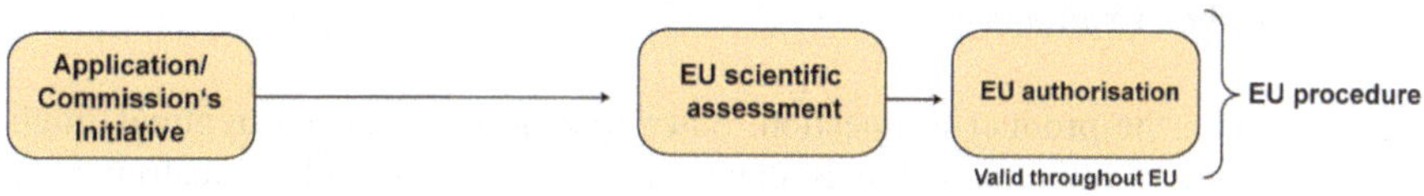

Fig. 5.8 Placing on the market—medicinal products: authorisation procedure

In addition, "other legitimate factors relevant to the matter under consideration" can be taken into account.[129]

5.4.3.3 Medicinal Products (Regulation (EC) No 726/2004)

To be placed on the European market,[130] medicinal products for human use developed through recombinant DNA technology require an authorisation following the centralised procedure laid down in Regulation (EC) No 726/2004 (Fig. 5.8).[131]

Companies submit their application to the European Medicines Agency (EMA). The EMA carries out a scientific assessment[132] and forwards an opinion to the European Commission as to whether the medicine should be authorised or not.[133] The latter[134] then grants or rejects the marketing authorisation based on the criteria quality, safety and efficacy.[135] The authorisation is valid throughout the EU. It lasts for 5 years and is renewable.[136]

In case the pharmaceutical or compounds of it are produced by plants (molecular farming), the uses of the GM production plant need to be authorised according to the general GMO authorisations, i.e. according to Directive 2009/41/EC for contained use and according to Directive 2001/18/EC for field trials as well as cultivation.[137]

---

[129]Regulation (EC) No 1829/2003, art. 7(1)/art. 19(1) as well as n. 157.

[130]Apart from the authorisation for the placing on the market, other authorisation requirements of the EU medical law apply, i.e. manufacturing and clinical trial authorisations, cf. Directive 2001/83/EC of the European Parliament and of the Council on the Community code relating to medicinal products for human use [2001] OJ L311/67, art. 40(1); Regulation (EU) No 536/2014 of the European Parliament and of the Council of 16 April 2014 on clinical trials on medicinal products for human use, and repealing Directive 2001/20/EC [2014] OJ L158/1, art. 4.

[131]Regulation (EC) No 726/2004, art. 3(1), annex; cf. the regulatory flowchart in European Food Safety Authority (2009), p. 11.

[132]The peculiarities of substances produced by transgenic plants are taken into account in a special EMA guideline on the "quality of biological active substances produced by stable transgene expression in higher plants", European Medicines Agency (2008).

[133]Regulation (EC) No 726/2004, arts. 5ff; cf. European Commission (n.d.-a).

[134]With the participation of the member states via a committee composed of member state representatives, Regulation (EC) No 726/2004, art. 10(2).

[135]Cf. Regulation (EC) No 726/2004, art. 12(1).

[136]Regulation (EC) No 726/2004, art. 14(1)-(3).

[137]Sparrow et al. (2013), pp. 4–6; the peculiarities of plant molecular farming are taken into account by the EFSA guidance document on the risk assessment of genetically modified plants used for non-food or non-feed purposes, cf. European Food Safety Authority (2009).

5.4.3.4 Other Products

Depending on the product in question, other specific authorisation regimes might apply. The most important case is plants producing plant-made industrial compounds (PMI)[138]: In case of the production of chemical elements or their compounds, the produced substance might need to be registered according to the REACH-Regulation.[139]

5.4.4 *General Principles and Concepts of All GMO Authorisation Regimes (Excluding Contained Use)*

5.4.4.1 Division Into Risk Assessment and Risk Management

As most GMO regulatory frameworks around the world,[140] the EU's authorisation procedures can be divided into a scientific assessment stage (risk assessment) and the political[141] authorisation decision (risk management).

[138]E.g. industrial enzymes or raw materials for the production of biopolymers, biofuels, paper and starch, European Food Safety Authority (2009), p. 11.

[139]Regulation (EC) No 1907/2006 of the European Parliament and of the Council of 18 December 2006 concerning the Registration, Evaluation, Authorisation and Restriction of Chemicals (REACH), establishing a European Chemicals Agency, amending Directive 1999/45/EC and repealing Council Regulation (EEC) No 793/93 and Commission Regulation (EC) No 1488/94 as well as Council Directive 76/769/EEC and Commission Directives 91/155/EEC, 93/67/EEC, 93/105/EC and 2000/21/EC [2006] OJ L396/1 (REACH), art. 5; applies to both manufacturing (in quantities of one tonne or more per year) as well as placing on the market, including import, REACH, arts. 5, 6. Many substances that are obtained from natural sources and have not been chemically modified are exempt from registration if the substance does not have adverse properties such as persistence or toxicity, cf. REACH, annex V No. 8, 9. A registration is not needed for substances that are already registered, provided their identity can be established, REACH, art. 10(a) (ii) in conjunction with annex VI(2) (establishing identity is challenging for biological materials, they belong to what is called the UVCB-substrates—substances of unknown or variable composition, complex reaction products or biological materials, cf. REACH, recital 45). For more information on registration of substances refer to European Chemicals Agency (2017).

[140]Craig et al. (2008), p. 853; cf. also international guidelines, e.g. United Nations Environment Programme (1995), p. 5, Codex Alimentarius Commission (2003b), Section 3; cf. text to n. 41.

[141]This means that the authorisation decision is not taken by scientific bodies but by political institutions. Risk management includes weighing of different options, deciding on the acceptability of risks and other value-based decisions, cf. Devos et al. (2010), p. 557.

5.4.4.2 General Principles and Concepts of Risk Assessment

The general principles of risk assessment are the same in the EU[142] as in most other countries.[143] The most important principles are:

The environmental and health safety assessment are based on a **comparative approach** (synonymous with the concept of substantial equivalence[144]).[145] The aim of the safety assessment is thus to show that the GMO is as safe as a comparable non-GM crop (relative safety). Following a molecular characterisation, the GMO is compared to its comparator, e.g. regarding compositional, phenotypic and agronomic characteristics. Statistically significant differences that fall outside the range of natural variation[146] are then assessed for their potential adverse effects on human health or the environment. Accordingly, both intended and unintended differences are identified and assessed.

Risk assessments are carried out on a **case-by-case** basis. This means that the required information may vary, depending e.g. "on the type of the GMOs concerned, their intended use and the potential receiving environment".[147]

Risk assessments maintain a **high scientific standard**.

The risk assessment may need to be **readdressed** if new information on the GMO and its effects on human health or the environment becomes available.

Regarding other concepts of EU risk assessments, the **data is collected and submitted by the applicant** and reviewed by national competent authorities and/or the EFSA. The **factors taken into consideration** during the environmental and health risk assessments are limited to risks of the individual GMO. Its benefits or overall risk/benefit considerations of GMOs are outside the scope of the scientific

[142]Cf. Directive 2001/18/EC, annex II(B).

[143]Cf. Tzotzos et al. (2009), pp. 52ff; Eckerstorfer et al. (2014), p. 62; Devos et al. (2010), pp. 562–565; cf. also the risk assessment principles in the Cartagena Protocol, which the EU risk assessments respect (Cartagena Protocol, annex III and Secretariat of the Convention on Biological Diversity (2016); Directive 2001/18/EC, recital 13); with respect to food cf. the Codex Alimentarius principles and guidelines and pertinent OECD guidelines, cf. text to n. 41, 126.

[144]OECD (1993), pp. 11ff: "existing organisms used as food, or as a source of food, can be used as the basis for comparison when assessing the safety of human consumption of a food or food component that has been modified or is new"; further FAO/WHO (2000).

[145]European Food Safety Authority (2011a), pp. 5–6; European Food Safety Authority (2010a), pp. 11–13; Schauzu (2011), pp. 73ff.

[146]In fact, this comprises two tests: the test of difference to verify whether the GM plant is different from its comparator (apart from the genetic modification) and the test of equivalence to verify whether the characteristics are within the range of natural variation, cf. European Food Safety Authority (2011a), pp. 5–6, 12; European Food Safety Authority (2010a), p. 28; regarding the determination of the range of natural variation cf. European Food Safety Authority (2010b) as well as e.g. the OECD Consensus Documents, OECD Working Group on the Safety of Novel Foods and Feeds (2018), or the ILSI database, International Life Sciences Institute (2016), Ridley et al. (2004).

[147]Directive 2001/18/EC, annex II(B); European Food Safety Authority (2010a), p. 3.

assessments.[148] **Scientific uncertainty** is identified by analysing the nature and level of uncertainty at various steps.[149] The competent authorities can request additional data if they are not satisfied that the GMO activity is safe.[150]

As opposed to the homogeneity of the basic risk assessment principles and concepts, there is considerable variation between the regulatory systems in the depth of information required for each part of the assessment dossier.[151] In comparison to the USA for instance, the EU requires quite large amounts of data.[152]

5.4.4.3 General Principles and Concepts of Risk Management

What are the general principles and concepts guiding the approval decision?

The **precautionary principle**[153] supplies general considerations on how to deal with uncertainty in the risk management decision.[154]

The **factors that can be taken into consideration** in the GMO approval decisions are predominantly the risks of the GMO to human health and the environment.[155] The legal space for considering non-safety aspects, in contrast, is very limited.[156]

---

[148]European Food Safety Authority (2010a), p. 10; UK House of Commons Science and Technology Committee (2015), p. 42; Baulcombe et al. (2014), p. 32.

[149]With respect to the uncertainty analyses to be provided by the applicant cf. the EFSA Guidance on Uncertainty Analysis in Scientific Assessments, European Food Safety Authority (2018a) as well as the EFSA document about the principles and methods behind EFSA's Guidance on Uncertainty Analysis in Scientific Assessment, European Food Safety Authority (2018b).

[150]E.g. Directive 2009/41/EC, art. 10(3)(a) (contained use); Directive 2001/18/EC, art. 6(6),(7) (field trials), art. 15(1) (cultivation); Regulation (EC) No 1829/2003, art. 6(2)/art. 18(2) (food/feed).

[151]Craig et al. (2008), p. 854, Eckerstorfer et al. (2014), p. 62; UK House of Commons Science and Technology Committee (2015), Q411.

[152]The Netherlands Commission on Genetic Modification (2009b), p. 21; European Food Safety Authority (2011a).

[153]TFEU, art. 191(2); with respect to the European understanding of the precautionary principle cf. European Commission (2000).

[154]For instance, measures adopted should be proportional, non-discriminatory and consistent. They should consider the benefits and costs of action and lack of action and be re-examined after some time, European Commission (2000), para. 6.3.

[155]Winter (2016b), p. 189.

[156]**Firstly**, ethical committees can be consulted, Directive 2001/18/EC, art. 29; Regulation (EC) No 1829/2003, art. 33, e.g. the European Group on Ethics in Science and New Technologies. However, this only plays a marginal role in the authorisation decision in practice, Scott (2005), p. 118; Lee (2008), pp. 81–82. **Secondly**, the authorisation regime for marketing GM food and feed allows to take into account "other [that is non-safety related] legitimate factors" in the authorisation decision, Regulation (EC) No 1829/2003, arts. 7(1), 19(1). However, no authorisation decision has referred to "other factors" yet and their scope and role are highly debated, European Commission (2015c), pp. 3–4; Spök (2010), p. 32; Jack (2009), p. 230; Lee (2008), pp. 83ff; Dederer (2016a), p. 106; Dederer (2016b), p. 149; Scott (2005), pp. 118–119. **Thirdly**, some member states have included an ethical and/or socio-economic impact assessment in their national legislations regarding field trials/cultivation. They are of little impact in practice, cf. Spök (2010), pp. 31ff; Winter (2016b), p. 189. **Fourthly**, environmental risks can be balanced against environmental benefits (health risks do not allow for such a risk-benefit analysis), Dederer (2016a), p. 102.

This marginal role of non-safety aspects in the legal framework does not mirror their actual political and social importance. This discrepancy has resulted in a politicisation of the decision-making processes.[157] Scientific concerns are oftentimes merely raised for political reasons.[158] The voting behaviour of the member states representatives in the approval decision[159] is motivated by general acceptance or rejection of GMOs in the respective member state rather than by the safety of the individual GMO in question.[160] There is a constant discussion of how to give non-safety aspects more room in the EU's regulatory framework for GMOs to stop them resulting in this paralysis of the authorisation procedure.[161]

5.4.5 *Problems Regarding the Authorisation of GEOs*

5.4.5.1 Detection and Identification Techniques Required for GMO Authorisation[162]

Application dossiers for the placing on the market of GMOs require methods for detection and identification of the transformation event.[163] Whereas the term transformation event is not defined in EU legislation, it is elsewhere understood as the site "where a conventional organism is 'transformed' through the introduction of a modified DNA sequence".[164] Regarding genome-edited plants, the genome-edited event is thus the altered sequence at a specific site in the genome.[165] In the case of GM food and feed, the methods have to be validated by the European Union Reference Laboratory for GM Food and Feed (EURL-GMFF) in cooperation with the European Network of GMO laboratories (ENGL).[166]

[157]Dederer (2016b), pp. 148ff.

[158]Dederer (2016b), p. 152.

[159]Cf. the comitology procedure, n. 100.

[160]European Commission (2015c), p. 3.

[161]Winter (2016b), p. 183.

[162]Cf. in detail the report of the European Network of GMO Laboratories, European Network of GMO Laboratories (2019).

[163]Cf. Directive 2001/18/EC, art. 13(2)(a) in conjunction with annex IV(A)(7) (cultivation); Regulation (EC) No 1829/2003, art. 5(3)(i)/art. 17(3)(i) (food/feed); cf. in detail Regulation 503/2013, annex III; Regulation (EC) No 641/2004, annex I as well as the guidelines available at European Union Reference Laboratory for GM Food and Feed (2018), especially the guideline "Definition of minimum performance requirements for analytical methods of GMO testing"; cf. further international standards on methods for detection and identification, e.g. Codex Alimentarius Commission (2010), International Organisation for Standardisation (2005a, b, c, 2006). According to Kalaitzandonakes et al. (2007), p. 510, the requirement of detection and identification methods for authorisation is EU-specific.

[164]European Commission (2001).

[165]European Network of GMO Laboratories (2019), p. 5.

[166]Cf. Regulation (EC) No 1829/2003, art. 6(3)(d), annex(3)(d); European Union Reference Laboratory for GM Food and Feed (2017).

It is still unclear how techniques that allow for unequivocal identification of SDN-1, SDN-2 and ODM events could be designed.[167] Therefore, placing GEOs containing these events on the market is currently not possible.[168]

As things stand now, a genetic modification introduced by SDN-1, SDN-2 or ODM can be detected[169] (provided that prior knowledge on the modification is available).[170] However, in general, it cannot be identified[171] as one that has been introduced by a genome editing technique.[172] The shorter the modification and the larger a plant's genome, the more likely it is that the DNA sequence already exists in the plant's genome or reoccurs naturally or as a result of traditional breeding.[173] According to some experts, for identification, information on a unique DNA sequence in the genome of at least 20 nucleotides is needed.[174]

This means that the identification of the *genome edited line* might be possible using the characteristic mutation sequence and the sequence in its direct vicinity plus possibly other sequences ("markers") of the genome edited line.[175] However, the identification of the genomic change itself and thus of the *event* remains challenging because the distinct DNA sequence of the genomic change itself is not long enough.[176] For organisms developed through SDN-3, including cisgenesis, the development of unequivocal detection and identification techniques is possible.[177]

---

[167]European Network of GMO Laboratories (2019), 7–13; 17; European Commission Scientific Advice Mechanism Group of Chief Scientific Advisors (2018), p. 4. Likely, the minimum performance requirements for detection and identification methods will have to be changed, European Network of GMO Laboratories (2019), p. 1.

[168]Kahrmann et al. (2017), p. 180; The Greens/European Free Alliance in the European Parliament (2018a).

[169]"Detection" of a genetic modification means "that it is possible to determine the existence of a change in the genetic material of an organism", Lusser et al. (2011), pp. 63–64.

[170]Lusser et al. (2011), pp. 69–70.

[171]Identification of a genetic modification means "that it is possible not only to detect the existence of a change in the genetic material of an organism [but also] to identify the genetic modification as one that has been intentionally introduced by a new technique", Lusser et al. (2011), p. 64.

[172]European Network of GMO Laboratories (2019), pp. 9–10; Lusser et al. (2011), pp. 69–70; Hilscher et al. (2017a), pp. 32–33.

[173]Plant genomes are very diverse and dynamic. That means that they change at a rapid pace and that plant genomes differ a lot between two varieties, European Food Safety Authority (2012a), p. 12; European Network of GMO Laboratories (2019), p. 11; regarding the maize genome Jiao et al. (2017), p. 526.

[174]Lusser et al. (2011), pp. 165, 169; more precise Grohmann et al. (2019), pp. 4–5.

[175]Hilscher et al. (2017a), pp. 32–33; however, the linkage between the mutation and the background markers is broken up if the variety is used in further breeding programmes, Hilscher et al. (2017a), p. 33.

[176]However, it might be deemend sufficient that the genome edited line can be identified, because this also clarifies the origin of the event, Hilscher et al. (2017a), pp. 32–33.

[177]Ribarits et al. (2014), p. 186; Lusser et al. (2011), pp. 69–70.

5.4.5.2 Risk Assessment: Safety Data Requirements Possibly Disproportionate

The basic risk assessment principles and concepts are adequate for the assessment of varieties and their products developed through genome editing[178] (notwithstanding the question whether a risk assessment is needed at all, which is a question of the scope of the framework[179]).

However, it would be disproportionate to require the same amount of safety data for transgene insertions and point mutations.

The risk assessment is already carried out on a case-by-case basis, which means that the required amount of data varies depending on the individual case.[180] The appropriateness of the risk assessments for organisms developed through **SDN-1, SDN-2 and ODM** thus depends on the flexibility of the application of data requirements in practice.[181] Possibly, the scientific information required by Directive 2001/18/EC and Regulation (EU) No 503/2013 as well as the EFSA guidance documents[182] need to be revised.[183] It might be of importance for applicants whether they can get pre-submission advice from EFSA on the required extent of studies to perform.

Risk assessments are appropriate for organisms developed through **SDN-3**. EFSA issued two opinions that its current guidance documents for GMO environmental and health risk assessments are adequate for plants developed through SDN-3 and cisgenesis.[184] On a case-by-case basis, less event-specific data will probably be necessary.[185] An update of the guidance documents to introduce further flexibility[186] should be considered.[187]

---

[178]Cf. Eckerstorfer et al. (2014), p. 62.

[179]Eckerstorfer et al. (2014), p. 62.

[180]Cf. text to n. 147.

[181]It is criticised that in practice, the amount and type of information is not sufficiently flexible, especially when it comes to *less* data, UK House of Commons Science and Technology Committee (2015), Q411, Q413 [Professor Hails]; different opinion (appropriate amount of data) e.g. UK House of Commons Science and Technology Committee (2015), GMC0016 [Professor Perry] paras. 6, 7.

[182]A revision of the EFSA guidance documents is even envisaged for SDN-3, cf. text to n. 188, and is therefore even more necessary for SDN-1 and SDN-2.

[183]Cf. also Ribarits et al. (2014), pp. 188–189, Wasmer (2019), pp. 8–9: Some elements, e.g. environmental risks caused by gene transfer, should be reconsidered.

[184]European Food Safety Authority (2012a), pp. 17ff; European Food Safety Authority (2012b), p. 21; cf. also (with respect to cisgenesis) Prins and Kok (2010), p. 23: "Existing knowledge [...] will, on a case-by-case basis, already be used within the current regulatory framework [...]. [...] no scientific basis for a general reduction of requirements".

[185]European Food Safety Authority (2012b), p. 21; European Food Safety Authority (2012a), p. 19.

[186]E.g. in the section molecular characterisation or with respect to scrutiny for unintended effects, European Food Safety Authority (2012a), p. 18; European Food Safety Authority (2012b), p. 21.

[187]European Food Safety Authority (2012a), p. 18; EFSA guidance documents are regularly updated anyway, European Food Safety Authority (n.d.).

5.5 Status Quo of Genome Edited Plants and Products Derived from Them

5.5.1 *Contained Use*

Public and private institutions in the EU are strong at research on genome edited plants. Regarding the articles published on the use of CRISPR systems in plant genome editing, the EU ranks third after China and the USA.[188] Much attention has been paid to the development of transgene-free plants.[189] However, it is expected that the GMO classification of all types of GEOs will lead to a significant decrease in public and industrial investments in research towards commercial applications of genome edited plants.[190]

Research projects often cover a large variety of species and traits.[191] An example is the French GENIUS project using targeted mutagenesis on a range of varieties to improve traits linked to crop culture (disease resistance, flowering time, plant architecture etc.) as well as to the quality for industrial purposes.[192] Nonetheless, Table 5.14 gives a few specific examples.

5.5.2 *Field Trials*

Initially, field trials of an oilseed rape developed using ODM ("Cibus SU Canola"[193]) were carried out in the UK and Sweden.[194] Competent authorities of six member states had decided that a GMO field trial authorisation is not required, in other terms that the ODM oilseed rape is not a GMO in their view.[195] Possibly, field trials were also carried out of other genome edited crops, e.g. a CRISPR/Cas9 mutated Arabidopsis.[196] However, in 2015, the European Commission asked the

---

[188]Ricroch et al. (2017), pp. 178, 179; Modrzejewski et al. (2018), p. 5.

[189]Ricroch et al. (2017), p. 180.

[190]Callaway (2018), and King (2018).

[191]OECD (2018a), p. 5.

[192]OECD (2018a), p. 5; Nogué et al. (2015).

[193]Cibus™ (2017, 2018), Lombardo et al. (2016), pp. 52–53.

[194]Abbott (2015).

[195]Swedish Board of Agriculture (2015) (Sweden); UK Advisory Committee on Releases to the Environment (2011) (UK); German Federal Office of Consumer Protection and Food Safety (2015) (Germany; withdrawn in 2018, German Federal Office of Consumer Protection and Food Safety (2018)); further Abbott (2015); Madre and Agostino (2017); Eriksson (2018), pp. 2–4; Corporate Europe Observatory (2016c).

[196]Not known for certain, cf. Eriksson et al. (2017), p. 226. Cf. also the decisions by the Swedish and Finnish authorities that a GMO field trial authorisation is not required, Swedish Board of Agriculture (2015); Finnish Centre of Excellence in Molecular Biology of Primary Producers (2016).

Table 5.14 Research examples of genome editing in crop plants in the EU[a]

Species	Trait(s)	Breeding technique	Country	Reference
Brassica oleracea	Shatter-resistant pods	SDN-1 (Gene knockout, CRISPR/Cas9)	UK	Lawrenson et al. (2015)
Camelina[b]	Altered oil quality	SDN-1 (Gene knockout, CRISPR/Cas9)	France, UK	Morineau et al. (2017)
Potato	Altered starch quality	SDN-1 (Gene knockout, CRISPR/Cas9)	Sweden	Andersson et al. (2017, 2018)
Tomato	Early flowering	SDN-1 (Gene knockout, CRISPR/Cas9)	Germany, France, USA	Soyk et al. (2017)
Tomato	Powdery mildew resistance	SDN-1 (Gene knockout, CRISPR/Cas9)	UK, Germany	Nekrasov et al. (2017)
Wheat	Low-gluten	SDN-1 (Gene knockout, CRISPR/Cas9)	Spain	Sánchez-León et al. (2017)
Cotton	Herbicide tolerance, trait stacking	SDN-3 (Gene insertion, meganuclease)	Belgium	D'Halluin et al. (2013)

[a]Data taken from Ricroch et al. (2017), pp. 171–176 and Modrzejewski et al. (2018), pp. 7–25, see there for further examples and further data. Cf. also Eriksson et al. (2018a), pp. 1–2 providing an exemplary country-by-country overview of the research projects on plant genome editing carried out by public research institutes

[b]At the time of writing, field trials are carried out, cf. text to n. 203

competent authorities of the member states not to allow field trials of plants obtained by new breeding techniques before the clarification of their GMO status at the EU level.[197] Consequently, field trials were suspended[198] and several other genome edited plants ready for field trials[199] were kept back to await the judgment of the European Court of Justice in Case C-528/16. One exception was a field trial of genome edited maize that started in Belgium in 2017.[200] Another was a field trial of

[197]European Commission Directorate-General for Health and Food Safety (2015a); cf. also Abbott (2015). Once a binding interpretation of the GMO definition is issued at the EU level by the European Court of Justice, national interpretations are void. They are de facto also void if the European Commission issues non-binding interpretation guidance. Member states decisions that no field trial authorisation is required have no impact on the necessity of a marketing authorisation because marketing authorisations are issued at the EU level.

[198]Abbott (2015).

[199]Eriksson et al. (2017), p. 226.

[200]Cf. Corporate Europe Observatory (2018), and Greenpeace EU (2018).

genome edited Camelina[201] that started in the UK in June 2018.[202] Both proceeded after the judgment of the European Court of Justice as they obtained a GM field trial authorisation in 2019.[203] All future field trials of genome edited plants will require a GMO field trial authorisation due to their GMO classification in the judgment in Case C-528/16.

5.5.3 Placing on the Market

It will take several years until the first genome edited plants and products derived from them enter the EU market due to the requirement of a GMO authorisation.[204]

5.6 Reform Efforts

Shortcomings and update proposals concerning only individual parts of the GMO regime are outlined in the respective sections, e.g. the shortcomings due to the impossibility to identify certain genetic alterations by genome editing.[205] This section, to the contrary, explores efforts to update the entire GMO regime with regard to the regulation of genome edited plants and products derived from them.

At the time of writing, there have been **no serious attempts to update the EU's regulatory framework for GMOs, so far** in reaction to the judgment of the European Court of Justice in Case C-528/16.[206] No reforms in the aftermath of the judgment are expected till the new European Commission takes office in November 2019.[207]

Nonetheless, the following section presents a few efforts, proposals and possibilities of reform that are discussed by stakeholders and legal scholars.

[201]Cf. "Morineau et al. (2017)"; UK Advisory Committee on Releases to the Environment (2018), p. 2.

[202]Faure and Napier (2018), p. 3.

[203]Faure and Napier (2018), pp. 3–4; European Commission Joint Research Centre (2019a); Rothamsted Research (2019); European Commission Joint Research Centre (2019b); VIB (2019).

[204]There are no pending applications for placing GMOs obtained by genome editing on the market, yet, cf. European Commission (n.d.-g).

[205]See e.g. Sect. 5.4 ("Regulatory Prerequisites for Activities Relating to Genome Edited Plants"), text to n. 163ff, 179ff; Sect. 5.7 ("Low Level Presence"), text to n. 275ff; Sect. 5.8 ("Labelling"), text to n. 329ff; Sect. 5.10 ("Liability"), text to n. 374.

[206]Cf. text to n. 50–59.

[207]Lappin (2018b), p. 3; Andriukaitis (2018).

5.6.1 Incentives and Disincentives for Reform

The EU **legislative organs are divided** regarding their position on genome edited organisms.[208] Therefore, strong external incentives are necessary to reach the required majorities for legislative amendments.

Stakeholder positions on the necessity and desirability of reforms are split between the sectors.[209] Environmental and consumer NGOs as well as the organic sector do not wish for reforms. Before the ruling of the European Court of Justice in Case C-528/16, they had demanded the regulation of GEOs under the GMO framework.[210] Accordingly, since the European Court of Justice classified all types of GEOs as GMOs, they are now satisfied with the current state.[211]

By contrast, scientists and large parts of the agricultural sector, i.e. the plant breeding and seed industry, the conventional agricultural sector,[212] the feed industry etc., support the idea of not regulating organisms developed through SDN-1, SDN-2 and ODM within the GMO framework but within the legal framework of traditional plant breeding,[213] possibly with some GEO-specific adjustments.[214] Therefore, they are predominantly dissatisfied with the European Court of Justice's judgment.[215] They expect that the GMO classification will considerably hinder the research, production and trade of GEOs.[216] Thus, they fear to miss out on advantages that GEOs promise or only obtain them at a high price[217] while, at the same time, they suffer competitive disadvantages compared to researchers and producers in non-EU countries. In consequence, these sectors favour deregulation now. The European Commission's Group of Chief Scientific Advisors recommended "revising the

---

[208]Cf. text to n. 395–418.

[209]Cf. also the stakeholders' positions on GEOs in general, n. 435–442.

[210]Laaninen (2016), p. 7; cf. the joint letter by environmental NGOs: EcoNexus et al. (2015); cf. the joint position paper by environmental NGOs and the organic sector: IFOAM EU Group (2017); cf. further GM Freeze (2016), Greenpeace (2015), GMWatch (2014), Steinbrecher (2015), and IFOAM EU Group (2015).

[211]Cf. Michalopoulos (2018), and IFOAM EU Group (2018).

[212]Note n. 441.

[213]Wolt et al. (2016), pp. 513–514; Laaninen (2016), p. 6; European Commission New Techniques Working Group (2011), para. 5.2.1.5 B; French Haut Conseil des Biotechnologies (2016a), pp. 95–96; European Seed Association (2017b), New Breeding Techniques Platform (2015b); cf. also the statements from the research and agricultural sectors listed in Stakeholder and Issue Mapping on New Breeding Techniques (2017). A frequent line of argumentation is: Organisms obtained by SDN-1, SDN-2 and ODM are similar to organisms resulting from traditional mutagenesis by radiation or chemical agents, which are not GMOs. Genome editing techniques even generate fewer unintended effects than traditional mutagenesis techniques, cf. European Commission New Techniques Working Group (2011), para. 5.2.1.5 B.

[214]Cf. Huang et al. (2016).

[215]Cf. e.g. the statements at Science Media Centre (2018), Clarke (2018), German Association of Biotechnology Industries (2018), and bioökonomie.de (2018).

[216]Cf. Lappin (2018a), pp. 4–5; European Seed Association (2017b), pp. 2–3.

[217]Cf. Michalopoulos (2018); cf. also text to n. 191.

existing GMO Directive to reflect current knowledge and scientific evidence, in particular on gene editing and established techniques of genetic modification."[218]

Most **consumers** would likely not support reforms aiming at *deregulation*. They would probably perceive deregulation of genome editing technologies in the way that food from "risky"[219] technologies ends up on their plates without adequate official risk controls. Some aspects are of particular importance to consumers, e.g. labelling. The abolition of the labelling obligation[220] for GEOs and their products would likely be perceived as depriving consumers of their choice, leaving them at the mercy of the highly unpopular agricultural industry.[221]

Apart from stakeholder calls for reform, reforms might be induced by **trade conflicts** with non-EU countries or even a **dispute before the WTO**.[222] The EU submits all GEOs to its GMO regime whereas most other countries do not.[223] Since the EU is the largest importer of agri-food products, third country exporters are compelled to comply with the EU's requirements.[224] This has significant **monetary consequences**. Examples are (1) costs and delayed market access because of the authorisation requirements, (2) poor marketability of products, namely foods and seeds, from genome edited plants in the EU[225] and (3) costs of segregation of non-GMOs from GMOs, now also including GEOs, on the field and during the subsequent production chain, which is necessary to fulfil the labelling obligation and to ensure that no unauthorised GEOs enter the EU market.[226] If importers do not comply with the EU's requirements, import bans or other **repressive measures** (fines etc.) are imposed on them (provided, an enforcement of the requirements is possible with regard to GEOs[227]). Such consequences are especially relevant for low-level

---

[218]European Commission Scientific Advice Mechanism Group of Chief Scientific Advisors (2018), p. 6.

[219]Regarding the risks perceived by consumers cf. n. 463.

[220]See Sect. 5.8 ("Labelling").

[221]With respect to the importance of consumer choice and personal control over exposure for the acceptance of new food technologies Frewer et al. (2011), p. 453; regarding the importance of labelling of GEOs to German focus groups participants Hopp et al. (2017), pp. 31–32; regarding British participants in a public dialogue van Mil et al. (2017), p. 87.

[222]Cf. also the past WTO dispute challenging the EU's GMO policies, WTO Panel Report, *European Communities - Measures Affecting the Approval and Marketing of Biotech Products (EC Biotech)* (2006) WT/DS291/R; WT/DS292/R; WT/DS293/R.

[223]Cf. e.g. chapter 7 (Country Report on the USA); chapter 1 (Country Report on Argentina). Cf. also third countries' disappointment regarding the European Court of Justice's decision in Case C-528/16, e.g. U.S. Department of Agriculture (2018), stating that they "encourage the European Union to seek input from the scientific and agricultural communities, as well as its trading partners, in determining the appropriate implementation of the ruling."

[224]Cf. Cantley (2007), p. 38.

[225]Cf. text to n. 328–329.

[226]Cf. Cantley (2007), p. 38; Paarlberg et al. (2004), p. 19; Beckmann et al. (2014), p. 385; Lezaun (2006), p. 502.

[227]See text to n. 275.

presence of GEOs not yet authorised in the EU. In the past, traces of unauthorised GMOs have already caused significant trade disruptions (import bans etc.).[228] Regarding GEOs, the problem is reinforced by the high number of genome edited varieties that will be developed in countries not submitting them to their GMO regime.[229] Furthermore, this high number would result in an equally high number of GMO import applications, which could push EFSA and the other authorities involved in the authorisation procedure to their capacity limits.[230] **Delays in the approvals** would be the consequence. Lastly, the EU's strict requirements regarding GEOs and consequences of failure to comply will have **knock-on effects**. Examples are (1) a general reluctance of non-EU countries, especially developing countries, to produce GEOs and (2) indirect protection of the EU agricultural market, as in the EU, GEO cultivation will likely be little,[231] and so the described costs do not arise there.[232]

Accordingly, the GMO classification of all types of GEOs could be the catalyst for countries exporting agri-food products to the EU to challenge (parts of) the EU's GMO regulatory framework before the WTO. As is known, the conformity of many aspects of the GMO framework with WTO rules is debated.[233] With regard to GEOs, the conformity is often even more questionable.[234] Several WTO members published a statement in political support of genome editing in agricultural applications shortly after the ruling of

---

[228]Regarding "asynchronous GMO approvals" and "isolated foreign approvals" Stein and Rodríguez-Cerezo (2009), pp. 19–21; Beckmann et al. (2014), pp. 385–386.

[229]Cronin and Stone (2018) ("The world already has experienced the economic impact of asynchronous approvals for GMOs, when China and Europe rejected U.S. grain imports because a GMO crop was not approved in that country. However, the impact could be several times greater this time because there are so many gene-edited crops in the commercial development pipeline.").

[230]Bruins (2018), p. 10.

[231]Regarding the situation of GM crop cultivation in the EU cf. text to n. 3–7.

[232]Cf. the experiences regarding GMOs, Paarlberg et al. (2004), pp. 5–6, 19.

[233]Lee (2008), pp. 234–240; regarding labelling and traceability e.g. Lezaun (2006), p. 501; Mansour and Key (2004), pp. 63–64; regarding the opt-out possibilities Dederer (2016b), pp. 163–164.

[234]In detail Dederer (2019); further Kahrmann et al. (2017), p. 182. Three examples:

(1) Many measures to restrict genome editing would qualify as measures to protect human, animal or plant life or health and thus as measures under the Agreement on the Application of Sanitary and Phytosanitary Measures (SPS Agreement), cf. SPS Agreement, art. 1.2 in connection with annex A. As such, they have to be based on a scientific risk assessment and need "sufficient scientific evidence" to be maintained, SPS Agreement, arts. 2.2, 5.1, 5.2; Alemanno (2013), pp. 292–293 (unless a risk assessment is not possible to perform, cf. SPS Agreement, art. 5.7). Do the EU risk regulation rules on GEOs withstand these procedural and material requirements? Is there "sufficient scientific evidence" that GEOs are risky? Are the extensive measures proportionate to the risk identified, cf. WTO Panel Report, *Japan – Measures Affecting the Importation of Apples* (2003) WT/DS245/R, para. 8.198? Is the huge discrepancy between the regulation of organisms developed through traditional breeding and the regulation of organisms developed through SDN-1, SDN-2 and ODM "inconsistent", SPS Agreement, art. 5.5?

(2) Even more than regarding GMOs, it could be claimed that the different treatment of GEOs (which will mostly be produced outside the EU) and traditionally bred organisms is a discrimination of "like products", cf. General Agreement on Tariffs and Trade 1994 (GATT 1994), Art.

the European Court of Justice, amongst them some of the EU's most important trading partners.[235] Furthermore, the USA, supported by Argentina and Paraguay, hinted that they consider the barrier to trade posed by the current EU's GMO regulation unjustified since the regulatory distinctions between random and targeted mutagenesis might lack a scientific basis.[236] A joint statement of several American countries supporting "risk- and science-based, predictable, consistent, and transparent" regulatory approaches[237] can be interpreted as a criticism of the EU's GMO regulation as well. A joint statement of several American countries supporting "risk- and science-based, predictable, consistent, and transparent" regulatory approaches240 can be interpreted as a criticism of the EU's GMO regulation as well. This indicates a growing conflict potential.

An important disincentive for any renegotiation of the EU's GMO regime is that amendments are **lengthy** and **burdensome.**[238] Deregulation attempts might possibly end up **counterproductive**. This means there is a risk that, instead of achieving deregulation, regulations get even more rigorous or complicated.[239]

Lastly, reforms do not necessarily have **practical implications**. This is especially true for deregulation: Even if the EU's regulatory framework is adapted to the needs of GEOs, e.g. by simplifying the application procedure, those simplifications are devoid of practical implications if member states block authorisations or hamper GEOs in other ways for political reasons.[240] The success of any change to the regulatory framework therefore does not only rely on legal details, but heavily depends on the public's and national authorities' attitude towards GEOs. In fact, the root of problems regarding GMO and GEO adoption in the EU is not the

III:4 as well as Agreement on Technical Barriers to Trade (TBT Agreement), art. 2.1 (as far as applicable, cf. also art. 1.5). Could it be justified?

(3) Is there a "legitimate objective" for traceability, labelling and coexistence regulations within the meaning of the TBT Agreement, art. 2.2? Is it a legitimate objective to fulfil the consumers' wish to know whether the product is genome edited/genetically modified or do consumers have to have a "good reason" for wanting to know it? Cf. Paarlberg et al. (2004), pp. 7, 14; Cheyne (2012), pp. 326–327, who further remarks that there is "great uncertainty about how a panel or the Appellate Body would resolve a dispute over existence and content of GMO labelling".

[235] World Trade Organization Committee on Sanitary and Phytosanitary Measures (2018).

[236] World Trade Organization Committee on Sanitary and Phytosanitary Measures (2019), 3.1.5.

[237] U.S. Department of Agriculture (2019).

[238] Phillips and Flach (2017), p. 3.

[239] Sikkema (2018) ("No proposal in the GMO file is promising – one always opens Pandora's box"); regarding the establishment of a trait-based system UK House of Commons Science and Technology Committee (2015), p. 41 ("very real possibility of ending up with the unsatisfactory GM regime simply being applied more generally to any novel crop"). The suspicion that attempts to deregulate genome editing could be counterproductive is underlined by calls for their tighter regulation, e.g. by the European parliamentary group The Greens/European Free Alliance, cf. the demand for an "enhanced" risk assessment for GEOs "to take into account the new set of risks linked to the gene editing techniques", The Greens/European Free Alliance in the European Parliament (2018a).

[240] Cf. UK House of Commons Science and Technology Committee (2015), p. 41, Q424; cf. the politicisation of the GMO authorisation process, n. 157–160.

regulatory framework but the political opposition towards GMOs of many member states and large parts of the European population.[241]

5.6.2 *Options for Deregulation*

The options for deregulation are presented according to the scale of the introduced changes: From a complete overhaul of the GMO regulatory framework and the establishment of a trait-based system to the revival of existing simplifications within the current framework.

5.6.2.1 Complete Overhaul of the EU's GMO Regulatory Framework

The emergence of new breeding techniques have given an impetus to criticise the EU's GMO regulatory framework as a whole. Firstly, the general, usual criticism was reignited ("politicised", unscientific,[242] "dysfunctional",[243] difficult to enforce[244] etc.). Secondly, the aptness of process-based frameworks to deal with technological progress was questioned in general. This was fostered by years of quarrels over how to fit the new processes into the existing GMO/non-GMO dichotomy.[245] Lastly, the huge gap between GMO-regulation and non-GMO regulation in the EU seems increasingly unjustified.[246] The reason is that the new breeding techniques fill the continuum between traditional breeding techniques and the established techniques of genetic modification.[247]

In consequence, the new breeding techniques revived the discussion about reforming the GMO regulatory framework as a whole.

---

[241] Cf. UK House of Commons Science and Technology Committee (2015), Q424.

[242] Cf. e.g. Baulcombe et al. (2014), pp. 36–38, 41.

[243] UK House of Commons Science and Technology Committee (2015), pp. 32–35; European Plant Science Organisation (2017), p. 3.

[244] Cf. European Parliament (2014c), para. 32: "The European Parliament [...] [n]otes that, with today's technique-based plant-breeding legislation, it has proven difficult, after the event, to define what technique was used at the time of plant-breeding, which serves to confirm the difficulties posed by technique-based laws".

[245] Marchant and Stevens (2015), pp. 236–237; UK House of Commons Science and Technology Committee (2015), p. 36. The initially foreseen mechanisms to deal with technological progress (update of the positive and negative lists of techniques in the GMO definition; simplified procedures and dossiers etc.) are (with very few exceptions) not used for political reasons, The Netherlands Government (2017), p. 2; text to n. 266.

[246] Tagliabue and Ammann (2018), p. 40 ("inconsistent"); Devos et al. (2012), p. 10771.

[247] Sprink et al. (2016), p. 1493.

Scientists, e.g. science academies or plant science organisations, often express the wish to move towards a trait-based system.[248] Frequently, the Canadian system is mentioned as a role model.[249] This wish was reinforced by the GMO classification of all types of GEOs.[250]

Yet, past experience shows that the fundamental characteristics of the EU's GMO framework are resilient.[251] They have never changed since the establishment of the framework in the 1990s despite major shocks (de facto moratorium, trade disputes, etc.).[252] At least, for the reasons mentioned above (consumers' lack of confidence in GMOs, some member states' opposition towards GMOs), a complete overhaul of the European GMO system is not a realistic option in the near future.[253] Nevertheless, a move towards a trait-based system, "little by little",[254] might be feasible or even necessary in the long term.[255]

5.6.2.2 Exclusion of Some Types of GEOs from the GMO Regulatory Framework

The most straightforward way to exclude some types of GEOs from the GMO regulatory framework is to amend annex I B Directive 2001/18/EC (the "grey list"). In September 2017, the Dutch government made such a proposal. The proposal exempted plants resulting from new breeding techniques from the GMO framework if certain conditions are met.[256] However, this amendment proposal was suggested before the European Court of Justice's decision on the GMO status of genome edited organisms. Therefore, it was premature and more of a position

---

[248]Cf. the overview of bodies wishing to move towards a trait-based system in UK House of Commons Science and Technology Committee (2015), p. 36 and Laaninen (2016), p. 5; examples of the statements in favour of a trait-based system are: European Academies Science Advisory Council (2015), p. 4; European Plant Science Organisation (2017); German National Academy of Sciences Leopoldina et al. (2015), p. 3; UK Advisory Committee on Releases to the Environment (2013a), p. 1; UK Biotechnology and Biological Sciences Research Council (2014), pp. 1, 4.

[249]Cf. UK House of Commons Science and Technology Committee (2015), p. 36; Tagliabue and Ammann (2018), p. 44.

[250]Cf. the statements in reaction to Case C-528/16, e.g. Prof Leyser or Prof Crute at Science Media Centre (2018).

[251]Pollack and Shaffer (2009), pp. 277–278; regarding the reasons for the continuance of the essential characteristics of GMO regulatory systems Pollack and Shaffer (2009), pp. 77–80.

[252]See n. 252.

[253]Cf. Dederer (2016b), p. 166; Hartung and Schiemann (2014), p. 750; UK House of Commons Science and Technology Committee (2015), Q543 [Lord de Mauley]; UK House of Commons Science and Technology Committee (2015), GMC0016 [Professor Perry] para. 8; The Netherlands Commission on Genetic Modification (2009b), p. 29.

[254]Eriksson and Ammann (2016), p. 3.

[255]Cf. UK House of Commons Science and Technology Committee (2015), p. 41; Marchant and Stevens (2015), p. 238; Hartung and Schiemann (2014), pp. 743–744.

[256]The Netherlands Government (2017); cf. further Phillips and Flach (2017).

statement than an earnest attempt to amend the EU legislation.[257] After the European Court of Justice's decision, stakeholders have started to raise demands to exclude some types of GEOs from the GMO regulatory framework.[258]

In fact, an adaptation of the lists of techniques in the GMO definition ("black list", "white list", "grey list") to technological progress has never taken place before.[259] One reason is that no specific procedure is foreseen for the update. Therefore, the normal legislative procedure with all its difficulties (lengthy, burdensome, necessary majorities cannot be reached) has to be applied.[260]

Regarding the exclusion of certain types of GEOs it is further problematic that according to the European Court of Justice, "the risks linked to the use of those new techniques/methods of mutagenesis might prove to be similar to those which result from the production and release of a GMO through transgenesis".[261] In order to comply with the precautionary principle,[262] the EU legislator should therefore carry out a risk evaluation, demonstrating the relative safety of GEOs compared to traditionally bred organisms if it wishes to exclude certain types of GEOs from the scope of the GMO regulatory framework.[263]

---

[257] Cf. Phillips and Flach (2017), p. 3.

[258] Cf. e.g. Bioökonomierat (2018) (in German) aiming at an exemption for plants with genetic alterations up to 20 base pairs (not clear which legal amendments are envisaged).

[259] Cf. The Netherlands Government (2017), p. 2.

[260] The Netherlands Government (2017), pp. 2–3.

[261] ECJ, Case C-528/16 *Confédération paysanne and Others* [2018] ECLI:EU:C:2018:583, paras. 48, 53.

[262] Cf. TFEU, art. 191(2); cf. also European Commission (2000), para. 5.1.2 ("An assessment of risk should be considered where feasible when deciding whether or not to invoke the precautionary principle."). Cf. also van Heezik and Tuinzing-Westerhuis (2018) recommending such a risk evaluation due to the current political climate.

[263] Possibly, the studies that have already been carried out on hazards associated with the application of the various new breeding techniques in plants, animals and micro-organisms would be sufficient; cf. the EFSA study on cisgenesis, which concludes "that similar hazards can be associated with cisgenic and conventionally bred plants, while novel hazards can be associated with intragenic and transgenic plants.", European Food Safety Authority (2012a), p. 20; cf. the studies analysing whether there are general differences in safety between organisms developed by the various new breeding techniques and traditionally bred organisms, e.g. European Commission Scientific Advice Mechanism High Level Group of Scientific Advisors (2017), pp. 17–19, 88–90 or the "Safety Issues" sections in Vogel (2012). The EU legislator enjoys a "wide measure of discretion" where complex technical assessments have to be made, cf. e.g. ECJ, Case T-177/13 *TestBioTech eV and Others v Commission* [2016] EU:T:2016:736, para. 77. It is therefore unlikely that the European Court of Justice would challenge the risk evaluation by the EU legislator. Cf. further Scott and Vos (2002), p. 254.

5.6.2.3 Reduction of Obligations Under the GMO Regulatory Framework for Some Types of GEOs

First, **existing possibilities for simplifications**[264] within the GMO regulatory framework might be used. Yet, the realisation of these simplifications might fail due to politically motivated opposition of the member states, which are involved in all simplification decisions.[265] At least, this conclusion may be drawn from past experiences.

Second, the EU's GMO regulatory framework could be reformed by **relaxing** it with respect to **certain types of GEOs**. Examples are a simplified authorisation procedure for certain types of GEOs,[266] a simplified risk assessment[267] or an exemption from the coexistence or labelling requirements.

Third, some **member states** favourable towards GEOs might **reduce the obligations** that are in their **national competence**. One example is simplifications regarding the contained use of certain types of GEOs. Another example is that some member states might not request opt-outs from GMO cultivation regarding certain genome edited crops.[268]

---

[264]**Contained use:** Directive 2009/41/EC provides the possibility to exempt GMMs with a high degree of familiarity and safety, Directive 2009/41/EC, art. 3(1)(b), annex II part B; Herdegen and Dederer (2009), para. 13, 22.

Field trials: Directive 2001/18/EC allows for the introduction of simplified procedures ("differentiated procedures") for GMOs with a high degree of familiarity and safety, Directive 2001/18/EC, art. 7; Di Fabio and Kreiner (2003), pp. 714ff; Herdegen and Dederer (2015), para. 76.

Placing on the market: There are two possibilities: In the individual case, the applicant can request to provide less information on the marketing of the GMO, Directive 2001/18/EC, art. 13(2)(2) in conjunction with annex IV sec. B (cf. Table 5.10 sec. "Additional Information"); Herdegen and Dederer (2015), para. 114. On a general basis, Directive 2001/18/EC allows for the introduction of simplified dossiers for certain types of GMOs with a high degree of familiarity and safety, Directive 2001/18/EC, art. 16; Herdegen and Dederer (2015), para. 115; Di Fabio and Kreiner (2003), para. 92.

[265]All decisions are made by way of comitology procedures and thus involve a committee composed of member state representatives; cf. Directive 2009/41/EC, annex II part B, art. 20(2) regarding exemptions from Directive 2009/41/EC; cf. Directive 2001/18/EC, arts. 7(3),(4), 30(2) regarding differentiated procedures; cf. Directive 2001/18/EC, arts. 16(2), 30(3) regarding simplified dossiers for placing certain types of GMOs on the market.

[266]E.g. notification instead of authorisation or fast-track authorisation procedures. Regarding the case of cisgenic plants, the European Parliament called on the European Commission to "differentiate between cisgenic and transgenic plants and to create a different approvals process for cisgenic plants" in 2014, European Parliament (2014a), para. 31; Laaninen (2016), p. 7.

[267]Cf. also text to n. 179ff.

[268]Cf. text to n. 116.

5.6.3 Options for More Regulation

The European Court of Justice decided as part of Case C-528/16 that member states are free to subject organisms that are exempted from the GMO regulatory framework, i.e. organisms developed through **traditional mutagenesis techniques**,[269] to the obligations applicable to GMOs or to other obligations on a national basis.[270] EU law, in particular the principle of free movement of goods, has to be respected.[271]

5.6.4 Summary

There are many options for reforming the EU's GMO regulatory framework with regard to the regulation of genome edited plants and products derived from them. However, for political reasons, it is unlikely or at least very doubtful whether any amendments will be introduced in the near future.

5.7 Low Level Presence

In general, there is **no tolerance level for unauthorised GMOs** and their products ("zero tolerance for unauthorised GMOs").[272] The only exception is unauthorised GM feed, which is tolerated up to a threshold of 0.1% in case the GM feed meets certain conditions.[273]

The **enforcement** of this zero tolerance policy for GMOs will be difficult or even impossible regarding plants and derived products obtained by SDN-1, SDN-2 and

[269]E.g. mutagenesis by chemicals or irradiation.

[270]ECJ, Case C-528/16 *Confédération paysanne and Others* [2018] ECLI:EU:C:2018:583, para. 82.

[271]ECJ, Case C-528/16 *Confédération paysanne and Others* [2018] ECLI:EU:C:2018:583, para. 82.

[272]European Commission EU Science Hub (2016). From time to time, it is discussed to introduce a tolerance level for unauthorised GMOs in seed, cf. IG Saatgut (2016), p. 1.

[273]Commission Regulation (EU) No 619/2011 of 24 June 2011 laying down the methods of sampling and analysis for the official control of feed as regards presence of genetically modified material for which an authorisation procedure is pending or the authorisation of which has expired [2011] OJ L166/9, art. 6, annex II(B)(2). Thereby, the EU wants to avoid problems of feed imports, Bergmann and Dederer (2012), para. 49. There has to be either an expired EU authorisation for the GM feed or both an authorisation in a third country and a pending application in the EU, Regulation (EU) No 619/2011, art. 2. This condition ensures that a risk assessment has been carried out. Cf. further Dederer (2016a), pp. 109–110.

ODM.[274] Genome edited plants and derived products will likely enter the EU market unnoticed.[275]

To give a scenario—which ways are there to identify unauthorised genome edited soybeans that are accidentally present in the cargo of a bulk shipment of soybean from a non-EU country?[276] Or how to identify traces of unauthorised genome edited grain in a food or feed composed of various ingredients?

To detect unauthorised GMOs or GM products, laboratory testing (typically PCR-based methods[277]) and controls using documentation are carried out.[278] While enforcement through documentation will not be efficient enough, analytical controls are difficult or even impossible, if relevant genetic changes could also occur naturally or be achieved through traditional breeding methods.

Two cases of analytical controls need to be distinguished.[279] First, the **targeted search for unauthorised "known GEOs"**: For some unauthorised GEOs, DNA sequence information[280] (or other information on the organism) might be available. Potential sources are pending applications in the EU, publications, patent specifications voluntary databases[281] and GMO authorisations or GMO status consultation procedures in other countries.[282] These characteristic DNA sequences of the GEO[283] can be detected. However, it is possible that the same DNA sequences

[274]Cf. in detail the report of the European Network of GMO Laboratories, European Network of GMO Laboratories (2019), pp. 14–16.

[275]Cf. European Network of GMO Laboratories (2019), p. 17. Cf. also the US company Cibus claiming that its ODM herbicide-tolerant canola might have already entered the European market unnoticed, Cibus Europe (2015); Corporate Europe Observatory (2016c).

[276]Cf. Jones (2015), pp. 2–3.

[277]European Network of GMO Laboratories (2017), p. 5; Lusser et al. (2011), p. 68.

[278]European Network of GMO Laboratories (2019), p. 14. Cf. also the EU Official Controls Regulation, Regulation (EU) 2017/625 of the European Parliament and of the Council of 15 March 2017 on official controls and other official activities performed to ensure the application of food and feed law, rules on animal health and welfare, plant health and plant protection products [2017] OJ L95/1.

[279]Cf. Bartsch et al. (2018), pp. 43–55; cf. also European Commission Scientific Advice Mechanism High Level Group of Scientific Advisors (2017), pp. 19–20; Lusser et al. (2011), pp. 63–71; Ribarits et al. (2014), pp. 185–186.

[280]I.e. sequence of the mutation site and possibly of other sites in the genome.

[281]E.g. the Plant Genome Editing Database on which plant breeders can voluntarily provide information about plants that have been generated using the CRISPR/Cas9 technology, Plant Genome Editing Database (2018).

[282]Chapter 1 (Country Report on Argentina), n. 42. Cf. also the central register mentioned in Regulation (EC) No 1830/2003, art. 9(3), which also contains, where available, relevant information concerning GMOs that are not authorised in the EU. Cf. calls for a global public register of released genome edited organisms as part of the existing registry of the Biosafety Clearing House on Living Modified Organisms (LMOs), European GMO-Free Regions Network (2018). Regarding the difficulties of establishing a register of all released GEOs worldwide European Network of GMO Laboratories (2019), p. 16.

[283]Or characteristic proteins, metabolites etc. However, GMO screening for enforcement purposes is currently mainly performed with DNA-based analysis, Ribarits et al. (2014), p. 186.

exist or reoccurred naturally or through traditional breeding.[284] Yet, absolute certainty is not required to proof the presence of an unauthorised GMO.[285] Taking the common principles of standard of proof (e.g. prima facie evidence[286]) into account, low probabilities can be disregarded.[287] The probability of a DNA sequence existing or reoccurring naturally or through traditional breeding depends on the characteristic sequence in question (particularly its length) and is, according to some experts, sufficiently low for characteristic DNA sequences with a length of at least 20 nucleotides. Therefore, detecting the characteristic DNA sequences might suffice to identify an unauthorised GEO and prove an infringement of the authorisation regime, provided that the sequence in question is sufficiently long.[288]

On the contrary, an **untargeted search for unauthorised "unknown GEOs"**, i.e GEOs whose existence is not known to the authorities in advance, is currently not possible.[289] It is unlikely that practically implementable methods to detect GEOs without prior knowledge on them will be developed in the near future. Even if small sequence changes are detected in plants compared to a reference genome,[290] it is equally possible that these changes were induced by traditional breeding or natural mutations.[291] Only in rare cases, it is almost certain that genome editing has been applied. An example is that all copies of a gene have been knocked out, which is currently not technically feasible using traditional methods.[292] In short, uncovering unauthorised "unknown GEOs" and proving an infringement of the authorisation regime will only seldom be possible. Thus, most "unknown GEOs" will probably enter the EU market undetected.

Moreover, unauthorised GEOs might be uncovered through **controls using documentation**, which has been prepared in the course of food traceability,[293] variety registration in the EU or a third country,[294] intellectual property protection,

---

[284]Cf. text to n. 172 (identification).

[285]The uniqueness of traditional GM transformation events is also only a probabilistic judgement (though, of course, one with a very high probability), Lezaun (2006), p. 511.

[286]Prima facie evidence in this case means that the authorities prove that it is improbable that the DNA sequence is not derived from genome editing. The developer can then prove the contrary, i.e. that it occurred naturally or through traditional breeding, e.g. by showing breeding protocols.

[287]For criminal sanctions, however, a higher standard of proof might apply.

[288]Cf. text to n. 175. The characteristic sequence can e.g. be composed of the sequence of the genetic modification and the sequence in its vicinity.

[289]European Network of GMO Laboratories (2019), pp. 1, 15, 17.

[290]As genomes differ between varieties of plant species (see n. 174), referring to a single reference genome is not possible. Pan-genome databases would have to be established, i.e. databases including variations between varieties.

[291]European Commission Scientific Advice Mechanism High Level Group of Scientific Advisors (2017), p. 19.

[292]Duensing et al. (2018), p. 4.

[293]Cf. Regulation (EC) No 178/2002, art. 3 no. 15, arts. 18, 11; IFS Food (2017), p. 75; Codex Alimentarius Commission (2006); International Organisation for Standardisation (2015), 8.5.2.

[294]Cf. Black et al. (2006), p. 368.

contractual agreements,[295] breeding programmes, etc.[296] However, this does not mean that the importer is required to show by documentation that the product is not a GMO. It is upon the authority to proof the existence of an unauthorised GMO.[297]

Enforcement deficiencies are especially pertinent with respect to **imported commodities**. Many countries outside the EU do *not* regulate some types of GEOs as GMOs. In these countries, GEOs are neither officially registered[298] nor (provided such GMO obligations exist) officially labelled, traceable and segregated. This increases the probability of unauthorised GEOs being unintentionally imported to the EU, e.g. due to accidental commingling of GEOs and non-GMOs or unintentional use of genome edited breeding material.[299] The efficiency of private party registration, segregation etc.[300] largely depends on the efficiency of the EU's enforcement of the zero tolerance regime. Private party measures will be lenient if the described enforcement deficiencies arise. The high number of GEOs that will be developed due to low regulatory burdens renders enforcement even more difficult.[301] In sum, the problem of unintentional import of unauthorised GMOs into the EU might be significantly greater regarding GEOs than regarding established GMOs.

Consequently, both loss of the EU's credibility[302] and trade disruptions[303] are feared.[304]

5.8 Labelling

As all types of GEOs are classified as GMOs, the GM labelling rules apply to them, which are described in the following (cf. also Table 5.15).[305]

---

[295]E.g. confirmations by producers that the product is GM-free or that segregation measures have been applied.

[296]Cf., however, Jones (2015), p. 3, pointing to the "inherent frailties and temptations for misuse" of documentation.

[297]Heinemann (2014), pp. 167–168.

[298]The Netherlands Commission on Genetic Modification and Health Council of the Netherlands (2016), p. 46.

[299]Cf. also the example in German Federal Office of Consumer Protection and Food Safety (2017), p. 6.

[300]Cf. Beckmann et al. (2014), p. 386.

[301]Bruins (2018), p. 10; Cronin and Stone (2018).

[302]Cf. Jones (2015), p. 3 presuming that the EU might "turn a blind eye to the potential presence of unauthorised products".

[303]Including import bans, Jones (2015), p. 3.

[304]Cf. Prof Huw Jones at Science Media Centre (2018); The Netherlands Commission on Genetic Modification and Health Council of the Netherlands (2016), pp. 46, 47; The Netherlands Commission on Genetic Modification (2009b), pp. 3–4. Cf. further text to n. 223ff.

[305]Regarding the EU's GM labelling rules cf. e.g. Plan and van den Eede (2010), p. 12; Devos et al. (2012), pp. 10775, 10777–10778; The group of FAS Biotechnology Specialists in the European Union (2018), pp. 29–31.

The EU maintains one of the most extensive GM labelling regimes in the world,[306] as a response to consumer demands for information.[307]

In the EU, labelling of GM products is **mandatory**. Regarding **food and feed**, any ingredient that is, contains or is produced from[308] a GMO needs to be indicated as such.[309] Eggs, milk, meat and dairy products from animals fed on GM feed do not have to be labelled as they are not produced from, but only with the help of GMOs.[310] Labelling requirements are fulfilled through writing "genetically modified", "contains genetically modified" or "produced from genetically modified" in brackets after the ingredient concerned in the ingredients list or in a footnote to it.[311] Additionally, special characteristics or properties have to be indicated.[312] The labelling obligation applies irrespective of the detectability of recombinant DNA or resulting proteins.[313] **Other products** need to be labelled if they are or contain GMOs, but not if they have been produced from GMOs (like clothes produced from GM cotton).[314] Again, the wording of the label is regulated in detail.[315]

Labelling obligations apply irrespective of whether the product originates in the EU or is imported from outside the EU.[316]

There is a **tolerance threshold** for traces of authorised GM material of **0.9%** (per ingredient) regarding food and feed as well as products intended for direct processing.[317] The condition is that the GM traces are adventitious or technically unavoidable. Regarding other products, e.g. seed, no tolerance threshold has been introduced so far.[318]

---

[306]Masip et al. (2013), p. 317.

[307]Cf. Regulation (EC) No 1829/2003, recital 21; Devos et al. (2006), p. 144.

[308]Regarding the term "produced from GMOs" see n. 118.

[309]Regulation (EC) No 1829/2003, arts. 12ff, 24ff in conjunction with art. 15(1). The labelling requirement thus mirrors the authorisation requirement—the same food and feed that have to be authorised need to be labelled, cf. text to n. 118–119.

[310]See n. 118.

[311]Regulation (EC) No 1829/2003, art. 13/art. 25 (food/feed). Regarding the labelling of compound ingredients cf. Regulation (EC) No 1829/2003, art. 13(1)(b); cf. further The group of FAS Biotechnology Specialists in the European Union (2018), p. 30.

[312]Regulation (EC) No 1829/2003, art. 13(2),(3)/art. 25(2)(c),(3) (food/feed).

[313]In oils, for example, no recombinant DNA is detectable as the DNA is destroyed during the refining process, cf. Aparicio (2017), p. 81.

[314]Regulation (EC) No 1830/2003, art. 4(6); Lee (2008), p. 144.

[315]Regulation (EC) No 1830/2003, art. 4(6).

[316]The group of FAS Biotechnology Specialists in the European Union (2018), p. 29.

[317]Regulation (EC) No 1829/2003, art. 12(2)/art. 24(2) (food/feed); Regulation (EC) No 1830/2003, arts. 4(7), (8), 5(4); Directive 2001/18/EC, art. 21(3). According to Masip et al. (2013), p. 317 "The adventitious presence thresholds in the EU are the strictest in the world".

[318]Cf. Regulation (EC) No 1830/2003, art. 4(7); Directive 2001/18/EC, art. 21(2); cf. further Devos et al. (2006), p. 142; Grossman (2010), pp. 141–146 and Norer and Preisig (2016), p. 54, also with respect to the ongoing discussion about the introduction of labelling thresholds. Cf. also European Seed Association (2012), pp. 1, 2, explaining that sampling and testing of seed differs widely between the member states ("patchwork of rules and practices").

Table 5.15 GM labelling in the EU

Labelling scheme	Pertinent legislation		
Mandatory positive labelling "GMO"	Food/feed:	Regulation (EC) No 1829/2003; Regulation (EC) No 1830/2003	Regulated at EU level
	Other products:	Regulation (EC) No 1830/2003	
Voluntary negative labelling	"GM-free" labels:	Respective laws and rules of member states and private operators	Regulated at member state level
	Organic schemes prohibiting use of GMOs:	Regulation (EC) No 834/2007 (EU organic scheme); member state organic schemes; regional organic schemes; private sector organic schemes	Regulated at EU level and member state level

In addition to the mandatory GM labelling, there are **voluntary "GM-free" labels** regulated at member state level.[319] Some member states have introduced legislation or guidelines to facilitate private GM-free labelling, others have prohibited it and yet others have no rules at all, leaving GM-free labelling entirely to private operators, e.g. private associations or supermarket chains.[320] Consequently, a variety of GM-free labels exist that differ significantly from each other in their product scope,[321] threshold levels for unintentional GM presence,[322] the admissibility of feeding animals on GM feed, certification, control and other aspects.[323] Beyond specific "GM-free" labels, "GM-free" is also a requirement of other quality labels, e.g. organic production labels.[324] The EU organic farming scheme[325] prohibits the use of GMOs, also in animal feed.[326] The threshold for

[319]The EU neither forbids nor regulates GMO-free labelling, European Commission Directorate-General for Health and Food Safety (2015b), p. 2.

[320]European Commission Directorate-General for Health and Food Safety (2015b), pp. 10ff.

[321]Most GM-free labels cover only food and feed, European Commission Directorate-General for Health and Food Safety (2015b), pp. 30ff; GM-free labels are mainly found on animal products, canned sweet corn and soybean products, The group of FAS Biotechnology Specialists in the European Union (2018), p. 31.

[322]Usually 0.9% or 0.1% for adventitious or technically unavoidable GM presence, European Commission Directorate-General for Health and Food Safety (2015b), pp. 33–34.

[323]European Commission Directorate-General for Health and Food Safety (2015b), pp. 10–47.

[324]European Commission Directorate-General for Health and Food Safety (2015b), pp. 2, 19–21.

[325]Council Regulation (EC) No 834/2007 of 28 June 2007 on organic production and labelling of organic products and repealing Regulation (EEC) No 2092/91 [2007] OJ L189/1. It is replaced from 2021 on by Regulation (EU) 2018/848 of the European Parliament and of the Council of 30 May 2018 on organic production and labelling of organic products and repealing Council Regulation (EC) No 834/2007 [2018] OJ L150/1. National, regional and private operator organic schemes partly apply stricter rules than the EU organic farming scheme with respect to GMOs, European Commission Directorate-General for Health and Food Safety (2015b), p. 19.

[326]Regulation (EC) No 834/2007, art. 9.

GM traces in organic products is 0.9%.[327] All GM-free labels have in common that they apply stricter standards than the mandatory GM labelling.

Regarding **genome edited plants and their products**, GM labelling will have far-reaching consequences: Experience shows that foods labelled as GM are barely found in European supermarkets due to **consumers' reluctance to buy them.**[328] Food producers will therefore avoid using GEOs in food production (unless it does not trigger the GM labelling obligation, e.g. if used as feed).[329]

Furthermore, it is feared that the GM labelling regime **cannot be enforced effectively** regarding plants obtained by SDN-1, SDN-2 and ODM and their products.[330] Labelling is based on mandatory traceability documentation. However, to control the labelling regime, additionally, analytical controls are applied.[331] Those might be difficult or even impossible regarding SDN-1, SDN-2 and ODM.[332] In particular, the labelling threshold (0.9%) will be difficult to measure because the development of unequivocal quantification methods for point mutations is challenging.[333] Even if such methods were developed they might not be suitable for routine testing. The likelihood that GEOs and products containing them or produced from them are illegally not labelled is particularly high regarding imported products, as many GEOs are not treated as GMOs outside the EU.[334] In any case, the question of how to enforce the labelling regime will have to be answered before the first GEOs can be authorised.

---

[327]Cf. Dederer (2016a), pp. 114–115; whether stricter thresholds for GM traces should be set in organic production is the subject of continuous debates, European Commission (2012), p. 9. However, the new Regulation (EU) 2018/848 also applies the 0.9% threshold, Regulation (EU) 2018/848, art. 30(4).

[328]European Commission (2016), p. 2; Sleenhoff and Osseweijer (2013), pp. 167, 168; Greenpeace (2005); The group of FAS Biotechnology Specialists in the European Union (2018), p. 29; except for consumers e.g. in the UK and Spain, The group of FAS Biotechnology Specialists in the European Union (2018), p. 41, below n. 443. Mandatory labelling is criticised for having a stigmatising effect, Sunstein (2017), pp. 1085–1087, Huffman and McCluskey (2014), p. 475, Peters and Lambert (2007), p. 159 ("message that genetic modification is harmful or bad"). Therefore, the reason why many consumers reject to buy GM labelled products is not only that they positively reject GMOs but also that some consumers are influenced by the label itself.

[329]Cf. The group of FAS Biotechnology Specialists in the European Union (2018), p. 41.

[330]Kahrmann et al. (2017), p. 180.

[331]Cf. e.g. Regulation (EC) No 1830/2003, art. 9(1). Admittedly, there are already products that do not allow for analytical controls, e.g. highly refined oil from GM plants, Davison and Ammann (2017), p. 22. Yet, the raw materials of the product can be tested.

[332]Currently, there are no unequivocal identification techniques for genome edited events, cf. the problem regarding the approval of these GEOs, text to 163ff. Control using several stretches of characteristic sequences in addition to the event might be possible, but could fail for progeny as progeny might have less or even none of these stretches due to crossing with other lines, cf. n. 176.

[333]European Network of GMO Laboratories (2019), p. 1; Brueller et al. (2012), pp. 126–127, 128; Ribarits et al. (2014), p. 187; Cotter et al. (2015), p. 12.

[334]The Netherlands Commission on Genetic Modification and Health Council of the Netherlands (2016), p. 46. Cf. also text to n. 299.

5.9 Identity Preservation System (Coexistence)

The **notion "coexistence"** describes the segregation between GM and non-GM, i.e. conventional and organic, production.[335]

Rules for coexistence of GM and non-GM plants and their products are set by the **member states.**[336] The **European Commission** issued a general guidance for the set-up of coexistence measures[337] as well as crop-specific guidance documents.[338]

The **aim** of these coexistence rules is to ensure that both producers and consumers can choose their preferred system of agricultural production, be it conventional, organic or agriculture using GMOs.[339]

The coexistence obligations are imposed on the **GM producer**. There are **two types of coexistence measures**[340]: The first type consists of ex-ante measures, i.e. measures to prevent GM traces in conventional and organic products. Examples of measures in place are (1) mandatory technical segregation measures such as isolation distances between GM and non-GM fields of the same species, pollen traps, crop rotation systems, cleaning of machinery and destruction of volunteers, (2) restrictions regarding the cultivation of GM crops, e.g. GM free areas and (3) mandatory information, registration and training procedures.[341] The second type consists of ex-post measures, i.e. measures to compensate non-GM producers for economic damages resulting from GMO admixture. Liability for GMO admixture and other compensation measures are described in Sect. 5.10 ("Liability").

The problem that certain genetic alterations by genome editing are **untraceable** is not relevant regarding ex-ante coexistence measures but only regarding liability and is therefore discussed in Sect. 5.10 ("Liability").

Coexistence rules **vary greatly** between the member states.[342] The coexistence rules of some member states are criticised as **politically motivated** and biased by an

---

[335]European Commission (n.d.-b).

[336]Cf. Directive 2001/18/EC, art. 26a: "Member States may take appropriate measures to avoid the unintended presence of GMOs in other products."; with respect to the coexistence measures adopted by the member states (as of 2009) European Commission (2009a), Beckmann et al. (2014), pp. 377–378.

[337]European Commission (2010b).

[338]Cf. the best practice documents by the European Coexistence Bureau, elaborated by technical working groups composed of member state experts, European Coexistence Bureau (n.d.).

[339]European Commission (2010b), annex 1.1.

[340]Cf. Beckmann et al. (2014), pp. 376–380; Dillen et al. (2016), pp. 64–65.

[341]Cf. the indicative catalogue of measures for coexistence at European Commission (2003), pp. 14–17; cf. European Commission (2009b), pp. 6–9; cf. also the overview of technical coexistence measures divided by production steps (seedbed preparation and start material—sowing—growing—harvest—post-harvest—storage, processing and transport) in Devos et al. (2012), p. 10778. In some countries, deviation from technical coexistence measures is possible provided that the neighbouring farmer consents, Schenkelaars and Wesseler (2016), p. 8. Registration and information responsibilities exist in almost all member states, Schenkelaars and Wesseler (2016), p. 7.

[342]Schenkelaars and Wesseler (2016), p. 7.

anti-GM attitude, as they are overly restrictive and thus preventing GMO cultivation in practice.[343] An example is Luxembourg that established an isolation distance of 600m for GM maize.[344] In Spain, in contrast, where 95% of the EU's GM crops are grown, GM-growers only have to adopt very few coexistence measures.[345]

Enterprises that acquire GM food or feed from **third countries** are required to set safety measures against GMO contamination by way of contract.[346]

5.10 Liability

Rules on liability are merely harmonised to a limited extent within the EU. While the field of product liability and to some extent environmental liability are harmonised, criminal, administrative[347] and civil liability[348] remain largely within national competences. Consequently, different national liability schemes applicable to GMOs have developed throughout Europe, so that this section can only illustrate their commonalities and differences.

5.10.1 Criminal and Administrative Liability

Directive 2001/18/EC (Deliberate Release) and Regulations (EC) No 1829/2003 (GM Food and Feed) and 1830/2003 (Traceability and Labelling) require member states to sanction breaches of their provisions with "effective, proportionate and dissuasive" penalties.[349] Accordingly, member states provide for fines and

[343]Masip et al. (2013), pp. 312, 317 ("de facto moratorium", "arbitrary"); Beckmann et al. (2014), pp. 384, 388.

[344]Schenkelaars and Wesseler (2016), p. 7.

[345]Spain has no national or regional coexistence regulations. GM farmers have to follow good agricultural practices and recommendations by the seed industry. The non-GM farmers are responsible themselves to adopt other necessary segregation measures, Schenkelaars and Wesseler (2016), p. 8; Guerrero (2017), p. 21; Asociación Nacional de Obtentores Vegetales (n.d.).

[346]Cf. Regulation (EC) No 1829/2003, art. 24(2),(3)—the labelling tolerance threshold for traces of authorised GM material of 0.9% only applies to traces that are adventitious or technically unavoidable (cf. text to n. 318–319). The operators must supply evidence that they have taken appropriate steps to avoid the presence of GM material. Contracts obliging suppliers from non-EU countries to take safety measures against GMO contamination is such an appropriate step, German Federal Ministry of Food and Agriculture and German Federal Office of Consumer Protection and Food Safety (2011), p. 5.

[347]E.g. liability for field trials or marketing without approval (as far as approval is required under Directive 2001/18/EC).

[348]E.g. liability for bodily injuries, damages to property and contamination of non-GMOs.

[349]Directive 2001/18/EC, art. 33; Regulation (EC) No 1829/2003, arts. 45(1), (2); Regulation (EC) No 1830/2003, arts. 11 (1), (2).

imprisonment for the deliberate release and/or placing on the market of a GM variety without approval and for breaches of labelling requirements: Italy and Germany for instance introduced imprisonment of up to 3 years for deliberate release without approval.[350]

Furthermore, an important number of countries provide for penalties for non-compliance with their respective national coexistence regulations, independent of the occurrence of damage through admixture.[351]

5.10.2 *Product Liability*

Civil liability for damages caused by defective products is fully harmonised within the EU under the Product Liability Directive (PLD), meaning that no member state may apply more stringent or more lenient rules.[352] Applying the general provisions of the PLD, both breeders and cultivators are strictly liable for death, personal injuries, and damage to privately used property if the variety or products processed from it are defective.[353] They are, however, exempt from liability according to the general development risk defence, Art. 7(e) PLD, if they can prove that the defect could not have been discovered when the GM variety was put into circulation.[354] Yet, with respect to GMOs some member states made use of the possibility to derogate from this defence.[355]

Moreover, the PLD does not cover damages sustained by cultivators as a result of admixture for two reasons: Firstly, because the PLD only covers damages to

[350]German Genetic Engineering Act: Gesetz zur Regelung der Gentechnik (GenTG), (BGBl. 1993 I p. 2066). Last amended by law of 17.07.2017 (BGBl. 2017 I p. 2421), Sec. 39 (2); Decreto legislativo of 8 July 2003. n. 222 (GU n. 194 of 22-8-2003), art. 35.1.

[351]E.g. Austria, Belgium, Denmark, France, Hungary, the Netherlands and the Czech Republic, Beckmann et al. (2014), p. 380; European Commission (2009b), p. 8.

[352]ECJ, Case C-52/00 *Commission v France* [2002] ECR I-3827, paras. 16–24; ECJ, Case C-154/00 *Commission v Greece* [2002] ECR I-2879, paras. 12–20; ECJ, Case C-183/00 *González Sánchez v Medicina Asturiana SA* [2002] ECR I-3901, paras. 25–32; Riehm (2017), para. 47.

[353]Council Directive 85/374/EEC of 25 July 1985 on the approximation of the laws, regulations and administrative provisions of the Member States concerning liability for defective products [1985] OJ L210/29 (Product Liability Directive), arts. 1ff; under the PLD three categories of 'defects' are recognised: Design defects, manufacturing defects, and warning defects, Brüggemeier (2015), para. 301; Hoffman and Hill-Arning (1994), pp. 6–7; Taschner and Frietsch (1990), Richtlinie Art. 6, para. 9. Accordingly, a GM variety is e.g. defective if it contains a potentially harmful trait, as is a food if it is erroneously processed from a GM variety not authorised for food use in the European Union, Koch (2010), para. 72.

[354]Product Liability Directive, art. 7(e); Brüggemeier (2015), paras. 324–325.

[355]Product Liability Directive, art. 15(1)(b); Luxembourg and Finland derogated from the development risk defence entirely, whereas Spain derogated with respect to food and pharmaceuticals, Koch (2010), para. 80, n.124; Giliker (2014), 334, n. 89. Moreover, Germany derogated from the defence with respect to GMOs, however, only concerning the liability of seed producers, GenTG. Sec. 37(2); Fedtke (2010), para. 40.

property that is used for private—not commercial—purposes or consumption,[356] and secondly, because it only applies to products put into circulation,[357] which is not the case for plants during their cultivation on the fields.[358]

5.10.3 Civil Liability

Beyond liability for defective products, civil liability for GMOs is regulated by the member states, predominantly in tort and property law. As a minimum protection, general tort law in all jurisdictions provides for compensation for **bodily injuries and damages to property** in cases of fault.[359] In detail, however, liability for GMOs is affected by general differences between European tort law systems. Examples for differences with an impact on liability for GMOs are the extent to which immaterial damages are compensated for,[360] the handling of alternative, cumulative and supervening causation,[361] and the applicable standard of proof.[362] Significant differences concerning periods of prescription are of particular importance in this context, since long-term damages of GMOs might only become apparent and materialise far in the future.[363] Moreover, certain countries introduced strict liability, i.e. liability without fault, for bodily injuries and damages to property caused by GMOs, though some limit this strict liability to the research and development stage of GMOs (e.g. Austria and Germany).[364]

In practice, questions of liability for **admixture of GM crops with non-GM crops**, e.g. through cross-pollination, are most relevant. While the European Commission issued guidelines for national measures ensuring the coexistence of GMOs

356 Product Liability Directive, art. 9(b)(i), (ii).

357 Product Liability Directive, art. 7(a); Brüggemeier (2015), paras. 319–320; Taschner and Frietsch (1990), Richtlinie Art. 7, para. 7.

358 Kohler (2005), p. 575; Koch (2013), p. 410.

359 European Commission (2009b), p. 5; Koch (2013), p. 407.

360 van Dam (2013), p. 346, Bar (2000), para. 145; Koch (2010), para. 8.

361 Koch (2010), paras. 45–47; regarding the different handling of alternative causation cf. Koziol (2007), pp. 387–389; Infantino and Zervogianni (2017), pp. 632–634; van Dam (2013), pp. 329–332; regarding cumulative causation, cf. Infantino and Zervogianni (2017), pp. 652–653; regarding supervening causation cf. Koch (2007), p. 501; Infantino and Zervogianni (2017), pp. 628–631.

362 Ranging from "more likely than not", e.g. in Italy, cf. Monti and Fusco (2010), para. 20 to almost "certainty", e.g. in Austria, cf. Weissenbacher (2010), para. 18; see Koch (2010), paras. 36–37.

363 Koch (2010), paras. 118–120, outlining that the length of objective periods of prescription, i.e. prescription commencing at the time of the harmful event independent of the victim's knowledge, varies from 3 years in the Czech Republic to 30 years e.g. in Austria, Germany, and the Netherlands.

364 Koch (2010), paras. 62–69; strict liability for classical damages in general applies in Finland, Hungary, Lichtenstein, Norway and Poland, strict liability only for the research and development stage in Austria (Sec. 79a Gentechnikgesetz) and Germany (GenTG, Sec. 32 (1), 37 (2)).

and non-GMOs, liability for admixture falls within the exclusive competence of the member states and is therefore not covered by the Commission's document.[365] As a consequence, liability provisions regarding admixture differ widely between the member states. About a third of the member states introduced specific liability provisions addressing damages arising from admixture.[366]

In contrast to other jurisdictions, it is generally the cultivator of non-GM crops, who can claim compensation for admixture. The damage sustained by cultivators generally is the price difference between organic, conventional and GM crops.[367] This is because admixture above the EU threshold of 0.9% requires non-GM farmers to label their products and prevents them from marketing them as organic.[368] Whether the non-GM farmer would also be indemnified for economic losses below the EU threshold, e.g. for losses resulting from contractual obligations requiring a higher degree of purity under private organic labelling, remains highly uncertain.[369] A slight tendency towards non-compensation of these damages can, however, be detected.[370] Moreover, depending on the legal doctrine, some jurisdictions even refrain entirely from providing for compensation where the farmer's loss does not arise as a consequence of an intermediary physical damage to the plants themselves ("pure economic loss").[371] Yet, certain countries already regard loss as consequence of an intermediary damage if it occurs as result of any physical alteration to the plant.[372]

The burden of proving causation between the presence of this genetically altered plant and an event in the defendant's sphere generally lies upon the claimant. However, an important number of member states provide for alleviations of the burden of proof: While few member states shift it entirely onto the defendant, some apply a (rebuttable or irrebuttable) presumption of causation in cases of GMO admixture, while again others rely on *prima facie* or circumstantial evidence.[373] Accordingly, in most jurisdictions it will probably not constitute a problem that it is not possible to identify with certainty a genetic alteration as one caused by genome

---

[365]European Commission (2010b), para. 2.5.

[366]Countries with such specific regulation are in particular Austria, Finland, France, Germany, Hungary, Norway, Poland, Slovakia and Switzerland, cf. Koch (2008a), paras. 156–183.

[367]Schenkelaars and Wesseler (2016), p. 8; Faure and Wibisana (2008), para. 75; Beckmann et al. (2014), p. 379; European Commission (2009b), p. 4.

[368]Cf. above Sect. 5.8 ("Labelling").

[369]Beckmann et al. (2014), p. 380; Koch (2008b), para. 4; see e.g. discussion in Germany regarding GenTG, Sec. 36a, which refers to statutory labelling requirements only, but is still interpreted as uncertain by some, cf. Palme (2005), p. 256; Fedtke (2008), para. 40.

[370]Schenkelaars and Wesseler (2016), p. 8.

[371]Koch (2008a), para. 33–34; Bar (2000) no. 25; Bussani and Palmer (2005), pp. 123–125.

[372]Bar (2000), para. 32; Koch (2008a), para. 34.

[373]Cf. analysis in Koch (2008a), paras. 47–50, France and Denmark only require proof of cultivation of GMOs in the vicinity (and presence of GMOs on the claimant's field) for compensation and thereby *de facto* apply an irrebuttable presumption of causation, cf. Ulfbeck (2008), para. 3; Taylor (2008), para. 6.

editing (SDN-1, SDN-2, ODM). Most likely, it suffices that the injured party can prove the presence of a characteristic sequence or a characteristic trait of the GEO. The theoretical and very unlikely possibility that a conventional plant has mutated exactly like the genome edited plant does therefore not hinder liability.

Liability for GMO admixture is fault-based in most European jurisdictions.[374] Non-compliance with coexistence legislation is regularly regarded as negligent.[375] Some jurisdictions even provide for strict liability, mostly in specific provisions concerning liability for GMOs.[376] France appears to be the only country that would also apply strict liability based on a more general norm, which prescribes strict liability for all hazardous activities.[377] Similar provisions do exist in other European jurisdictions. However, it is not yet clear and largely left at the discretion of their national courts whether they would also regard the cultivation of GMOs as hazardous activity entailing strict liability.[378]

In addition to the law of torts described hitherto, almost all European jurisdictions developed rules governing the liability between adjacent landowners (neighbour law), which are not GMO-specific but also apply to GMO admixture caused by pollen drift.[379] Most of these rules are independent of fault on the part of the disturbing landowner, i.e. GMO cultivating farmer.[380] Despite major doctrinal differences,[381] all those rules focus on whether a particular land use is customary in a place and whether the interference with the use of the adjoining land is significant and unreasonable.[382] Liability under neighbour law can therefore essentially depend on whether GM farming has become a common agricultural practice (in a certain area), making GMO admixture a reasonable minor interference, which adjacent landowners have to tolerate.[383]

---

[374]European Commission (2009b), p. 5.

[375]Koch (2008a), para. 56.

[376]European Commission (2009b), p. 5; cf. the presentation of special liability regimes for GMO admixture in Europe by Koch (2008a), paras. 156–173: Strict liability is provided for in Austria, France (also within a special liability system, cf. art. 8 du Loi n°2008-595 du 25 juin 2008 relative aux organismes génétiquement modifiés, JORF n°0148 du 26 juin 2008, 10218), Finland (if qualified as environmental harm), Germany, Hungary, Luxemburg (Goergen (2010), para. 4), Norway, Poland, Slovakia.

[377]Koch (2008a), para. 59, drawing attention to the fact that the wording of the Belgium civil code is identical, but interpreted differently.

[378]Koch (2013), p. 409; Koch (2008a), para. 59.

[379]Koch (2008a), para. 67; Bar (1998), para. 531.

[380]Gordley (2010), p. 24; Bar (1998), paras. 531, 535–544; Koch (2008a), para. 69; except for the Netherlands which made liability between neighbours dependent on fault, Castillo and van Boom (2008), para. 32.

[381]Gordley (2010), pp. 23–24; Bar (1998), para. 533; Koch (2008a), para. 68.

[382]Bar (1998), para. 534; Koch (2008a), para. 69; Koch (2008b), para. 14.

[383]Koch (2008a), para. 69; Bar (1998), para. 534; different in Germany, where no distinction is made between GM and conventional cultivation as regards the customary character of farming (GenTG, Sec. 36a (3)), while any GMO admixture is legally defined as significant interference, GenTG, Sec. 36a (1).

Finally, several member states set up **mandatory compensation funds** covering losses resulting from adventitious presence of GM plants in conventional or organic crops, funded by GM farmers (e.g. Belgium and Denmark), or require **mandatory insurance** for GM farmers (e.g. Austria and France).[384]

5.10.4 Environmental Liability

According to Art. 3(1)(a) Environmental Liability Directive (ELD) in conjunction with Annex III (11), both breeders and cultivators are strictly liable for environmental damage arising from the deliberate release, transporting, or placing on the market of GMOs.[385] Thereby the ELD implements the "polluter-pays-principle".[386] It covers environmental damage, i.e. damages to protected species and natural habitats (biodiversity), water damage and land contamination creating significant risks for human health.[387] Given the reference to "species" only, the Directive does, however, not encompass injury to the genetic variability among species, and accordingly does not consider plant-to-plant gene flow as such as an environmental damage.[388] Furthermore, losses that are only secondary to environmental harm, such as personal injuries or damages to property, are not covered.[389]

Exceptions from strict liability may be provided for by the member states for emissions that had been authorised by the competent authority and for emissions or products which were not considered likely to cause environmental damage according to the state of scientific and technical knowledge at the time they were released ("*development risk defence*").[390] Several member states implemented these defences,[391] yet producers and cultivators may only invoke them if they were not at fault.[392]

[384]Beckmann et al. (2014), p. 380; European Commission (2009b), p. 5.

[385]Directive 2004/35/CE of the European Parliament and of the Council of 21 April 2004 on environmental liability with regard to the prevention and remedying of environmental damage [2004] OJ L143/56 (Environmental Liability Directive), art. 3(1)(a) in conjunction with annex III (11); Barns (2018), p. 5.

[386]Environmental Liability Directive, art. 1; Hinteregger (2008), para. 5/6.

[387]Environmental Liability Directive, art. 2(1)(a)-(c).

[388]European Commission (2002), pp. 19f; Duikers (2006), pp. 626–627; Meßerschmidt (2011), para. 16; Hinteregger (2008), para. 5/9.

[389]Environmental Liability Directive, art. 3(3).

[390]Environmental Liability Directive, art. 8(4)(a), (b); Bergkamp and van Bergeijk (2013), paras. 4.38–4.46; Barns (2018), pp. 6–7.

[391]e.g. Czech Republic, Greece, Italy, Malta, Portugal, Spain and the UK (with the exception of Scotland and Wales), cf. Goldsmith and Lockhart-Mummery (2013), para. 7.26; Bergkamp and van Bergeijk (2013), para. 4.47.

[392]Environmental Liability Directive, art. 8(4); Barns (2018), pp. 6–7.

Table 5.16 Liability

	GMOs (incl. GEOs)
Criminal & administrative liability	*Monetary penalties* and/or *imprisonment* for unapproved **deliberate release/marketing** and breaches of **labelling** requirements
Civil liability	– Liability for bodily injuries and damages to property: fault-based liability; in certain countries: strict liability – Liability for admixture of GM-crops with non-GM crops: general tort law (fault-based); neighbour law; in certain countries: special liability regimes (strict liability); damage = admixture beyond 0.9% threshold
Environmental liability	**Strict liability** for damages to protected species, natural habitats, water, land.
Product liability	**Strict liability** for damages caused by defective varieties or processed products. Damages sustained by cultivators as result of admixture are not covered.

Ultimately, the ELD only sets out minimum requirements for environmental liability, which may be exceeded by nature conversation legislation in the member states.[393]

5.10.5 Summary

With respect to GM varieties and their products, the general liability schemes are often tightened and/or supplemented by specific liability schemes. This is the case both within the Environmental Liability Directive of the European Union and as a matter of national legislation (Table 5.16). There is a suspicion that member states have used their competences as a political margin of manoeuvre, either to facilitate coexistence or to discourage the cultivation of GM crops through the creation of increased risks of liability. In particular, comparatively strict liability schemes governing admixture of GM-crops with non-GM crops are apt to deter farmers from cultivating GM-crops in a number of European countries.

[393]Environmental Liability Directive, art. 16(1); Hinteregger (2008), para. 5/7; Meßerschmidt (2011), para. 8.

5.11 Perception of Genome Editing

5.11.1 Position of Public Authorities

5.11.1.1 EU Authorities

As regards genome editing, the positions of the European Commission,[394] the European Parliament[395] and the Council of the European Union[396] are the most important. These EU institutions are involved when it comes to reforms of the GMO framework regarding genome editing technologies: In the ordinary legislative procedure, the European Commission submits legislative proposals whilst the European Parliament and the Council of the European Union jointly adopt legislation.[397] As the European Commission is the executive body of the EU, it also plays an important role in the implementation of the EU's GMO law.[398]

5.11.1.1.1 European Commission

Generally, the Commission is innovation- and biotech-friendly.[399] However, regarding genome editing, the Commission has been inactive so far, probably due to political pressures from different sides—civil society organisations have called for a strict regulation of new breeding techniques, the trading partners of the CETA (and formerly TTIP) trade agreements as well as the biotechnology industry have been

[394]The European Commission is composed of one commissioner from each member state. Examples of functions are: Proposal of new laws, management of EU policies and allocation of EU funding, enforcement of EU law and international representation of the EU, cf. Consolidated version of the Treaty on European Union [2012] OJ C326/1 (TEU), art. 17; European Union (2018a), and European Commission (2018b).

[395]The European Parliament is directly elected. Examples of functions are: Adoption of EU laws together with the Council of the European Union, decision on international agreements, supervisory and budgetary functions, cf. TEU, art. 14; European Union (2018b).

[396]The Council of the European Union is composed of government ministers from each member state. Examples of functions are: Adoption of EU laws together with the European Parliament, coordination of EU policies, conclusion of agreements between the EU and other countries or international organisations and budgetary functions, cf. TEU, art. 16; European Union (2017).

[397]TEU, arts. 14(1)(1), 16(1)(1), 17(2); TFEU, arts. 289(1), 294; European Parliament (2017), pp. 11ff.

[398]Examples are: the strong role of the Commission in the GMO approval process—de facto, it takes the final authorisation decisions, see n. 100; the Commission's guidance documents, e.g. on coexistence, text to n. 338–339; collection and dissemination of information and data on GMOs, cf. e.g. the GMO register, European Commission (n.d.-g) or various reports and studies; the enforcement of EU law through infringement procedures (TFEU, art. 258), e.g. when several member states did not implement Directive 2001/18/EC promptly, Poli (2006), p. 390.

[399]Bernauer and Aerni (2008), p. 188. From the various research projects on GMOs the Commission funded, it drew the conclusion that "GMOs [...] are not per se more risky than e.g. conventional plant breeding technologies", European Commission (2010a), p. 16.

seeking to avoid their regulation altogether and the European Parliament is split down in the middle.[400]

Since 2007, the Commission has sought scientific advice on various matters concerning new breeding techniques, e.g. their risks and potential,[401] the legal classification of organisms obtained by them as GMOs[402] as well as similarities and differences between new breeding techniques, traditional breeding techniques and established techniques of genetic modification.[403,404] In addition, the Commission initiated a stakeholder dialogue.[405]

In 2015, the Commission announced a (non-binding) interpretation of the GMO definition, i.e. a legal classification of organisms obtained by new breeding techniques as GMOs.[406] To date (June 2019), the legal opinion has not yet been produced because the Commission had awaited the ruling of the European Court of Justice in Case C-528/16.[407] It is not clear whether the Commission still intends to produce a legal opinion, e.g. to cover the techniques Case C-528/16 did not deal with,[408] or not.[409]

In Case C-528/16, the Commission supported the interpretation that new mutagenesis techniques are encompassed by the legal term "mutagenesis" and thus exempted from the GMO regulatory framework.[410]

In November 2019, a new president of the European Commission and a new team of Commissioners will take office. The Commission hinted that the new Commission might then think about a legislative proposal amending the EU genetic engineering law.[411]

400Cf. Valavanidis (2016), p. 30; Leroux (2016), Corporate Europe Observatory (2016a), p. 4; Fladung (2016), p. 474; text to n. 206, 209, 427–433.

401Lusser et al. (2011), and European Food Safety Authority (2012a, b).

402European Commission New Techniques Working Group (2011).

403European Commission Scientific Advice Mechanism High Level Group of Scientific Advisors (2017).

404Cf. European Commission (n.d.-h).

405High-level conference "Modern Biotechnologies in agriculture – Paving the way for responsible innovation", 28 September 2017, Brussels, cf. European Commission (2017).

406Laaninen (2016), pp. 1, 2; Gene editing in legal limbo in Europe (2017); Corporate Europe Observatory (2016a), p. 13.

407Tagliabue and Ammann (2018), p. 51; O'Reilly (2017). This inaction despite the urgent need for regulatory clarification has been heavily criticised, cf. European Seed Association (2017b), p. 2.

408Cf. text to n. 486–488.

409Optimistic Purnhagen et al. (2018), p. 573. Cf. also Duensing et al. (2018), p. 2 ("during the hearing [. . .] in the Case C-528/16 [. . .], the Commission vaguely stated that they were preparing something about this 'new' problem.").

410Bobek (2018), paras. 76–78.

411Fortuna (2019).

5.11.1.1.2 European Parliament

The European Parliament is divided regarding its position on GEOs, which is not surprising as it is composed of multiple political groups.[412] However, what *is* surprising is that many of these groups either have no position at all or no coherent position on GMOs, including GEOs.[413]

The diverging opinions and indecision are reflected by the resolutions issued by the European Parliament. On one hand, the resolutions support the development, funding and use of new breeding techniques and stress the opportunities in ensuring food security and food quality, responding to climate change and reducing the environmental impact of agriculture.[414] On the other hand, they point to the precautionary principle and are concerned about the techniques being "safe and proven".[415] The current European Parliament was elected in May 2019. Its positions were not yet fully foreseeable at the time of writing.

5.11.1.1.3 Council of the European Union

The Council of the European Union represents the governments of the member states. Therefore, it does not have a unanimous position but mirrors the positions of the various member states and their ministries, which are analysed in the following subsection. The view of the Council is often unpredictable.[416] It can be important which country is holding presidency of the Council.[417]

---

[412]The two largest groups, the **Group of the European People's Party (Christian Democrats)** and the **Group of the Progressive Alliance of Socialists and Democrats**, do not have an official position on GEOs. Regarding established GMOs, the European People's Party is conflicted. The Socialists and Democrats Group tends to reject GMOs, cf. e.g. Martin de la Torre, Victoria (2016). Members of the **Group of the Alliance of Liberals and Democrats for Europe+Renaissance +USR PLUS** and the **European Conservatives and Reformists Group**, i.e. the 3rd and 5th largest groups, initiated parliamentary reports stressing the potential of the new breeding techniques (see n. 415), cf. further Bruins (2017), McIntyre (2016). The **Group of the Greens/European Free Alliance**, i.e. the 4th largest group, reject GEOs. They call to be prudent and have always favoured the regulation of all GEOs as GMOs, The Greens/European Free Alliance in the European Parliament (2018b); The Greens/European Free Alliance in the European Parliament (2017). Regarding the composition of the EU Parliament 2019–2024 see European Parliament (2019). Regarding the position of the groups towards established GMOs cf. further the voting patterns of the European Parliament groups in a resolution concerning the placing on the market of the GMO Pioneer 1507 (European Parliament (2014b)), Mühlböck and Tosun (2015). Regarding the positions on established GMOs of the member state parties forming the European Parliament groups Wortmann (2015), pp. 14–23.

[413]Wortmann (2015), p. 3; n. 413.

[414]European Parliament (2016a), para. 32; European Parliament (2016b), paras. 27–28; European Parliament (2014c), paras. 7–8.

[415]European Parliament (2016a), para. 32; European Parliament (2016b), para. 27.

[416]Holland (2004), p. 5.

[417]The presidency rotates among the member states and is held for six months, TEU, art. 16(9).

Regarding genome editing, in an Agriculture and Fisheries Council meeting in May 2019, 14 EU member states requested the Commission to review the current GMO legislation.[418]

5.11.1.2 Member States Authorities[419]

There is large variation among the 28 EU member states with respect to their attitudes towards GEOs.

England,[420] Spain, and the Netherlands,[421] for example, are open to the new techniques and would have preferred that organisms having only genetic changes that could be obtained by traditional breeding methods are not regulated under the GMO regime.[422] Other EU countries, such as Germany, France or the Czech Republic, are divided in the sense that supporting and rejecting forces are pulling into opposite directions.[423] Others again, e.g. Austria,[424] reject genome editing in agriculture and welcome the regulation through the GMO regime. Most member states have not yet officially positioned themselves.[425]

In the end, the attitudes of the member states towards genome editing in agriculture will likely mirror those towards traditional forms of genetic engineering.[426] The

---

[418]Council of the European Union (2019), p. 7; Fortuna (2019).

[419]Cf. the country overview in The group of FAS Biotechnology Specialists in the European Union (2017), pp. 50–51 as well as the "Innovative Biotechnologies" sections in the USDA Foreign Agricultural Service reports about the individual member states, The group of FAS Biotechnology Specialists in the European Union (2018), p. 53. Cf. further Madre and Agostino (2017); with respect to the Scandinavian countries Denmark, Finland, Norway and Sweden Eriksson et al. (2017); Stakeholder and Issue Mapping on New Breeding Techniques (2017) gives an overview of the political climate, positions of ministries and regulatory agencies in selected countries; cf. also the positions held by the national governments in Case C-528/16, Bobek (2018), paras. 71–75.

[420]UK House of Commons Science and Technology Committee (2015), p. 16; UK Government (2016); differing Wales, Scotland and Northern Ireland, cf. UK House of Commons Science and Technology Committee (2015), p. 17; The group of FAS Biotechnology Specialists in the European Union (2017), p. 50.

[421]Cf. The Netherlands Government (2017).

[422]The group of FAS Biotechnology Specialists in the European Union (2017), p. 50; cf. further the UK parliamentary written answer Eustice (2018) ("the Government's view is that specific regulation of this technology is not required where the induced genetic change could have occurred naturally or been achieved through traditional breeding methods") as well as the Netherlands' proposal to exempt certain new plant breeding techniques from the GMO framework, The Netherlands Government (2017); Corporate Europe Observatory (2016a), p. 14.

[423]The group of FAS Biotechnology Specialists in the European Union (2017), pp. 50–51.

[424]GMWatch (2016).

[425]The group of FAS Biotechnology Specialists in the European Union (2017), p. 51; Eriksson et al. (2017), p. 229 (regarding Denmark and Norway, which is a member of the European Economic Area).

[426]With respect to the attitudes of the EU member states towards GMOs cf. the classification into adopters—conflicted—opposed in The group of FAS Biotechnology Specialists in the European

reason is that the underlying political, economic and social interests are similar. Economic interests are e.g. on the one hand the commercial importance of plant research and breeding[427] or of the export of agricultural products and on the other hand the importance of the organic sector or local food cultures.[428]

The UK's case is exceptional as it leaves the EU at the end of October 2019 ("Brexit").[429] At the time of writing, there is considerable uncertainty regarding the content of a possible withdrawal agreement as political discussions are still ongoing.[430] The UK will apply the GMO regulatory framework in the short term whether or not a withdrawal agreement is reached.[431] Whether the UK has an interest in relaxing the framework in the long term depends on the potential biotechnology market demand and size on the one hand and the economic importance of trading relationships with the EU on the other hand.[432] At present, the EU is the UK's most important trading partner for agricultural products.[433]

5.11.2 Stakeholders' Opinions

The various stakeholders have already positioned themselves according to their interests[434] and become active (Table 5.17). On one side, civil society organisations

Union (2018), pp. 41–42 and International Service for the Acquisition of Agri-biotech Applications (2017), p. 97; cf. also the member states' positions in the Council of the European Union regarding GMO authorisation requests (pro GMOs—abstain—against GMOs), see the figure in Mühlböck and Tosun (2017) (before 2014, the Council was involved in the GMO approval process). Cf. further the overview of selected EU countries' stance towards GMOs (UK, Germany, Poland, Spain) in Clancy (2017), pp. 28–29; 33–34; 36–40; 41–44.

[427]Cf. the strong plant research and breeding sectors in the Netherlands, Phillips and Flach (2017), Corporate Europe Observatory (2016a), p. 5; in Sweden, Eriksson et al. (2017), pp. 219, 225–226, Corporate Europe Observatory (2016a), p. 5; or in the UK, International Service for the Acquisition of Agri-biotech Applications (2016), p. 80 ("world leadership in plant science"), Sciencewise and Nuffield Council on Bioethics (2016), p. 11 ("global leadership in genome editing").

[428]E.g. in Greece and Italy, Lee (2017), p. 1222; cf. also Austria aiming at the protection of Alpine biodiversity, Lee (2017), p. 1222.

[429]Cf. European Commission Directorate-General for Health and Food Safety (2018).

[430]European Commission Directorate-General for Health and Food Safety (2018).

[431]Cf. UK Department for Environment, Food and Rural Affairs (2019) and Downing and Coe (2018), pp. 71–72 regarding the 'no deal' scenario; cf. BBC News (2018) regarding the transition period in case a withdrawal agreement is reached.

[432]International Service for the Acquisition of Agri-biotech Applications (2016), p. 80. Regarding the UK's future regulatory options for the regulation of GMOs and organisms obtained by new plant breeding techniques in general Border and Walker (2017), pp. 3–4 and Brookes (2018). In reaction to the ruling of the European Court of Justice in Case C-528/16 voices were raised in favour of a distinct regulation or at least interpretation of the GMO definition in the UK that is more favourable towards GEOs, cf. the open letter by the British research and agricultural sectors calling for a "science-based approach to regulation", John Innes Centre et al. (2018).

[433]Potton and Webb (2017), pp. 6–9.

[434]Regarding the interests of the various stakeholders Hamburger (2018).

Table 5.17 Some EU stakeholder positions on new plant breeding techniques

Positively disposed towards new plant breeding techniques	Negatively disposed towards new plant breeding techniques
• Most plant scientists • Plant breeding industry, seed industry • Large parts of the conventional agricultural sector	• Most environmental and consumer civil society organisations (e.g. Friends of the Earth Europe, Greenpeace EU, Trans Atlantic Consumer Dialogue) • Organic sector

opposing GMOs[435] as well as the organic sector[436] are negatively disposed towards GEOs. On the other side, scientists and their representations[437] as well as the breeding/seed sector[438] try to influence the debate into a direction favourable to GEOs.[439] The conventional agricultural sector is also predominantly in favour of genome editing techniques.[440] Thus, the stakeholders' positions towards GEOs are identical to those towards GMOs.[441]

[435]Cf. GM Freeze (2016), Greenpeace (2015), GMWatch (2014), Steinbrecher (2015), Trans Atlantic Consumer Dialogue (2016); cf. also the joint letter by environmental NGOs: EcoNexus et al. (2015) and the joint position paper by environmental NGOs and other associations, IFOAM EU Group et al. (2017).

[436]Examples of position papers are IFOAM EU Group (2015), German Organic Food Production Alliance (2017), Agir pour l'Environnement et al. (2017).

[437]Cf. the position statements by European and national scientific organisations, e.g. European Plant Science Organisation (2017), European Academies Science Advisory Council (2015), German National Academy of Sciences Leopoldina et al. (2015), UK Biotechnology and Biological Sciences Research Council (2014). Plant scientists from the Netherlands, the UK and Sweden are amongst the most active in the debate and also receive high media coverage, Stakeholder and Issue Mapping on New Breeding Techniques (2017), executive summary. An example is a Swedish researcher who grows genome edited cabbage in his garden and served it to a journalist as the alleged world's first CRISPR meal, Lawrence (2016), Callaway (2018).

[438]Cf. the communication initiative "Embracing the Power of Nature", European Seed Association (n.d.); European Seed Association (2017a); cf. the New Breeding Techniques Platform, New Breeding Techniques Platform (2015a). Differing Helliwell et al. (2017), p. 2090, speaking of a "wait-and-see" strategy regarding the breeding/seed sector.

[439]Helliwell et al. (2017), p. 2090.

[440]Cf. e.g. Copa-Cogeca (2017) (European umbrella organisation of agricultural cooperatives "Copa-Cogeca"); Morgan (2017) (National Farmers Union, UK); French Haut Conseil des Biotechnologies (2016b), Annex 1, pp. 23–24, 102 (various agricultural cooperatives, France). There are also some farming cooperatives and associations sceptical or negatively disposed towards organisms developed through new breeding techniques, cf. e.g. French Haut Conseil des Biotechnologies (2016b), Annex 2, pp. 37ff, 102; Eriksson et al. (2017), p. 232 (regarding farmers associations in Norway, which is a member of the European Economic Area); German Syndicate of Traditional Agriculture (2016). Cf. further the sections on national associations in Stakeholder and Issue Mapping on New Breeding Techniques (2017).

[441]Cf. the overview of stakeholders' opinions towards GMOs in The group of FAS Biotechnology Specialists in the European Union (2018), pp. 39–41.

5.11.3 Public Opinion

On the basis of public opinions on the "old", established GMOs in the EU, which are presented at (1.), some aspects relevant for the evolvement of public opinions on the "new" GMOs, i.e. GEOs, are outlined (2.).

5.11.3.1 Public Opinion on GMOs

In the EU, there is a high level of awareness regarding GM food.[442] Whereas GMOs are widely accepted in the medical sector, European consumers are, in an international comparison, **sceptical[443] towards GMOs in agriculture**, especially in foodstuffs.[444] The main issue is food safety.[445] According to surveys, the support for GM food has kept declining for more than a decade.[446] Looking at real-world purchasing decisions, by contrast, acceptance of GMOs is in fact increasing.[447] Furthermore, consumer attitudes towards GMOs differ widely between the member states[448] and between traits and uses[449] of GMOs.

[442] TNS Opinion & Social (2010a), pp. 13–17 (84% of the respondents have heard about GM food; 38% of those who have heard about GM food actively search for information on it).

[443] Including feelings of wariness, unease and uncertainty, Rollin et al. (2011), p. 100.

[444] Gaskell (2010), pp. 37–38; on average, the opponent/supporter-ratio regarding GM food was 3:1 in 2010, Gaskell (2010), p. 7. According to Rollin et al. (2011), p. 100 the "majority [...] are undecided or feel that they don't know enough to form a view". There are "minorities with strongly positive or negative opinions".

[445] Gaskell (2010), p. 38; TNS Opinion & Social (2010a), pp. 18; 28–29 (59% of the respondents do not think that GM food is safe for their health and that of their family). In spontaneous responses to the questions "What are all the things that come to your mind when thinking about possible problems or risks associated with food and eating", 8% mentioned GMOs, which puts GMOs to the 7th place); in prompted responses, 66% of the respondents indicated that they are "very worried" or "fairly worried" about GMOs in food or drinks (putting GMOs to the 5th place), TNS Opinion & Social (2010b), pp. 19, 21.

[446] Gaskell (2010), pp. 39–40.

[447] European Commission Directorate-General for Health and Food Safety (2015b), pp. 48ff; Sleenhoff and Osseweijer (2013), pp. 166, 168–169; Lucht (2015), pp. 4258–4259; International Service for the Acquisition of Agri-biotech Applications (2017), pp. 97–98; European Academies Science Advisory Council and German National Academy of Sciences Leopoldina (2013), p. 30; The group of FAS Biotechnology Specialists in the European Union (2018), p. 40. This is particularly true for the UK, cf. UK House of Commons Science and Technology Committee (2015), Q437 [Professor Poppy].

[448] TNS Opinion & Social (2010a), pp. 18–32. Gaskell (2010), p. 40 shows that relatively high support of GM food amongst the population is found e.g. in the UK, the Czech Republic, Portugal and Spain (with more than 35% of respondents agreeing that GM food should be encouraged, UK at the top with 44%); countries with low support are e.g. Greece, France, Germany or Austria (23% or below agree that GM food should be encouraged).

[449] Traits offering consumer or environmental benefits and non-food use of GM crops are better accepted, The group of FAS Biotechnology Specialists in the European Union (2018), p. 40.

Main **causes** of the cautious attitude towards GMOs in agriculture in the EU were general cultural and economic grounds,[450] food scandals, above all BSE, in the 1990s[451] and media-effective protests by civil society organisations like Greenpeace[452] such as communication campaigns and destructions of field trials.[453] Often, GMOs are a **symbol for deeper concerns**, e.g. the lack of trust in regulators and the feeling of not being taken seriously by them, discontent with the productivity-based agricultural economy, ethical objections regarding the violation of the "integrity of nature", concerns over imbalances of power in society and scepticism towards technical progress or globalisation in general.[454]

5.11.3.2 Public Opinion on GEOs

Most consumers in the EU have **not yet heard of genome editing**.[455] Therefore, there is no strong consumer position on genome editing, yet. The low level of public awareness has also the consequence that genome edited foods are currently of very little concern to consumers.[456]

Both the mass press and specialised press frequently report on genome editing, also in agriculture.[457] Therefore, some experts expect a public debate about genome editing to break out in the near future.[458]

What could the public opinion on genome editing be like in the **future**?

It is likely that a **high level of unawareness** of genome editing techniques or at least of the differences between them and traditional forms of genetic engineering remains since these differences are highly technical.[459] It is also likely that even people who know the technical differences still **regard genome editing techniques**

---

[450]Cf. Pollack and Shaffer (2009), pp. 73–74; Burchardi (2007), pp. 35–37.

[451]Pollack and Shaffer (2009), pp. 75–76; Devos et al. (2012), p. 10771; Paarlberg et al. (2004), p. 6.

[452]The group of FAS Biotechnology Specialists in the European Union (2018), pp. 39; Lucht (2015), pp. 4257–4258.

[453]Cf. Kuntz (2012).

[454]The group of FAS Biotechnology Specialists in the European Union (2018), pp. 39; Scholderer (2004), pp. 169, 220–221.

[455]The group of FAS Biotechnology Specialists in the European Union (2018), p. 43; examples: 86% of a random sample of German consumers have never heard of genome editing, German Federal Institute for Risk Assessment (2019), p. 7; 90% of the French people polled have not heard of CRISPR-Cas9, Institut d'études Opinion et Marketing (2016), p. 7.

[456]For Germany: German Federal Institute for Risk Assessment (2019), p. 9.

[457]Cf. Le Déaut and Procaccia (2017), p. 286. However, the media coverage differs widely between the member states, cf. the country sections in Stakeholder and Issue Mapping on New Breeding Techniques (2017).

[458]Sciencewise and Nuffield Council on Bioethics (2016), pp. 10–11; Malyska et al. (2016), pp. 530–531.

[459]Hamburger (2018), p. 6; Ishii and Araki (2016), p. 1508.

as forms of GM techniques,[460] not least because GEOs are "officially" classed as GMOs in the EU. This means they include GEOs into their attitude towards GMOs.[461]

In addition, many consumers might reject genome editing techniques for the same reasons that they reject established GM techniques, e.g. because they perceive these techniques as risky[462] or because most GEOs authorised in the EU will be produced by large agrochemical companies like Monsanto/Bayer[463] and can thus be regarded as a symbol for large corporations controlling the agricultural sector.[464]

In short, the **public opinion towards GEOs** will probably be the **same as towards established GMOs**.

Yet, **individual applications offering clear consumer or environmental benefits** might find acceptance.[465] EU consumers often criticise the lack of need for GMOs because in their view the introduced traits do not have such benefits.[466]

---

[460]This is also a result of focus groups on public opinions towards genome editing in Germany, Hopp et al. (2017), p. 21 (in German).

[461]Regarding the activation of existing attitudes in case of similarity of new technologies to existing ones in general Frewer et al. (2011), pp. 453, 454; with respect to genome editing Bruce (2017), p. 390.

[462]Cf. Cardello (2003), p. 218, explaining that new food technologies typically possess many factors that foster consumers' perception of risk. This also applies to genome editing techniques: Consumers cannot judge (without a label) whether a food derives from genome edited plants by inspecting or using it—the perceived risk is unobservable; the technologies are new; the risks are—in the eyes of consumers and as propagated by some NGOs—uncertain and delayed, etc. On top, European consumers tend to be risk adverse regarding new technologies for food production, Pollack and Shaffer (2009), p. 73; Rollin et al. (2011), p. 100. Cf. also Hopp et al. (2017), pp. 27–29 showing that the participants in the focus groups tend to regard not only GM food but also genome edited food as unhealthy.

[463]Cf. the high financial expenditure and regulatory expertise that is necessary to go through the GMO authorisation process; cf. further Nuffield Council on Bioethics (2016), pp. 118–119; European Academies Science Advisory Council and German National Academy of Sciences Leopoldina (2013), p. 29; Malyska et al. (2016), p. 532; Stokstad (2018).

[464]Cf. the concern, mainly by NGOs and the political left, that the control over agriculture and foods is getting more and more into the hands of large agricultural companies ("corporate control"), instead of farmers and consumers ("food sovereignty"), Harriss and Stewart (2015), pp. 45, 54–55; Helliwell et al. (2017), pp. 2091–2092.

[465]As is the case for established GM crops, cf. n. 450. Cf. the acceptance criteria for GM applications (including genome editing applications) analysed in van Mil et al. (2017), pp. ii–iii. Examples of applications that found support amongst the participants in the public dialogue are the use of genome editing to produce cheaper medicines, to produce more nutritious crops to supplement dietary insufficiencies or to protect crops from damage, e.g. through late blight, van Mil et al. (2017), pp. 58, 81–90.

[466]Gaskell (2010), pp. 37–38; Gaskell et al. (2004), p. 193 ("the 'Achilles heel' of GM foods is not so much the misperception of the scientific risks, but rather the perceived absence of benefit for the consumer"); Lucht (2015), p. 4260. With respect to the importance of perceived benefits regarding the acceptance of a new technology in general Rollin et al. (2011), p. 100; Ronteltap et al. (2007), pp. 6–9. Stressing the importance of individual benefits (as distinguished from broader societal and environmental benefits) for individual acceptance Rollin et al. (2011), p. 105.

There are indications that more EU consumers would buy GM products if they delivered consumer benefits other than price reductions, European Commission Directorate-General for

Research into genome editing applications in agriculture, by contrast, is often designed to offer consumer benefits.[467] Examples are oils with reduced trans-fat[468] or low-gluten wheat for people suffering from coeliac disease.[469]

Furthermore, there is a slight possibility that some consumers will distinguish between **transgenic** and **non-transgenic** products in their opinions. According to focus groups and surveys, there is a tendency for consumers to accept cisgenic foods better than transgenic ones.[470] One reason is that plants with genetic changes that could occur naturally are perceived as more natural by some consumers.[471] Thus, GM products using additional voluntary labelling to indicate they are non-transgenic might be better accepted.

Lastly, there is an—even slighter—possibility that a **more positive general view**[472] on genome editing is achieved jointly by characteristics of the genome editing technologies themselves (more naturalness), acceptable first applications[473] (including success stories from research on genome editing in the generally better

Health and Food Safety (2015b), pp. 49–50; King's College London (2008), para. 5–6; with respect to quality being the leading purchasing factor for EU consumers when buying dairy products and meat, far ahead of price TNS Opinion & Social (2014), p. 58.

Cf. also the fear of NGOs that genome editing will only be used "to make rich people richer, not to make the world less hungry or more bio-diverse or more resilient to climate change", Helliwell et al. (2017), p. 2092.

[467]Cf. Ricroch et al. (2017), pp. 170, 177 and Modrzejewski et al. (2018), p. 6 (in German) showing the share of the different breeding goals regarding genome editing applications in agriculture. Food quality plays an important role.

[468]Hilscher et al. (2017b), p. 8; APHIS Deputy Administrator Michael J. Firko (2015), Haun et al. (2014).

[469]Sánchez-León et al. (2017)

[470]Gaskell (2010), pp. 46–50; Lombardo and Zelasco (2016), p. 497.

[471]Gaskell (2010), p. 48 (72% of the respondents agree or tend to agree that transgenic crops are unnatural compared to 52% with respect to cisgenic crops). "Unnaturalness" is one of the main concerns associated with GM food in the EU, Gaskell (2010), pp. 7, 38, 46; Frewer et al. (2011), p. 448. However, there are different types of "unnaturalness" in consumer perception. Crossing species boundaries or more generally biological similarity is just one of them, Andersen et al. (2015), p. 431; Mielby et al. (2013), p. 478. In other terms, genome edited plants could still be perceived as unnatural because of their "unnatural" *production process*, Lucht (2015), p. 4270. Especially consumer and environmental NGOs reject the *production process*, Trans Atlantic Consumer Dialogue (2016), pp. 2, 4 Greenpeace (2015); cf. also Hopp et al. (2017), pp. 21–23 showing that the participants in the focus groups regard GEOs as equally unnatural as GMOs due to their unnatural production process.

[472]Cf. surveys indicating that EU consumers might accept genome edited products a bit better than traditional GM products, e.g. Shew et al. (2018), p. 74 (Belgium, France); Hopp et al. (2017), p. 38 (Germany).

[473]Cf. Bruce (2017), p. 388, explaining that early applications of genome editing might shape opinions towards the technique in general.

accepted medical sector[474]) and public participation[475] as well as "advertisement" for the new technologies by governments[476] and the industry.[477]

5.12 Treatment of Other New Breeding Technologies

5.12.1 Regulatory Status

Organisms developed through **cisgenesis** and **intragenesis** are GMOs, as explained in Sect. 5.3 ("Regulatory Status of Genome Edited Plants").[478] This applies irrespective of whether SDN-3 or traditional forms of genetic engineering (e.g. agrobacterium-mediated transformation) are applied to insert the cisgene. Several field trials on cisgenic and intragenic crops (traditional forms of genetic engineering, i.e. not genome edited) have already been carried out in Europe from 2002 on, e.g. on high-amylopectin potatoes, late blight-resistant potatoes or scab-resistant apples.[479] An application for the placing on the market has been withdrawn, though.[480]

No regulatory decision has yet been made at the EU or national level[481] at the time of writing regarding the GMO classification of organisms developed through

474 E.g. Schwank et al. (2013), Osborn et al. (2015), Kaminski et al. (2016), Cyranoski (2016), van Diemen and Lebbink (2017); cf. also The Netherlands Commission on Genetic Modification (2014), p. 13.

475 At the time of writing, the extent and orientation of societal involvement in genome editing issues differs between the EU member states, cf. the overview of selected countries in Le Déaut and Procaccia (2017), pp. 295ff (in French); cf. also Sciencewise and Nuffield Council on Bioethics (2016). Le Déaut and Procaccia (2017), p. 286 warn that inaction in terms of societal involvement bears the risk that the debate is taken over by the opponents of biotechnology as was the case in the GMO debate.

476 Especially national governments favouring GEOs, e.g. of the UK and the Netherlands, Corporate Europe Observatory (2016b).

477 Cf. the attempts to frame genome editing in a positive "innovation" or "naturalness" context instead of the GMO context, e.g. the communication initiative "Embracing the Power of Nature" or the term "plant breeding innovation" by the European Seed Association, European Seed Association (n.d.).

478 Cf. text to n. 73.

479 European Commission Joint Research Centre (n.d.); Holme et al. (2013), p. 404; Hou et al. (2014), p. 2; Halterman et al. (2016), p. 4.

480 The intragenic potato "Modena", European Food Safety Authority (2018c), Holme et al. (2013), p. 403.

481 The German competent authorities have been asked whether apple trees obtained by accelerated breeding are GMOs but have not yet decided on that issue, German Federal Ministry of Food and Agriculture (2018) (in German).

other new breeding techniques, notably **grafting,**[482] **agro-infiltration,**[483] **RNA-dependent DNA methylation (RdDM), accelerated breeding**[484] **and reverse breeding.**[485] It is possible that the European Commission will issue a guidance document[486] on the scope of the GMO legislation, which might also cover organisms developed through these techniques.[487]

In Detail The main **issues of interpretation** that are still unsettled are (cf. also Table 5.18):

First, are epigenetic alterations[488] alterations of the genetic material within the meaning of the GMO definition?[489] This is mainly relevant for RdDM, agro-infiltration,[490] grafting,[491] and the use of SDNs to make epigenetic changes.[492] Many experts agree that an alteration of the genetic material only means a change to the nucleotide sequence.[493]

Second, is it sufficient for a finding of an alteration of the genetic material that unintended genetic alterations may have been induced?[494] Most new breeding

---

482 Grafting a non-GM scion onto a GM rootstock.

483 Agro-infiltration sensu stricto, i.e. "non-germline tissues [...] are agro-infiltrated in order to obtain localised expression", European Commission New Techniques Working Group (2011), para. 5.5.2.

484 Here defined as a technique, "in which an intermediate plant contains a transgene to accelerate the breeding process, but which is subsequently crossed out and only the null-segregants are used for further breeding"; also referred to as "fast breeding" or "rapid crop cycle breeding", Schiemann and Hartung (2014), pp. 207–208.

485 The understanding of Lusser et al. (2011), pp. 23–27 is adopted with regard to these techniques if not defined separately.

486 Not legally binding, Craig and de Búrca (2015), pp. 109–110; Laaninen (2016), p. 2; Commission guidance documents nevertheless have practical effects or even a de facto binding effect, Snyder (1993), p. 32.

487 Cf. text to n. 407–410.

488 Defined as "changes in gene function that are mitotically and/or meiotically heritable and do not entail a change in DNA sequence", cf. Nap and van Kessel (2011), p. 17; Riggs et al. (1996), p. 1.

489 Cf. European Commission New Techniques Working Group (2011), para. 5.6.4 A; UK Advisory Committee on Releases to the Environment (2013b), pp. 25–26; Vogel (2012), pp. 27–28; The Netherlands Commission on Genetic Modification (2009a), pp. 4, 25; European Food Safety Authority (2015), p. 3.

490 Cf. UK Advisory Committee on Releases to the Environment (2013b), p. 19.

491 Cf. European Commission New Techniques Working Group (2011), para. 5.4.4.

492 Cf. The Netherlands Commission on Genetic Modification (2009a), p. 25. However, most applications of SDNs for targeted gene expression regulation make use of transgenic plants, which are of course GMOs, cf. Hilscher et al. (2017a), pp. 21–22. The applications that are potentially non-GMOs are those inducing changes that are inherited (epigenetic changes), which allows the elimination of the inducer of the change, e.g. the transgene, cf. Nap and van Kessel (2011), p. 12.

493 European Food Safety Authority (2015), p. 3; European Commission New Techniques Working Group (2011), para. 5.6.4 A; more cautious UK Advisory Committee on Releases to the Environment (2013b), p. 26; The Netherlands Commission on Genetic Modification (2006), p. 19.

494 Cf. Vogel (2012), pp. 28, 34.

techniques can induce unintended genetic alterations.[495] Most experts agree that unintended alterations can be disregarded. At least, according to them, this holds true for techniques inducing unintended alterations that are comparable to such induced naturally or by traditional breeding methods.[496] Yet, it is doubtful whether this line of argumentation is still valid[497] in the light of the process-based interpretation of the European Court of Justice in Case C-528/16: The question whether the genetic material is altered through an unnatural process[498] does not depend on whether this alteration is intentional or not.

Third, are organisms necessarily GMOs if use was made of genetic modification in their development (*strictly* process-based interpretation)?[499] In other terms, is it sufficient to use a GM technique in order to classify the resulting organism and all its progeny as GMOs no matter whether they still have the genetically engineered modification? According to experts, "this aspect is not considered in the Directives".[500] Yet, it is relevant for all of the aforementioned techniques. During an intermediate step in the development, they use genetic modification. The end product, i.e. the commercial variety, lacks the genetically engineered modification (i.e. the GM event). It is for example a negative segregant.[501] The interpretational

---

[495]It could be further distinguished whether the unintended alterations are induced by an "unnatural" breeding step, e.g. the intermediate integration of a transgene, or by a traditional breeding step, e.g. tissue culture.

[496]European Commission New Techniques Working Group (2011), para. 4.4.

[497]At the time of writing, no updated expert opinions on the GMO classification were available, yet.

[498]Process-based interpretation of the GMO definition, Directive 2001/18/EC, art. 2(2).

[499]Cf. European Commission New Techniques Working Group (2011), paras. 4.4, 4.5, 5.6.4 A; UK Advisory Committee on Releases to the Environment (2013b), pp. 16–17; The Netherlands Commission on Genetic Modification (2006), p. 14. In fact, there are several cases that might entail different legal treatment, e.g.

- stable integration of a transgene into the plant genome and subsequent crossing out (or other way of removal) (e.g. reverse breeding, accelerated breeding)
- transient introduction of a transgene into the cell, i.e. no integration into the genome and no capacity of autonomous replication (e.g. RdDM, Agro-infiltration)
- unnatural techniques not making use of DNA/RNA at all, e.g. delivery of zinc finger nucleases for epigenetic changes in protein form

cf. Rodriguez-Cerezo (2014), p. 9; German Central Committee on Biological Safety (2012), p. 6; European Commission New Techniques Working Group (2011), paras. 4.4, 4.5; The Netherlands Commission on Genetic Modification (2009a), p. 25.

[500]European Commission New Techniques Working Group (2011), para. 4.4. In fact, two lines of interpretation could be followed regarding the GMO definition: The **strictly process-based interpretation**, according to which any use of genetic engineering techniques results in GMOs and the **less stringent process-based interpretation**, according to which the end product has to have a genetic alteration (1) which has been induced by an unnatural technique (2), cf. Vogel (2012), p. 94. The wording of the GMO definition ("altered...in a way that does not occur naturally") suggests the second interpretation. The judgment of the European Court of Justice in Case C-528/16 does not discuss this subject.

[501]"Plants that are negative segregants lack the transgenic event and can be produced, for example, by self-fertilisation of hemizygous GM plants, or from crosses between hemizygous GM plants and non-GM plants.", European Food Safety Authority (2011b), p. 9 Examples of "negative segregant techniques" are reverse breeding and accelerated breeding.

Table 5.18 Main issues of interpretation regarding other new breeding techniques

Issue	Legal wording	Main opinion on interpretation[a]
Epigenetic alterations	'Genetic material [. . .] altered'	Only change to the nucleotide sequence is an alteration of the genetic material
Possible unintended alterations	'Altered'	Can be disregarded
Use of genetic modification in the development process	'Altered in a way that does not occur naturally'	No GMO if absence of genetically engineered modification can be demonstrated

[a]However, not necessarily followed by the European Court of Justice should a Court case be initiated regarding the issues of interpretation, cf. also the European Court of Justice's deviation from most expert opinions in Case C-528/16, text to n. 55–58

issue is not only relevant for breeding but also e.g. for the question whether negative segregants resulting from commingling on the field are GMOs. Many expert statements express the opinion that the final organism/offspring resulting from the use of GM techniques is not a GMO provided that the GM event, e.g. the foreign DNA, is no longer present.[502] However, the absence of the GM event, e.g. the foreign DNA, has to be demonstrated.[503] Of course, this view, furthermore presupposes that unintentional genetic alterations can be disregarded (see second issue) (Table 5.18).[504]

This leads to the following result (Table 5.19):

Should organisms developed by RdDM, grafting, reverse breeding, accelerated breeding and agro-infiltration be classified as GMOs, the enforcement of the GMO regime will rely solely on documentation as there is generally no way to recognise or detect them.[505]

In sum, some interpretational issues are still unsettled. Therefore, the regulatory status of organisms developed through the various new breeding techniques will likely continue to occupy the EU for some time in the future.

---

[502]European Commission New Techniques Working Group (2011), paras. 4.4, 4.5, 5.6.4 A; German Central Committee on Biological Safety (2012), p. 6; more cautious French Haut Conseil des Biotechnologies (2016a), p. 98 ("should be exempt from risk assessment and could be considered to be a plant obtained by conventional breeding"); cf. also Vogel (2012), pp. 11–12.

[503]In legal terms, the use of a GM technique leads to a rebuttable GMO presumption. According to European Commission New Techniques Working Group (2011), para. 4.4 "[c]lear criteria would be needed to establish whether the 'foreign' genetic material is no longer present in the resulting organism".

[504]Cf. also European Commission New Techniques Working Group (2011), para. 4.4.

[505]Lusser et al. (2011), p. 70; Ribarits et al. (2014), p. 186; Rodriguez-Cerezo (2014), p. 20; The Netherlands Commission on Genetic Modification (2006), p. 14.

Table 5.19 Regulatory classification of organisms developed through other new breeding techniques

	Regulatory classification of intermediate and resulting organism
Grafting on GM-rootstock	Plant: GMO[a] Offspring (e.g. fruit): ?[b]
Agro-infiltration	Agro-infiltrated plant: contains GMMs[c] Offspring[d]: ?[e]
RdDM	Intermediate organisms: GMO[f] Resulting organism: ?[g]
Accelerated breeding	Intermediate organisms: GMO Resulting organism: ?[h]
Reverse breeding	Intermediate organisms: GMO[i] Resulting organism: ?[j]
Mutation by chemicals or radiation	Exempted GMO[k]

[a]European Commission New Techniques Working Group (2011), para. 5.4.5

[b]Most expert opinions: not a GMO, European Commission New Techniques Working Group (2011), para. 5.4.5; French Haut Conseil des Biotechnologies (2016a), p. 97; German Central Committee on Biological Safety (2012), p. 10; indecided UK Advisory Committee on Releases to the Environment (2013b), p. 23 and The Netherlands Commission on Genetic Modification (2006), pp. 22–23

[c]Thus regulated under Directive 2009/41/EC, European Commission New Techniques Working Group (2011), p. 31

[d]In practice not very relevant as most applications make use of the expressed protein, not the progeny of the plant, cf. European Commission New Techniques Working Group (2011), para. 5.5.3

[e]Most expert opinions: not a GMO, European Commission New Techniques Working Group (2011), para. 5.5.4; French Haut Conseil des Biotechnologies (2016a), p. 97; German Central Committee on Biological Safety (2012), p. 11; UK Advisory Committee on Releases to the Environment (2013b), pp. 19–20

[f]Differing European Commission New Techniques Working Group (2011), para. 5.6.4 A if the RNA is directly delivered into the cell without being able to replicate; in this case it could be debated whether the genetic material is "altered" within the meaning of Directive 2001/18/EC, art. 2 (2)

[g]Most expert opinions: not a GMO, European Commission New Techniques Working Group (2011), para. 5.6.4 A; French Haut Conseil des Biotechnologies (2016a), p. 97; German Central Committee on Biological Safety (2012), pp. 12–13

[h]Most expert opinions: not a GMO, Schiemann and Hartung (2014), pp. 207–208

[i]European Commission New Techniques Working Group (2011), para. 5.7.4 A

[j]Most expert opinions: not a GMO, European Commission New Techniques Working Group (2011), para. 5.7.4; German Central Committee on Biological Safety (2012), pp. 13–14; UK Advisory Committee on Releases to the Environment (2013b), p. 17; The Netherlands Commission on Genetic Modification (2006), p. 14

[k]See Directive 2001/18/EC, annex I B(1)

5.12.2 *The Regulatory Framework for Non-GMOs: An Overview*

As it is possible that some plants obtained by other new breeding techniques are not classified as GMOs, in the following, an overview of the general laws applicable to any new plant variety and products obtained from it is provided ("Non-GMO Regime").

5.12.2.1 Contained Use, Field Trials

No authorisation for contained use or field trials of non-GM plants is required. Of course, if GMOs are used during an intermediate step in the breeding process, a GMO contained use/field trial authorisation is required during that step regardless of whether the resulting organism is classified as a GMO or not.

5.12.2.2 Placing on the Market

5.12.2.2.1 Cultivation

In the EU, **variety registration** is mandatory for new varieties.[506] Registration takes place at member state level in compliance with EU harmonised requirements and procedures. Those requirements are, firstly, distinctness,[507] uniformity[508] and stability[509] of the variety (DUS criteria).[510] Secondly, a variety of an agricultural plant

---

[506]For example:

- Varieties of agricultural plant species: Directive 2002/53/EC, art. 3(1); Directive 2003/90/EC of 6 October 2003 setting out implementing measures for the purposes of Article 7 of Council Directive 2002/53/EC as regards the characteristics to be covered as a minimum by the examination and the minimum conditions for examining certain varieties of agricultural plant species [2003] OJ L254/7, art. 1(1)
- Vegetable seed: Directive 2002/55/EC, art. 3(1);
- Fruit: Council Directive 2008/90/EC of 29 September 2008 on the marketing of fruit plant propagating material and fruit plants intended for fruit production [2008] OJ L267/8, art. 7(2), (4);
- Vine: Council Directive 68/193/EEC of 9 April 1968 on the marketing of material for the vegetative propagation of the vine [1968] OJ L93/15, art. 5.

[507]The variety is clearly distinguishable on one or more important characteristics from any other registered variety, cf. Directive 2002/53/EC, art. 5(1).

[508]The plants of which the variety is composed are composed of identical plants, cf. Directive 2002/53/EC, art. 5(3).

[509]The variety maintains its essential characteristics over successive generations, cf. Directive 2002/53/EC, art. 5(2).

[510]For example: Varieties of agricultural plant species: Directive 2002/53/EC, art. 4(1); vegetable seed: Directive 2002/55/EC, art. 4(1). The same criteria are internationally relevant for variety protection, cf. e.g. International Union for the Protection of New Varieties of Plants (1991), art. 5ff.

species has to be of a satisfactory value for cultivation (VCU criterion).[511] Testing is carried out by official variety testing institutions in the respective member state[512] and takes 2–3 years.[513] Costs are low and mostly shared or fully paid by public authorities.[514] Member states register approved varieties in their national catalogues.[515] The variety is then also included in the corresponding European Common Variety Catalogue and, from that point on, marketable throughout the EU.[516]

Furthermore, seed and plant propagating material can only be marketed after is has been **officially certified**.[517] Certification shows that the identity, health and quality standards set in the seed Directives are met.

Moreover, **plant health requirements** have to be met[518]: For the introduction of living plant material (fruit, seeds, etc.) into the EU from non-EU countries, phytosanitary certificates[519] are generally required.[520] For movements of plants within the EU, plant passports[521] might be required.[522] Prohibitions, restrictions and special provisions apply for plants that are listed as pests[523] or as plants likely to host pests.[524]

---

[511]"Its qualities [. . .] offer [. . .] a clear improvement [. . .] for cultivation or [. . .] uses which can be made of the crops or the products derived therefrom", Directive 2002/53/EC, arts. 4(1), 5(4); criteria the value is based on are yield, resistance to harmful organisms, response to the environment and quality characteristics, cf. Directive 2003/90/EC, annex III.

[512]Cf. e.g. regarding varieties of agricultural plant species Directive 2002/53/EC, art. 7(1).

[513]German Federal Plant Variety Office (2017), p. 24; Madre and Agostino (2017).

[514]Cf. Madre and Agostino (2017).

[515]For example: Varieties of agricultural plant species: Directive 2003/90/EC, art. 3(1); Directive 2002/53/EC, art. 3(1); vegetable seed: Directive 2002/55/EC, art. 3(2).

[516]For example: Varieties of agricultural plant species: Directive 2002/53/EC, arts. 16(1), 17; vegetable seed: Directive 2002/55/EC, arts. 3(3), 16(1), 17. Regarding the plant variety catalogues, databases and information systems see European Commission (n.d.-i).

[517]Example: Cereal seed: Council Directive 66/402/EEC of 14 June 1966 on the marketing of cereal seed [1966] OJ 125/2309, art. 3(1). Further European Commission (n.d.-d); Black et al. (2006), p. 368. With respect to the requirements for equivalence of seed produced in non-EU countries cf. European Commission (n.d.-c).

[518]Cf. Regulation (EU) 2016/2031 of the European Parliament and of the Council of 26 October 2016 on protective measures against pests of plants, amending Regulations (EU) No 228/2013, (EU) No 652/2014 and (EU) No 1143/2014 of the European Parliament and of the Council and repealing Council Directives 69/464/EEC, 74/647/EEC, 93/85/EEC, 98/57/EC, 2000/29/EC, 2006/91/EC and 2007/33/EC [2016] OJ L317/4. Comes into effect in December 2019, replacing the current Directive 2000/29/EC. Cf. the overview of the provisions of Regulation (EU) 2016/2031 in Schiffers (2017). Cf. further de Jong et al. (2018), pp. 255–256 describing the provisions of the EU's Plant Health Regulation in the context of plants developed through new breeding techniques.

[519]Regulation (EU) 2016/2031, art. 71: "document, issued by a third country, which [. . .] certifies that the plant, plant product or other object concerned complies with all of the following requirements [in the following specified]".

[520]Regulation (EU) 2016/2031, arts. 71–74.

[521]Regulation (EU) 2016/2031, art. 78: "official label for movement of plants, plant products and other objects within the Union territory [. . .] which attests compliance with all requirements [. . .]".

[522]Regulation (EU) 2016/2031, arts. 78ff.

[523]Regarding the different categories of pests see Regulation (EU) 2016/2031, arts. 3, 4, 6, 32, 36.

[524]Regulation (EU) 2016/2031, chapter II–V.

Table 5.20 Summary: regulation and market acceptance of GMOs in the EU

Commercial production; exports; imports	Hardly any GMO cultivation (only 1 GM line grown—GM maize MON810, almost exclusively in Spain; GMO cultivation area amounts to 0.07% of total utilised agricultural area in the EU; 2/3 of the EU member states restricted or banned cultivation of authorised GM crops in their territories); no exports of GMOs/GM products; import of large amounts of GMOs/GM products for feed, mainly GM soybean/soybean meal
Regulatory prerequisites for uses	Authorisation requirements for **contained use**, **field trials** and any **placing on the market**. Placing on the market takes 4–7 years and costs 7–15 million euros (appr. 8.5–18.5 million US dollar); politicised authorisation process (deadlocks, politically motivated objections) causes delays and unpredictable outcomes
Low level presence	No tolerance level for unauthorised GMOs ("zero tolerance" policy)
Labelling	Labelling of products consisting of or containing the GM plant as well as food and animal feed produced from it
Identity preservation (Coexistence)	Detailed and often burdensome rules on coexistence introduced by the member states
Liability	Tightened liability for environmental damages; liability for GM admixture as provided for by the member states
Perception	EU population predominantly sceptical towards GMOs in agriculture (differences between EU member states); only a few GM labelled products in the supermarkets

5.12.2.2.2 Food and Feed

Both food and feed have to comply with general safety requirements.[525] Notably, food must not be injurious to health or unfit for human consumption.[526] In rare cases, the Novel Food authorisation regime[527] might be applicable. This

[525]Food: Regulation (EC) No 178/2002, art. 14; feed: Regulation (EC) No 178/2002, art. 15, Regulation (EC) No 767/2009 of the European Parliament and of the Council of 13 July 2009 on the placing on the market and use of feed, amending European Parliament and Council Regulation (EC) No 1831/2003 and repealing Council Directive 79/373/EEC, Commission Directive 80/511/EEC, Council Directives 82/471/EEC, 83/228/EEC, 93/74/EEC, 93/113/EC and 96/25/EC and Commission Decision 2004/217/EC [2009] OJ L229/1, art. 4. Compliance is ensured by official controls, cf. also the EU Official Controls Regulation, Regulation (EU) 2017/625, harmonising the organisation of official controls.

[526]Regulation (EC) No 178/2002, art. 14(2).

[527]Regulation (EU) 2015/2283 of the European Parliament and of the Council of 25 November 2015 on novel foods, amending Regulation (EU) No 1169/2011 of the European Parliament and of the Council and repealing Regulation (EC) No 258/97 of the European Parliament and of the Council and Commission Regulation (EC) No 1852/2001 [2015] OJ L327/1.

might be[528] the case if, firstly, the plant has been obtained by a non-traditional breeding practice[529] (i.e. a breeding practice not used before 15 May 1997) and, secondly, if the food is significantly changed in composition or structure affecting its nutritional value, metabolism or level of undesirable substances.[530] Furthermore, in specific cases, specific authorisation regimes, e.g. for food or feed additives, or specific compositional requirements[531] apply. In general, no authorisation is required for feed.

5.13 Conclusion

All types of GEOs are GMOs in the EU. The European Court of Justice's judgment in Case C-528/16 in 2018 put an end to more than a decade of discussions about their GMO status. This legal certainty was welcomed by everybody. Legal uncertainty, however, remains for organisms developed through other new breeding techniques, e.g. grafting on a GM rootstock, RdDM or reverse breeding.

The GMO classification of all types of GEOs is likely to considerably hinder research, production and trade of genome edited crops and products obtained from them[532]: Plants or derived products classified as GMOs are not only subject to the lengthy and costly approval procedure for placing GMOs on the market, which is an important, in some cases even prohibitive barrier.[533] They also face other obstacles such as the EU's zero tolerance policy for unauthorised GMOs, member states' cultivation bans, strict rules on labelling, coexistence and liability and, last but not least, societal and political rejection of GMOs.[534] Indeed, it is not an exaggeration to say that the applicability of the regulatory framework for GMOs might prevent the successful adoption of genome editing in agriculture in the EU (Table 5.20).[535]

Acknowledgment The research for this article was conducted as part of the project "Genome editing in plant biotechnology – a science based legal analysis of regulatory problems" funded by

[528]So far, no decision has been made in this regard. The Novel Food definition includes several ambiguities, thus its interpretation is not straightforward.

[529]The legal term "propagating practice" probably includes breeding practices, cf. Voigt and Klima (2017), pp. 324–325.

[530]Regulation (EU) 2015/2283, art. 3(2)(a)(iv).

[531]E.g. foods for specific groups, cf. Regulation (EU) No 609/2013 of the European Parliament and of the Council of 12 June 2013 on food intended for infants and young children, food for special medical purposes, and total diet replacement for weight control and repealing Council Directive 92/52/EEC, Commission Directives 96/8/EC, 1999/21/EC, 2006/125/EC and 2006/141/EC, Directive 2009/39/EC of the European Parliament and of the Council and Commission Regulations (EC) No 41/2009 and (EC) No 953/2009 [2013] OJ L181/35; European Commission (n.d.-e).

[532]Cf. Lappin (2018a), pp. 4–5; European Seed Association (2017b), pp. 2–3.

[533]Kalaitzandonakes et al. (2007), pp. 509, 511; Baulcombe et al. (2014), p. 34.

[534]The group of FAS Biotechnology Specialists in the European Union (2017), pp. 22–25.

[535]Hartung and Schiemann (2014), p. 745.

the German Federal Ministry of Education and Research; principal investigator Professor Dr. Hans-Georg Dederer, University of Passau, Head of Chair of Constitutional and Administrative Law, Public International Law, European and International Economic Law.

The authors would like to thank their colleague Katharina Schreiber for her valuable research for Sect. 5.1 ("Introduction"). They would also like to thank the other team members, Sabrina Brzezinski, Clemens Dienstbier, Sebastian Graup and David Hamburger as well as the project coordinator Professor Dr. Dederer for their support.

References[536]

Abbott A (2015) Europe's genetically edited plants stuck in legal limbo. Nature 528:319–320. https://doi.org/10.1038/528319a

Agir pour l'Environnement, Les Amis de la Confédération paysanne, Branche Inra de SUD-Recherche EPST, Collectif vigilance OGM et Pesticides 16, Comité de recherche et d'information indépendantes sur le génie génétique, Comité de soutien aux Faucheurs Volontaires 49, Conféderation Paysanne, Fédération nationale d'agriculture biologique des régions de France, Foll'avoine, France Nature Environnement, Générations Futures, Groupe international d'études transdisciplinaires, Nature & Progrès, OGM Dangers, Réseau Semences Paysannes, Sciences Citoyennes, Union nationale de l'apiculture française (2017) Pas d'OGM dans nos champs ni dans nos assiettes, ni anciens, ni nouveaux !: Lettre ouverte à Monsieur Nicolas Hulot, ministre de la Transition écologique et solidaire, par un collectif de scientifiques et d'organisations. http://www.fnab.org/actualites/actualites-fnab/993-pas-d-ogm-dans-nos-champs-ni-dans-nos-assiettes-ni-anciens-ni-nouveaux. Accessed 13 July 2018

Alemanno A (2013) Public perception of risks under WTO law: a normative perspective. In: van Calster G, Prévost D (eds) Research handbook on environment, health and the WTO. Edward Elgar, Cheltenham, pp 270–303

Andersen MM, Landes X, Xiang W, Anyshchenko A, Falhof J, Østerberg JT, Olsen LI, Edenbrandt AK, Vedel SE, Thorsen BJ, Sandøe P, Gamborg C, Kappel K, Palmgren MG (2015) Feasibility of new breeding techniques for organic farming. Trends Plant Sci 20:426–434. https://doi.org/10.1016/j.tplants.2015.04.011

Andersson M, Turesson H, Nicolia A, Fält A-S, Samuelsson M, Hofvander P (2017) Efficient targeted multiallelic mutagenesis in tetraploid potato (Solanum tuberosum) by transient CRISPR-Cas9 expression in protoplasts. Plant Cell Rep 36:117–128. https://doi.org/10.1007/s00299-016-2062-3

Andersson M, Turesson H, Olsson N, Fält A-S, Ohlsson P, Gonzalez MN, Samuelsson M, Hofvander P (2018) Genome editing in potato via CRISPR-Cas9 ribonucleoprotein delivery. Physiol Plant. https://doi.org/10.1111/ppl.12731

Andriukaitis V (2018) Reply of commissioner Andriukaitis to open letter: l'application aux nouvelles techniques de mutagenèse de la législation européenne sur les OGM. https://cdn.website-editor.net/ed25e686182040aeb41d3b3d05cc2cd2/files/uploaded/Letter%2520to%2520WGG_AFBV.pdf. Accessed 1 Dec 2018

Aparicio A (2017) Labelling requirements of GM food as compared to those of other foods. In: Escajedo San-Epifanio L (ed) Towards a new regulatory framework for GM crops in the European Union: scientific, ethical, social and legal issues and the challenges ahead. Wageningen Academic Publishers, Wageningen, pp 77–85

APHIS Deputy Administrator Michael J. Firko (2015) Confirmation that FAD3KO Soybean is not a regulated article. https://www.aphis.usda.gov/biotechnology/downloads/reg_loi/15-071-01air.pdf. Accessed 22 July 2018

[536]Legislation and court cases are cited in footnotes and not included in the bibliography.

Asociación Nacional de Obtentores Vegetales (n.d.) Guía Técnica y de Buenas Prácticas para el Cultivo de maíz Bt. http://web.anove.es/media/Guia-maiz-Bt-2016.pdf. Accessed 22 Aug 2018
Bar Cv (1998) The common European law of torts: Volume I: The core areas of tort law, its approximation in Europe, and its accommodation in the legal system. The common European law of torts, by Christian von Bar; vol 1. Clarendon Press, Oxford
Bar Cv (2000) The common European law of torts: Volume II: damage and damages, liability for and without personal misconduct, causality, and defences. The common European law of torts, by Christian von Bar; vol 2. Beck, München
Barns E (2018) The environmental liability directive: legal background and requirements. In: Lipton J, Özdemiroğlu E, Chapman D, Peers J (eds) Equivalency methods for environmental liability: assessing damage and compensation under the european environmental liability directive. Springer Netherlands, Dordrecht, pp 3–20
Bartsch D, Bendiek J, Braeuning A, Ehlers U, Dagand E, Duensing N, Fladung M, Franz C, Groeneveld E, Grohmann L, Habermann D, Hartung F, Keilwagen J, Leggewie G, Matthies A, Middelhoff U, Niemann H, Petersen B, Scheepers A, Schenkel W, Sprink T, Stolz A, Tebbe C, Wahler D, Wilhelm R (2018) Wissenschaftlicher Bericht zu den neuen Techniken in der Pflanzenzüchtung und der Tierzucht und ihren Verwendungen im Bereich der Ernährung und Landwirtschaft: - überarbeitete Fassung vom 23.02.2018. https://www.bmel.de/SharedDocs/Downloads/Landwirtschaft/Pflanze/GrueneGentechnik/Bericht_Neue_Zuechtungstechniken.pdf;jsessionid=F87980BA2DA64034A1B22375A55361D5.2_cid288?__blob=publicationFile. Accessed 7 Aug 2018
Baulcombe D, Dunwell J, Jones J, Pickett J, Puigdomenech P (2014) GM Science Update: A report to the Council for Science and Technology. https://www.gov.uk/government/uploads/system/uploads/attachment_data/file/292174/cst-14-634a-gm-science-update.pdf. Accessed 14 Mar 2018
BBC News (2018) The UK and EU agree terms for Brexit transition period. https://www.bbc.co.uk/news/uk-politics-43456502. Accessed 1 May 2018
Beckmann V, Soregaroli C, Wesseler J (2014) Coexistence. In: Smyth SJ, Phillips PWB, Castle D (eds) Handbook on agriculture, biotechnology and development. Edward Elgar, Cheltenham, pp 372–391
Bergkamp L, van Bergeijk A (2013) Exceptions and defences. In: Bergkamp L, Goldsmith B (eds) EU environmental liability directive: a commentary. Oxford University Press, pp 80–94
Bergmann N, Dederer H-G (2012) Verordnung (EG) Nr. 1829/2003. In: Herdegen M, Dederer H-G (eds) Internationales Biotechnologierecht: Gentechnik, Biopatente, genetische Ressourcen, vol 42. Müller, Heidelberg
Bernauer T, Aerni P (2008) Trade conflict over genetically modified organisms. In: Gallagher KP (ed) Handbook on trade and the environment. Elgar, Cheltenham, pp 183–193
Bioökonomie.de (2018) Surprise ruling by ECJ. https://biooekonomie.de/en/nachrichten/surprise-ruling-ecj. Accessed 31 Aug 2018
Bioökonomierat (2018) Genome Editing: Europa benötigt ein neues Gentechnikrecht: Vorläufige Version. http://biooekonomierat.de/fileadmin/Publikationen/berichte/BOER-Memo_Genome-Editing.pdf. Accessed 31 Aug 2018
Black M, Bewley JD, Halmer P (eds) (2006) The encyclopedia of seeds: science, technology and uses. CABI, Wallingford
Bobek M (2018) Opinion of Advocate General Bobek: Case C-528/16, Confédération paysanne and Others. ECLI:EU:C:2018:20
Border P, Walker J (2017) New Plant Breeding Techniques: POSTnote Number 548. https://researchbriefings.parliament.uk/ResearchBriefing/Summary/POST-PN-0548#fullreport. Accessed 27 Jan 2019
Brookes G (2018) UK plant genetics: a regulatory environment to maximise advantage to the UK economy post Brexit: Briefing paper. https://pgeconomics.co.uk/pdf/abc%20brexit%20report%20final.pdf. Accessed 27 Jan 2019
Bruce A (2017) Genome edited animals: Learning from GM crops? Transgenic Res 26:385–398. https://doi.org/10.1007/s11248-017-0017-2

Brueller W, Hartmann J, Hochegger R, Leonhardt C, Mechtler K, Peterseil V, Ribarits A, Söllinger J, Stepanek W, Widhalm I, Woegerbauer M (2012) Cisgenesis: A report on the practical consequences of the application of novel techniques in plant breeding. https://www.bmgf.gv.at/cms/home/attachments/6/6/0/CH1052/CMS1352183689337/cisgenesis_20121105.pdf. Accessed 31 Aug 2018
Brüggemeier G (2015) Tort law in the European Union. Kluwer Law International
Bruins M (2017) Innovation in EU Agriculture. Europeanseed 4:38–39
Bruins M (2018) A great loss for Europe: the impact of the recent ECJ ruling on plant breeding innovation. Europeanseed 5:8
Burchardi J-E (2007) Die Vereinbarkeit der europäischen Vorschriften zur Kennzeichnung gentechnisch veränderter Lebensmittel mit dem Welthandelsrecht, 1st edn. Rechtsfragen der Globalisierung. Duncker & Humblot GmbH, Berlin
Bussani M, Palmer VV (2005) The liability regimes of Europe - their facades and interiors. In: Bussani M, Palmer VV (eds) Pure economic loss in Europe. Cambridge University Press, Cambridge, pp 120–159
Callaway E (2018) CRISPR plants now subject to tough GM laws in European Union. Nature. https://doi.org/10.1038/d41586-018-05814-6
Callebaut S (2015) New developments in modern biotechnology: a survey and analysis of the regulatory status of plants produced through new breeding techniques: Master Thesis. Faculty of Law Ghent University, Ghent
Canadian Biotechnology Action Network (2015) Where in the world are GM crops and foods? The reality of GM crops in the ground and on our plates. Report 1. https://gmoinquiry.ca/wp-content/uploads/2015/03/where-in-the-world-gm-crops-foods.pdf. Accessed 10 March 2018
Cantley M (2007) An overview of regulatory tools and frameworks for modern biotechnology: a focus on agro -food: OECD international futures project on "The bioeconomy to 2030: designing a policy agenda". https://www.oecd.org/futures/longtermtechnologicalsocietalchallenges/40926623.pdf. Accessed 2 March 2018
Cardello AV (2003) Consumer concerns and expectations about novel food processing technologies: effects on product liking. Appetite 40:217–233. https://doi.org/10.1016/S0195-6663(03)00008-4
Castillo MM, van Boom WH (2008) Economic loss caused by GMOs in the Netherlands. In: Koch BA (ed) Economic loss caused by genetically modified organisms: liability and redress for the adventitious presence of GMOs in Non-GM crops. Springer, Vienna, pp 347–360
Cheyne I (2012) Consumer labelling in EU and WTO law. In: Gaines SE (ed) Liberalising trade in the EU and the WTO: a legal comparison. Cambridge University Press, Cambridge, pp 309–332
Cibus Europe (2015) Letter to DG SANTE. https://corporateeurope.org/sites/default/files/attachments/12.pdf. Accessed 24 Feb 2018
Cibus™ (2017) Cibus SU CanolaTM. http://www.cibuscanola.com/. Accessed 2 Aug 2018
Cibus™ (2018) Cibus crops. https://www.cibus.com/crops.php
Clancy KA (2017) The politics of genetically modified organisms in the United States and Europe. Springer, Cham
Clarke M (2018) A major upset for the EU plant breeding community. Europeanseed 5:6
Codex Alimentarius Commission (2003a) Guideline for the Conduct of Food Safety Assessment of Foods Derived from Recombinant-DNA Plants: CAC/GL 45-2003. http://www.fao.org/fao-who-codexalimentarius/codex-texts/guidelines/en/. Accessed 31 Aug 2018
Codex Alimentarius Commission (2003b) Principles for the Risk Analysis of Foods Derived from Modern Biotechnology: CAC/GL 44-2003. http://www.fao.org/fao-who-codexalimentarius/codex-texts/guidelines/en/. Accessed 12 Mar 2018
Codex Alimentarius Commission (2006) Principles for Traceability/Product Tracing as a Tool Within a Food Inspection and Certification System: CAC/GL 60-2006. http://www.fao.org/fao-who-codexalimentarius/codex-texts/guidelines/en/. Accessed 4 Aug 2018
Codex Alimentarius Commission (2010) Guidelines on performance criteria and validation of methods for detection, identification and quantification of specific DNA sequences and specific

proteins in foods: CAC/GL 74-2010. http://www.fao.org/fao-who-codexalimentarius/codex-texts/guidelines/en/. Accessed 16 Aug 2018

Copa-Cogeca (2017) Press Release: Copa and Cogeca underline urgent need to develop new and better plant varieties using new breeding techniques to deal with increasing challenges like climate change. http://copa-cogeca.be/Main.aspx?page=Archive&lang=en. Accessed 23 Feb 2018

Corporate Europe Observatory (2016a) Biotech lobby's push for new GMOs to escape regulation. https://corporateeurope.org/sites/default/files/attachments/biotechlobbies.pdf. Accessed 30 Jan 2018

Corporate Europe Observatory (2016b) Of apples and potatoes: the Dutch lobby for the deregulation of cisgenesis. https://corporateeurope.org/food-and-agriculture/2016/02/apples-and-pota toes-dutch-lobby-deregulation-cisgenesis. Accessed 22 July 2018

Corporate Europe Observatory (2016c) US company railroads EU decision-making on new GM. https://corporateeurope.org/food-and-agriculture/2016/02/us-company-railroads-eu-deci sion-making-new-gm#footnote18_butybpp. Accessed 15 Feb 2018

Corporate Europe Observatory (2018) ECJ ruling on gene editing products: victory for consumers, farmers, environment. https://corporateeurope.org/pressreleases/2018/07/ecj-ruling-gene-editing-products-victory-consumers-farmers-environment. Accessed 31 Aug 2018

Cotter J, Zimmermann D, van Bekkem H (2015) Application of the EU and Cartagena definitions of a GMO to the classification of plants developed by cisgenesis and gene-editing techniques: Greenpeace Research Laboratories Technical Report (Review) 07-2015. http://www.greenpeace.to/greenpeace/wp-content/uploads/2015/11/Application-of-GMO-definitions-to-plants-developed-by-cisgenesis-and-gene-editing-techniques.pdf. Accessed 29 Aug 2017

Council of the European Union (2019) Outcome of the council meeting: 3689th council meeting. Agriculture and Fisheries. Brussels, 14 May 2019. https://www.consilium.europa.eu/media/39365/st09271-en19.pdf. Accessed 7 June 2019

Court of Justice of the European Union (2018) Organisms obtained by mutagenesis are GMOs and are, in principle, subject to the obligations laid down by the GMO Directive: Press Release No 111/18. https://curia.europa.eu/jcms/upload/docs/application/pdf/2018-07/cp180111en.pdf. Accessed 3 Aug 2018

Craig PP, de Búrca G (2015) EU law: text, cases, and materials, 6th edn. Oxford University Press, Oxford

Craig W, Tepfer M, Degrassi G, Ripandelli D (2008) An overview of general features of risk assessments of genetically modified crops. Euphytica 164:853–880. https://doi.org/10.1007/s10681-007-9643-8

Cronin J, Stone A (2018) European Union Issues Crucial Ruling on Regulating Gene-Edited Organisms as GMOs. https://cspinet.org/news/european-union-issues-crucial-ruling-regulat ing-gene-edited-organisms-gmos-20180726. Accessed 23 Aug 2018

Custers R (2016) When is an organism subject to the provisions of the EU GMO legislation? An in-depth analysis. http://www.vib.be/en/about-vib/organization/Documents/rc_bvl_2016_00533%20GMO%20definition%20legal%20analysis_final.pdf. Accessed 29 Aug 2018

Cyranoski D (2016) CRISPR gene-editing tested in a person for the first time. Nature 539:479. https://doi.org/10.1038/nature.2016.20988

D'Halluin K, Vanderstraeten C, van Hulle J, Rosolowska J, van den Brande I, Pennewaert A, D'Hont K, Bossut M, Jantz D, Ruiter R, Broadhvest J (2013) Targeted molecular trait stacking in cotton through targeted double-strand break induction. Plant Biotechnol J 11:933–941. https://doi.org/10.1111/pbi.12085

Davison J, Ammann K (2017) New GMO regulations for old: determining a new future for EU crop biotechnology. GM Crops Food 8:13–34. https://doi.org/10.1080/21645698.2017.1289305

de Jong P, Bertolotto E, de Seze I (2018) From farm to fork: the regulatory status of Non-GMO plant innovations under current EU law. Bio-Sci Law Rev 16:251–266

de Visser CLM, Schreuder R, Stoddard F (2014) The EU's dependency on soya bean import for the animal feed industry and potential for EU produced alternatives. OCL 21:D407. https://doi.org/10.1051/ocl/2014021

Dederer H-G (2016a) Genetic technology and food security: country report – Germany. In: Norer R (ed) Genetic technology and food safety. Springer, Heidelberg, pp 73–124

Dederer H-G (2016b) The challenge of regulating genetically modified organisms in the European Union: trends and issues. In: Nakanishi Y (ed) Contemporary issues in environmental law: the EU and Japan. Springer, Heidelberg, pp 139–168

Dederer H-G (2019) WTO law and International trade in agricultural genome edited products. In: Spranger TM (ed) Genome editing under gene technology law (forthcoming)

Devos Y, Reheul D, de Waele D, van Speybroeck L (2006) The interplay between societal concerns and the regulatory frame on GM crops in the European Union. Environ Biosaf Res 5:127–149. https://doi.org/10.1051/ebr:2007002

Devos Y, Lheureux K, Schiemann J (2010) Regulatory oversight and safety assessment of plants with novel traits. In: Kempken F, Jung C (eds) Genetic modification of plants: agriculture, horticulture and forestry. Springer, Berlin, pp 553–574

Devos Y, Craig W, Schiemann J (2012) Transgenic crops, risk assessment and regulatory framework in the European Union. In: Meyers RA (ed) Encyclopedia of sustainability science and technology. Springer, New York, pp 10765–10796

Di Fabio U, Kreiner S (2003) § 63 Bio- und Gentechnik. In: Badura P, Rengeling H-W (eds) Handbuch zum europäischen und deutschen Umweltrecht: Eine systematische Darstellung des europäischen Umweltrechts mit seinen Auswirkungen auf das deutsche Recht und mit rechtspolitischen Perspektiven, 2nd edn. Heymanns, Köln, pp 678–762

Dillen K, Rizov I, Rodriguez-Cerezo E (2016) Developing solutions for coexistence in the EU — legal, technical, and economic issues. In: Kalaitzandonakes NG, Phillips PWB, Wesseler J, Smyth SJ (eds) The coexistence of genetically modified, organic and conventional foods: Government policies and market practices. Springer, New York, pp 63–70

Downing E, Coe S (2018) Brexit: Future UK Agriculture Policy: Briefing Paper Number CBP 8218. http://researchbriefings.files.parliament.uk/documents/CBP-8218/CBP-8218.pdf. Accessed 27 Jan 2019

Duensing N, Sprink T, Parrott WA, Fedorova M, Lema MA, Wolt JD, Bartsch D (2018) Novel features and considerations for ERA and regulation of crops produced by genome editing. Front Bioeng Biotechnol, 6. https://doi.org/10.3389/fbioe.2018.00079

Duikers J (2006) EG-Umwelthaftungsrichtlinie und deutsches Recht: Struktur der Richtlinie und Hinweise für die Umsetzung. Natur und Recht 28:623–631. https://doi.org/10.1007/s10357-006-1117-6

Eckerstorfer M, Miklau M, Gaugitsch H (2014) New plant breeding techniques and risks associated with their application. REP-0477. Report commissioned by the Swiss Federal Ethics Committee on Non-Human Biotechnology (ECNH). http://www.umweltbundesamt.at/fileadmin/site/publikationen/REP0477.pdf

EcoNexus, European Beekeeping Coordination, CEO, Friends of the Earth Europe, GeneWatch UK, Greenpeace, Test Biotech, European Coordination Via Campesina (2015) Joint Open Letter: Open Letter to the Commission on New Genetic Engineering Methods. http://www.genewatch.org/uploads/f03c6d66a9b354535738483c1c3d49e4/New_Breeding_Techniques___Open_Letter_27_Jan_2015.pdf. Accessed 24 Sept 2018

Eriksson D (2018) The Swedish policy approach to directed mutagenesis in a European context. Physiologia Plantarum. https://doi.org/10.1111/ppl.12740

Eriksson D, Ammann KH (2016) A universally acceptable view on the adoption of improved plant breeding techniques. Front Plant Sci 7:1. https://doi.org/10.3389/fpls.2016.01999

Eriksson D, Brinch-Pedersen H, Chawade A, Holme IB, Hvoslef-Eide TAK, Ritala A, Teeri TH, Thorstensen T (2017) Scandinavian perspectives on plant gene technology: applications, policies and progress. Physiol Plant 162:219–238. https://doi.org/10.1111/ppl.12661

Eriksson D, Harwood W, Hofvander P, Jones H, Rogowsky P, Stöger E, Visser RGF (2018a) A welcome proposal to amend the GMO legislation of the EU. Trends Biotechnol. https://doi.org/10.1016/j.tibtech.2018.05.001

Eriksson D, de Andrade E, Bohanec B, Chatzopolou S, Defez R, Eriksson NL, van der Meer P, van der Meulen B, Ritala A, Sági L, Schiemann J, Twardowski T, Vaněk T (2018b) Why the European Union needs a national GMO opt-in mechanism. Nat Biotechnol 36:18–19. https://doi.org/10.1038/nbt.4051

European Academies Science Advisory Council (2015) New breeding techniques: statement. https://easac.eu/fileadmin/PDF_s/reports_statements/Easac_14_NBT.pdf. Accessed 24 July 2018

European Academies Science Advisory Council; German National Academy of Sciences Leopoldina (2013) Planting the future: opportunities and challenges for using crop genetic improvement technologies for sustainable agriculture. EASAC policy report, vol 21. EASAC Secretariat; German National Academy of Sciences Leopoldina, Halle (Saale)

European Coexistence Bureau (n.d.) Best practice documents. http://ecob.jrc.ec.europa.eu/documents.html. Accessed 28 Feb 2018

European Commission (2000) Communication from the Commission on the precautionary principle: COM(2000) 1 final. https://eur-lex.europa.eu/legal-content/EN/TXT/PDF/?uri=CELEX:52000DC0001&rid=1. Accessed 29 Aug 2018

European Commission (2001) Proposal for a Regulation of the European Parliament and of the Council concerning traceability and labelling of genetically modified organisms and traceability of food and feed products produced from genetically modified organisms and amending Directive 2001/18/EC, Explanatory Memorandum. OJ C304E/327

European Commission (2002) Proposal for a Directive of the European Parliament and of the Council on environmental liability with regard to the prevention and remedying of environmental damage, Explanatory Memorandum: COM(2002) 17 final. https://eur-lex.europa.eu/legal-content/EN/TXT/PDF/?uri=CELEX:52002PC0017&from=EN. Accessed 30 July 2018

European Commission (2003) Commission Recommendation 2003/556 EC of 23 July 2003 on guidelines for the development of national strategies and best practices to ensure the co-existence of genetically modified crops with conventional and organic farming: C(2003) 2624. OJ L189/36

European Commission (2009a) Commission Staff Working Document accompanying Report from the Commission to the Council and the European Parliament on the coexistence of genetically modified crops with conventional and organic farming. Implementation of national measures on the coexistence of GM crops with conventional and organic farming {COM(2009) XXX final}: SEC(2009) 408 final. https://ec.europa.eu/agriculture/sites/agriculture/files/gmo/coexistence/com2009_153_annex_en.pdf. Accessed 28 Feb 2018

European Commission (2009b) Report from the Commission to the Council and the European Parliament on the coexistence of genetically modified crops with conventional and organic farming: COM(2009) 153 final. https://eur-lex.europa.eu/legal-content/EN/TXT/PDF/?uri=CELEX:52009DC0153&rid=1. Accessed 29 Aug 2018

European Commission (2010a) A decade of EU-funded GMO research: (2001–2010). Project Information. Publ. Office, Luxemburg

European Commission (2010b) Commission Recommendation of 13 July 2010 on guidelines for the development of national co-existence measures to avoid the unintended presence of GMOs in conventional and organic crops. OJ C200/1

European Commission (2015a) Fact Sheet: Questions and Answers on EU's policies on GMOs. http://europa.eu/rapid/press-release_MEMO-15-4778_en.htm. Accessed 10 Mar 2018

European Commission (2015b) GMOs: EU decision-making process explained. https://ec.europa.eu/food/sites/food/files/plant/docs/gmo_auth_decision-making-process.pdf. Accessed 7 Mar 2018

European Commission (2015c) Proposal for a Regulation of the European Parliament and of the Council amending Regulation (EC) No 1829/2003 as regards the possibility for the Member States to restrict or prohibit the use of genetically modified food and feed on their territory: COM(2015) 177 final. http://eur-lex.europa.eu/legal-content/EN/TXT/PDF/?uri=CELEX:52015PC0177&from=EN. Accessed 22 Jan 2018

European Commission (2016) Genetically modified commodities in the EU: Commission Staff Working Document. SWD(2016) 61 final. https://ec.europa.eu/transparency/regdoc/rep/10102/2016/EN/10102-2016-61-EN-F1-1.PDF. Accessed 15 Aug 2018
European Commission (2017) Conference Booklet: Modern Biotechnologies in Agriculture: Paving the way for responsible innovation. 28 September 2017. https://ec.europa.eu/food/sites/food/files/safety/docs/20170928_modern_biotech_conference-booklet.pdf. Accessed 9 July 2018
European Chemicals Agency (2017) Registration. Version 3.0. https://echa.europa.eu/documents/10162/23036412/nutshell_guidance_registration_en.pdf/9fda505f-972d-4bbb-b22c-bd71b6941571. Accessed 6 March 2018
European Commission (2018a) Agri-food trade in 2017: another record year for EU agri-food trade: Monitoring Agri-trade Policy MAP 2018-1. https://ec.europa.eu/info/sites/info/files/food-farming-fisheries/news/documents/agricultural-trade-report_map2018-1_en.pdf. Accessed 26 Aug 2018
European Commission (2018b) Fact Sheets on the European Union: The European Commission. http://www.europarl.europa.eu/factsheets/en/sheet/25/the-european-commission. Accessed 31 Aug 2018
European Commission Joint Research Centre (2019a) Notification report B/GB/19/R08/01. http://gmoinfo.jrc.ec.europa.eu/gmp_report.aspx?CurNot=B/GB/19/R08/01. Accessed 26 May 2019
European Commission Joint Research Centre (2019b) Notification report B/BE/19/V1. http://gmoinfo.jrc.ec.europa.eu/gmp_report.aspx?CurNot=B/BE/19/V1. Accessed 26 May 2019
European Commission (n.d.-a) Authorisation procedures - The centralised procedure. https://ec.europa.eu/health/authorisation-procedures-centralised_en
European Commission (n.d.-b) Coexistence of genetically modified crops with conventional and organic agriculture. https://ec.europa.eu/agriculture/gmo/coexistence_en. Accessed 28 Feb 2018
European Commission (n.d.-c) Equivalence requirements for non-EU countries. https://ec.europa.eu/food/plant/plant_propagation_material/equivalence_requirements_non-eu_en. Accessed 31 Aug 2018
European Commission (n.d.-d) EU marketing requirements. https://ec.europa.eu/food/plant/plant_propagation_material/legislation/eu_marketing_requirements_en. Accessed 31 Aug 2018
European Commission (n.d.-e) Foods for specific groups. https://ec.europa.eu/food/safety/labelling_nutrition/special_groups_food_en. Accessed 21 Feb 2018
European Commission (n.d.-f) GMO authorisations for cultivation. https://ec.europa.eu/food/plant/gmo/authorisation/cultivation_en. Accessed 29 Aug 2018
European Commission (n.d.-g) GMO Register. https://ec.europa.eu/food/plant/gmo/eu_register_en. Accessed 29 Aug 2018
European Commission (n.d.-h) Modern Biotechnologies in the Agri-food Sector. https://ec.europa.eu/food/plant/gmo/modern_biotech_en. Accessed 10 Mar 2018
European Commission (n.d.-i) Plant variety catalogues, databases & information systems. https://ec.europa.eu/food/plant/plant_propagation_material/plant_variety_catalogues_databases_en. Accessed 9 July 2018
European Commission (n.d.-j) Restrictions of geographical scope of GMO applications/authorisations: Member States demands and outcomes. https://ec.europa.eu/food/plant/gmo/authorisation/cultivation/geographical_scope_en. Accessed 29 Aug 2018
European Commission (2012) Report from the commission to the European Parliament and the council on the application of council regulation (EC) No 834/2007 on organic production and labelling of organic products: COM(2012) 212 final. http://ec.europa.eu/transparency/regdoc/rep/1/2012/EN/1-2012-212-EN-F1-1.Pdf. Accessed 27 May 2018
European Commission Directorate-General for Health and Food Safety (2015a) Letter to Competent Authorities. https://corporateeurope.org/sites/default/files/attachments/18._2015.06.15_lettre_autorites_competentes_redacted.pdf. Accessed 15 Feb 2018
European Commission Directorate-General for Health and Food Safety (2015b) State of play in the EU on GM-free food labelling schemes and assessment of the need for possible harmonisation: final report. Publications Office, Luxembourg
European Commission Directorate-General for Health and Food Safety (2018) Notice to stakeholders: withdrawal of the United Kingdom and EU rules on genetically modified food and feed

and the deliberate release of genetically modified organisms into the environment. https://ec.europa.eu/food/sites/food/files/plant/docs/notice_brexit_gmo.pdf. Accessed 27 Jan 2019

European Commission EU Science Hub (2016) GMOs. https://ec.europa.eu/jrc/en/research-topic/gmos. Accessed 25 Feb 2018

European Commission Joint Research Centre (n.d.) Deliberate Release and Placing on the EU Market of GMOs - GMO Register: Deliberate release into the environment of plants GMOs for any other purposes than placing on the market (experimental releases). http://gmoinfo.jrc.ec.europa.eu/gmp_browse.aspx. Accessed 15 Aug 2018

European Commission New Techniques Working Group (2011) Final Report. Unofficial final report, available as a leaked document. http://www.keinegentechnik.de/fileadmin/files/Infodienst/Dokumente/2011_Report_of_the_Working_Group_on_New_Techniques_Final.pdf. Accessed 24 Sept 2018

European Commission Scientific Advice Mechanism Group of Chief Scientific Advisors (2018) Statement by the Group of Chief Scientific Advisors: A Scientific Perspective on the Regulatory Status of Products Derived from Gene Editing and the Implications for the GMO Directive. https://ec.europa.eu/info/sites/info/files/2018_11_gcsa_statement_gene_editing_2.pdf. Accessed 18 Dec 2018

European Commission Scientific Advice Mechanism High Level Group of Scientific Advisors (2017) New techniques in Agricultural Biotechnology: Explanatory Note 02. https://ec.europa.eu/research/sam/pdf/topics/explanatory_note_new_techniques_agricultural_biotechnology.pdf#view=fit&pagemode=none. Accessed 14 Dec 2018

European Food Safety Authority (2009) Scientific opinion on guidance for the risk assessment of genetically modified plants used for non-food or non-feed purposes. EFSA J 7:1164. https://doi.org/10.2903/j.efsa.2009.1164

European Food Safety Authority (2010a) Guidance on the environmental risk assessment of genetically modified plants. EFSA J 8:1879. https://doi.org/10.2903/j.efsa.2010.1879

European Food Safety Authority (2010b) Scientific opinion on statistical considerations for the safety evaluation of GMOs. EFSA J 8:1250. https://doi.org/10.2903/j.efsa.2010.1250

European Food Safety Authority (2011a) Guidance for risk assessment of food and feed from genetically modified plants. EFSA J 9:2150. https://doi.org/10.2903/j.efsa.2011.2150

European Food Safety Authority (2011b) Guidance on selection of comparators for the risk assessment of genetically modified plants and derived food and feed. EFSA J 9:2149. https://doi.org/10.2903/j.efsa.2011.2149

European Food Safety Authority (2012a) Scientific opinion addressing the safety assessment of plants developed through cisgenesis and intragenesis. EFSA J 10:2561–2593. https://doi.org/10.2903/j.efsa.2012.2561

European Food Safety Authority (2012b) Scientific opinion addressing the safety assessment of plants developed using Zinc Finger Nuclease 3 and other Site-Directed Nucleases with similar function. EFSA J 10:2943. https://doi.org/10.2903/j.efsa.2012.2943

European Food Safety Authority (2013) EFSA guidance on the submission of applications for authorisation of genetically modified plants under Regulation (EC) No 1829/2003. EFSA J 11:3491. https://doi.org/10.2903/j.efsa.2013.3491

European Food Safety Authority (2015) Response to Mandate M-2015-0183 ("Request for EFSA to provide technical assistance with regard to issues related to the legal analysis of new plant breeding techniques"). http://registerofquestions.efsa.europa.eu/roqFrontend/questionDocumentsLoader?question=EFSA-Q-2015-00525. Accessed 26 Feb 2018

European Food Safety Authority (2018a) Guidance on uncertainty analysis in scientific assessments. EFSA J 16:54. https://doi.org/10.2903/j.efsa.2018.5123

European Food Safety Authority (2018b) The principles and methods behind EFSA's guidance on uncertainty analysis in scientific assessment. EFSA J 16:1141. https://doi.org/10.2903/j.efsa.2018.5122

European Food Safety Authority (2018c) Register of Questions: EFSA-Q-2009-00552. http://registerofquestions.efsa.europa.eu/roqFrontend/questionLoader?question=EFSA-Q-2009-00552. Accessed 6 Aug 2018

European Food Safety Authority (n.d.) GMO applications: regulations and guidance. https://www.efsa.europa.eu/en/applications/gmo/regulationsandguidance. Accessed 25 Feb 2018

European GMO-Free Regions Network (2018) Berlin declaration: continue with the precautionary approach to GMOs. https://gmofree-eu-network.hessen.de/sites/gmofree-eu-network.hessen.de/files/Deklaration%20Gentechnikfreier%20Regionen%20_0.pdf. Accessed 10 Sept 2018

European Medicines Agency (2008) Guideline on the quality of biological active substances produced by stable transgene expression in higher plants(EMEA/CHMP/BWP/48316/2006). http://www.ema.europa.eu/docs/en_GB/document_library/Scientific_guideline/2009/09/WC500003154.pdf. Accessed 5 Sept 2017

European Network of GMO Laboratories (2017) Detection, Interpretation and Reporting on the presence of authorised and unauthorised genetically modified materials

European Parliament (2014a) European Parliament resolution of 11 March 2014 on the future of Europe's horticulture sector — strategies for growth (2013/2100(INI)): P7_TA(2014)0205. OJ [2017] C378/44

European Network of GMO Laboratories (2019) Detection of food and feed plant products obtained by new mutagenesis techniques. http://gmo-crl.jrc.ec.europa.eu/doc/JRC116289-GE-report-ENGL.pdf. Accessed 1 May 2019

European Parliament (2014b) European Parliament resolution of 16 January 2014 on the proposal for a Council decision concerning the placing on the market for cultivation, in accordance with Directive 2001/18/EC of the European Parliament and of the Council, of a maize product (Zea mays L., line 1507) genetically modified for resistance to certain lepidopteran pests (2013/2974 (RSP)): P7_TA(2014)00. OJ [2016] C482/110

European Parliament (2014c) European Parliament resolution of 25 February 2014 on plant breeding: what options to increase quality and yields? (2013/2099(INI)): P7_TA(2014)0131. OJ [2017] C285/47

European Parliament (2016a) European Parliament resolution of 7 June 2016 on enhancing innovation and economic development in future European farm management (2015/2227 (INI)): P8_TA(2016)0252. OJ [2018] C86/62

European Parliament (2016b) European Parliament resolution of 7 June 2016 on technological solutions for sustainable agriculture in the EU (2015/2225(INI)): P8_TA(2016)0251. OJ [2018] C86/51

European Parliament (2017) Handbook on the ordinary legislative procedure: a guide to how the European Parliament co-legislates. PE 608:827

European Parliamentary Research Service (2017) Preliminary reference procedure: Briefing. http://www.europarl.europa.eu/RegData/etudes/BRIE/2017/608628/EPRS_BRI(2017)608628_EN.pdf. Accessed 9 Mar 2018

European Parliament (2019) European elections results. http://www.europarl.europa.eu/news/en/press-room/elections-press-kit/8/european-elections-results. Accessed 7 June 2019

European Plant Science Organisation (2017) Crop Genetic Improvement Technologies. http://www.epsoweb.org/webfm_send/2263. Accessed 6 Sept 2017

European Seed Association (2012) Position - Presence of EU-approved GMOs in seed. https://www.euroseeds.eu/system/files/publications/files/esa_12.0231_0.pdf. Accessed 4 Mar 2018

European Seed Association (2017a) Embracing the Power of Nature. plantbreeding.eu. Accessed 11 July 2018

European Seed Association (2017b) Position - Plant Breeding Innovation: Applying the latest Plant Breeding Methods for the benefit of sustainable Agriculture, Consumers and Society. https://www.euroseeds.eu/system/files/publications/files/esa_17.0510.pdf. Accessed 16 Mar 2018

European Seed Association (n.d.) Embracing the Power of Nature. https://www.euroseeds.eu/embracing-power-nature. Accessed 11 July 2018

European Union (2017) Council of the European Union. https://europa.eu/european-union/about-eu/institutions-bodies/council-eu_en. Accessed 31 Aug 2018

European Union (2018a) European Commission. https://europa.eu/european-union/about-eu/institutions-bodies/european-commission_en. Accessed 31 Aug 2018

European Union (2018b) European Parliament. https://europa.eu/european-union/about-eu/institutions-bodies/european-parliament_en. Accessed 31 Aug 2018

European Union Reference Laboratory for GM Food and Feed (2017) Description of the EURL GMFF Validation Process. http://gmo-crl.jrc.ec.europa.eu/doc/Description%20of%20the%20EURL%20GMFF%20Validation%20Process.pdf. Accessed 25 Feb 2018

European Union Reference Laboratory for GM Food and Feed (2018). http://gmo-crl.jrc.ec.europa.eu/guidancedocs.htm. Accessed 25 Feb 2018

Eurostat (2018) Extra-EU trade in agricultural goods. http://ec.europa.eu/eurostat/statistics-explained/index.php/Extra-EU_trade_in_agricultural_goods. Accessed 26 Aug 2018

Eustice G (2018) Agriculture: Genetic Engineering: Written question - 131586. https://www.parliament.uk/business/publications/written-questions-answers-statements/written-question/Commons/2018-03-07/131586. Accessed 10 July 2018

FAO/WHO (2000) Safety aspects of genetically modified foods of plant origin: Report of a Joint FAO/WHO Expert Consultation on Foods Derived from Biotechnology, Geneva, Switzerland, 29 May-2 June 2000. http://www.fao.org/fileadmin/templates/agns/pdf/topics/ec_june2000_en.pdf. Accessed 15 Jan 2018

Faure J-D, Napier JA (2018) Europe's first and last field trial of gene-edited plants? Elife 7. https://doi.org/10.7554/eLife.42379

Faure M, Wibisana A (2008) Liability in cases of damage resulting from GMOs: an economic perspective. In: Koch BA (ed) Economic loss caused by genetically modified organisms: liability and redress for the adventitious presence of GMOs in Non-GM crops. Springer, Vienna, pp 531–575

Fedtke J (2008) Economic loss caused by GMOs in Germany. In: Koch BA (ed) Economic loss caused by genetically modified organisms: liability and redress for the adventitious presence of GMOs in Non-GM crops. Springer, Vienna, pp 213–232

Fedtke J (2010) Damage caused by GMOs under German Law. In: Koch BA (ed) Damage caused by genetically modified organisms: comparative survey of redress options for harm to persons, property or the environment. de Gruyter, Berlin, pp 212–246

Ferer BS (2016) The European Commission's GMO opt-out for Member States: a WTO perspective. Eur J Risk Regul 7:187–190. https://doi.org/10.1017/S1867299X00005481

Finnish Centre of Excellence in Molecular Biology of Primary Producers (2016) Arabidopsis genome editing. http://www.primaryproducers.fi/pages/CRISPR_GTLK_EvaMarille.pdf. Accessed 22 July 2018

Fladung M (2016) Cibus' Herbicide-resistant canola in European limbo. Nat Biotechnol:473–474

French Haut Conseil des Biotechnologies (2016a) "Nouvelles Techniques" – "New Plant Breeding Techniques". Première étape de la réflexion du HCB - Introduction générale. Document No 1: Comité scientifique: Note sur les "nouvelles techniques". http://www.hautconseildesbiotechnologies.fr/sites/www.hautconseildesbiotechnologies.fr/files/file_fields/2016/02/04/160204hcb-note-csnpbt.pdf. Accessed 13 July 2018

French Haut Conseil des Biotechnologies (2016b) "Nouvelles Techniques" – "New Plant Breeding Techniques". Première étape de la réflexion du HCB - Introduction générale. Document No 2: Comité économique, éthique et social: Synthèse des contributions et débats relativs aux "nouvelles techniques". http://www.hautconseildesbiotechnologies.fr/fr/system/files/file_fields/2016/03/30/cees_1.pdf. Accessed 13 July 2018

Frewer LJ, Bergmann K, Brennan M, Lion R, Meertens R, Rowe G, Siegrist M, Vereijken C (2011) Consumer response to novel agri-food technologies: implications for predicting consumer acceptance of emerging food technologies. Trends Food Sci Technol 22:442–456. https://doi.org/10.1016/j.tifs.2011.05.005

Fortuna G (2019) 14 EU countries call for 'unified approach' to gene editing in plants. https://www.euractiv.com/section/agriculture-food/news/14-eu-countries-call-for-unified-approach-to-gene-editing-in-plants/. Accessed 7 June 2019

Friant-Perrot M (2010) The European Union regulatory regime for genetically modified organisms and its integration into community food law and policy. In: Bodiguel L, Cardwell M (eds) The regulation of genetically modified organisms: comparative approaches. Oxford University Press, Oxford, pp 79–100

Gaskell G (2010) Europeans and biotechnology in 2010: winds of change?; A report to the European Commission's Directorate-General for Research. Studies and reports. European Commission, Brussels

Gaskell G, Allum N, Wagner W, Kronberger N, Torgersen H, Hampel J, Bardes J (2004) GM foods and the misperception of risk perception. Risk Anal 24:185–194. https://doi.org/10.1111/j.0272-4332.2004.00421.x

Gene Editing in Legal Limbo in Europe (2017) Nature 542:392. https://doi.org/10.1038/542392a

German Association of Biotechnology Industries (2018) Biotechnology industry criticizes the undifferentiated ECJ definition in genetic engineering. https://www.vci.de/langfassungen/langfassungen-pdf/2018-07-25-ecj-judgment-on-genome-editing-blocks-innovations-dib.pdf. Accessed 6 Aug 2018

German Central Committee on Biological Safety (2012) Position statement of the ZKBS on new plant breeding techniques. https://www.bvl.bund.de/ZKBS/EN/04_Allgemeine%20Stellungnahmen/allgemeine_stellungnahmen_node.html. Accessed 25 Aug 2018

German Federal Institute for Risk Assessment (2019) BfR Consumer Monitor: 02/2019. https://www.bfr.bund.de/cm/364/bfr-consumer-monitor-02-2019.pdf. Accessed 01 May 2019

German Federal Ministry of Food and Agriculture (2018) Fragen und Antworten: Neue Züchtungstechniken. https://www.bmel.de/DE/Landwirtschaft/Pflanzenbau/Gentechnik/_Texte/FAQ-NeueZuechtungstechnologien.html;jsessionid=9957E097C8F0F455813B2E15D74236AA.2_cid296#doc11284300bodyText12. Accessed 31 Aug 2018

German Federal Ministry of Food and Agriculture, German Federal Office of Consumer Protection and Food Safety (2011) Leitfaden zur Kontrolle von GVO in Futtermitteln. https://www.bvl.bund.de/SharedDocs/Downloads/02_Futtermittel/fm_leitfaden_kontrolle_GVO.pdf?__blob=publicationFile&v=2. Accessed 31 Aug 2018

German Federal Office of Consumer Protection and Food Safety (2015) Bescheid: Antrag der Firma Cibus auf Feststellung, dass mittels des Rapid Trait Development Systems (RTDSTM) hergestellte herbizidresistente Rapslinien keine gentechnisch veränderten Organismen i.S.d. Gentechnikgesetzes darstellen(AZ 42050). http://www.umweltinstitut.org/fileadmin/Mediapool/Bilder/01_Themen/04_Gentechnik/Neue_Gentech-Methoden/BVL_Cibus.pdf. Accessed 17 Aug 2018

German Federal Plant Variety Office (2017) Plant breeders' rights and national listing. https://www.bundessortenamt.de/internet30/fileadmin/Files/PDF/BroschuereBSA_engl.pdf. Accessed 1 March 2018

German Federal Office of Consumer Protection and Food Safety (2017) Opinion on the legal classification of New Plant Breeding Techniques, in particular ODM and CRISPR-Cas9. http://www.bvl.bund.de/SharedDocs/Downloads/06_Gentechnik/Opinion_on_the_legal_classification_of_New_Plant_Breeding_Techniques.pdf?__blob=publicationFile&v=4. Accessed 25 Aug 2018

German Federal Office of Consumer Protection and Food Safety (2018) Cibus Raps-Bescheid vom BVL zurückgenommen. https://www.bvl.bund.de/DE/06_Gentechnik/04_Fachmeldungen/2018/2018_08_17_Fa_Cibus_Raps_Bescheid.html?nn=1471850. Accessed 17 Aug 2018

German National Academy of Sciences Leopoldina, Acatech - German National Academy of Science and Engineering, German Research Foundation, Union of the German Academies of Sciences and Humanities (2015) The opportunities and limits of genome editing: Statement. http://www.dfg.de/download/pdf/dfg_im_profil/reden_stellungnahmen/2015/stellungnahme_genome_editing_2015.pdf. Accessed 24 July 2018

German National Academy of Sciences Leopoldina, Acatech, Union der deutschen Akademien der Wissenschaften (2015) Academies issue statement on progress in molecular breeding and on the possible national ban on cultivation of genetically modified plants. http://www.leopoldina.org/

uploads/tx_leopublication/2015_03_26_Statement_on_Molecular_Breeding_final.pdf. Accessed 23 Feb 2018

German Organic Food Production Alliance (2017) BÖLW-Position zu neuen Gentechnologien: Vorsorgend handeln bei neuartigen Gentechnikverfahren! Das EU-Gentechnikrecht muss konsequent umgesetzt werden. https://www.boelw.de/fileadmin/media/pdf/Themen/Gentechnik/170209_BOELW_Position_neue_Gentechnik.pdf. Accessed 31 Aug 2018

German Syndicate of Traditional Agriculture (2016) Keine Gentechnik durch die Hintertür!: AbL-Position & Hintergründe zu neuen Gentechnik-Verfahren. Vorsorgeprinzip stärken und konsequent anwenden. http://www.abl-ev.de/fileadmin/Dokumente/AbL_ev/Gentechnikfrei/Hintergrund/AbL-Positionsapier_neue_GenT-Verfahren_Febr_2016_a.pdf. Accessed 22 Oct 2017

Giliker P (2014) Tort, 5th edn. Sweet & Maxwell

GM Freeze (2016) The case for regulating Gene Edited crops. http://www.gmfreeze.org/news-releases/266/. Accessed 24 Sept 2018

GMWatch (2014) "Genome editing": GM by another name. http://www.gmwatch.org/en/news/archive/2014/15546-genome-editing-gm-by-another-name. Accessed 24 Sept 2018

GMWatch (2016) New GMO techniques do give rise to GMOs – Austrian govt official. https://www.gmwatch.org/en/news/latest-news/17230-new-gmo-techniques-do-give-rise-to-gmos-austrian-govt-official. Accessed 11 July 2018

Goergen P (2010) Damage caused by GMOs under Luxembourg law. In: Koch BA (ed) Damage caused by genetically modified organisms: comparative survey of redress options for harm to persons, property or the environment. de Gruyter, Berlin, pp 350–380

Goldsmith B, Lockhart-Mummery E (2013) The ELD's national transposition. In: Bergkamp L, Goldsmith B (eds) EU Environmental liability directive: a commentary. Oxford University Press, pp 139–159

Gordley J (2010) Disturbances among neighbours: an introduction. In: Gordley J (ed) The development of liability between neighbours. Cambridge University Press, pp 1–28

Greenpeace (2005) No Market for GM Labelled Food in Europe. http://www.greenpeace.nl/Global/nederland/report/2008/7/eu-markets-no-market-for-gm-l.pdf. Accessed 9 Mar 2018

Greenpeace (2015) Policy Briefing: Gene-editing of plants - GM through the back door? http://www.greenpeace.org/eu-unit/Global/eu-unit/reports-briefings/2015/Greenpeace_Gene-editing_30112015%20-%202.pdf. Accessed 21 July 2018

Greenpeace EU (2018) Unauthorised GMO field trial exposed as EU takes hands-off approach. https://www.greenpeace.org/eu-unit/issues/nature-food/1260/unauthorised-gmo-field-trial-exposed-as-eu-takes-hands-off-approach-greenpeace/. Accessed 31 Aug 2018

Gross D (2006) Das gemeinschaftsrechtliche Genehmigungsverfahren bei der Freisetzung und dem Inverkehrbringen gentechnisch veränderter Organismen. Zugl.: Freiburg, Univ., Diss., 2006. Arbeiten aus dem Iuristischen Seminar der Universität Freiburg, vol 255. Schulthess, Zürich

Grossman MR (2010) Coexistence of genetically modified, conventional and organic crops in the European Union: the community framework. In: Bodiguel L, Cardwell M (eds) The regulation of genetically modified organisms: comparative approaches. Oxford University Press, Oxford, pp 123–162

Grohmann L, Keilwagen J, Duensing N, Dagand E, Hartung F, Wilhelm R, Bendiek J, Sprink T (2019) Detection and identification of genome editing in plants: challenges and opportunities. Front Plant Sci 10. https://doi.org/10.3389/fpls.2019.00236

Guerrero M (2017) Agricultural Biotechnology Annual Spain: USDA Foreign Agricultural Service - Gain Report. https://gain.fas.usda.gov/Recent%20GAIN%20Publications/Agricultural%20Biotechnology%20Annual_Madrid_Spain_12-12-2017.pdf. Accessed 22 Aug 2018

Halterman D, Guenthner J, Collinge S, Butler N, Douches D (2016) Biotech Potatoes in the 21st century: 20 years since the first biotech Potato. Am J Potato Res 93:1–20. https://doi.org/10.1007/s12230-015-9485-1

Hamburger DJS (2018) Normative criteria and their inclusion in a regulatory framework for new plant varieties derived from genome editing. Front Bioeng Biotechnol 6:176

Harriss J, Stewart D (2015) Science, politics, and the framing of modern agricultural technologies. In: Herring RJ (ed) The Oxford handbook of food, politics, and society. Oxford University Press, Oxford, pp 43–64

Hartung F, Schiemann J (2014) Precise plant breeding using new genome editing techniques: opportunities, safety and regulation in the EU. Plant J 78:742–752. https://doi.org/10.1111/tpj.12413

Haun W, Coffman A, Clasen BM, Demorest ZL, Lowy A, Ray E, Retterath A, Stoddard T, Juillerat A, Cedrone F, Mathis L, Voytas DF, Zhang F (2014) Improved soybean oil quality by targeted mutagenesis of the fatty acid desaturase 2 gene family. Plant Biotechnol J 12:934–940. https://doi.org/10.1111/pbi.12201

Heinemann A (2014) Access to evidence and presumptions – communicating vessels in procedural law. In: Hüschelrath K, Schweitzer H (eds) Public and private enforcement of competition law in Europe, vol 48. Springer, Berlin, pp 167–191

Helliwell R, Hartley S, Pearce W, O'Neill L (2017) Why are NGOs sceptical of genome editing? NGOs' opposition to agricultural biotechnologies is rooted in scepticism about the framing of problems and solutions, rather than just emotion and dogma. EMBO Reports 18:2090–2093. https://doi.org/10.15252/embr.201744385

Herdegen M, Dederer H-G (2009) Richtlinie 2009/41/EG. In: Herdegen M, Dederer H-G (eds) Internationales Biotechnologierecht: Gentechnik, Biopatente, genetische Ressourcen, vol 36. Müller, Heidelberg

Herdegen M, Dederer H-G (2015) Richtlinie 2001/18/EG. In: Herdegen M, Dederer H-G (eds) Internationales Biotechnologierecht: Gentechnik, Biopatente, genetische Ressourcen, vol 49. Müller, Heidelberg

Hilscher J, Bürstmayr H, Stöger E (2017a) RNAi-based techniques, accelerated breeding and CRISPR-Cas: basics and application in plant breeding. https://www.bmgf.gv.at/cms/home/attachments/2/5/0/CH1052/CMS1493814297724/techniken_pflanzenzuechtung_engl_20170323.pdf. Accessed 5 July 2017

Hilscher J, Bürstmayr H, Stoger E (2017b) Targeted modification of plant genomes for precision crop breeding. Biotechnol J:12. https://doi.org/10.1002/biot.201600173

Hinteregger M (2008) Environmental liability. In: Koziol H, Schulze R (eds) Tort law of the European community. Springer, Heidelberg, pp 103–119

Hoffman WC, Hill-Arning S (1994) Guide to product liability in Europe: the new strict product liability laws, pre-existing remedies, procedure and costs in the European Union and the European Free Trade Association. Kluwer Law and Taxation Publication, Deventer

Holland N (2004) Power Struggles over Biotech in Brussels. http://archive.corporateeurope.org/biotechbrussels.pdf. Accessed 1 Oct 2018

Holme IB, Wendt T, Holm PB (2013) Intragenesis and cisgenesis as alternatives to transgenic crop development. Plant Biotechnol J 11:395–407. https://doi.org/10.1111/pbi.12055

Hopp M, Lange S, Epp A, Lohmann M, Böl G-F (2017) Durchführung von Fokusgruppen zur Wahrnehmung des Genome Editings (CRISPR/Cas9): Abschlussbericht. BfR-Wissenschaft, 2017, 04. Bundesinstitut für Risikobewertung, Berlin

Hou H, Atlihan N, Lu Z-X (2014) New biotechnology enhances the application of cisgenesis in plant breeding. Front Plant Sci 5:389. https://doi.org/10.3389/fpls.2014.00389

Huang S, Weigel D, Beachy RN, Li J (2016) A proposed regulatory framework for genome-edited crops. Nat Genet 48:109–111. https://doi.org/10.1038/ng.3484

Huffman WE, McCluskey JJ (2014) Labeling of genetically modified foods. In: Smyth SJ, Phillips PWB, Castle D (eds) Handbook on agriculture, biotechnology and development. Edward Elgar, Cheltenham, pp 467–487

Hunt J (2011) Ploughing their own furrow: subnational regions and the regulation of GM crop cultivation. Camb Yearb Eur Legal Stud 13:135–159. https://doi.org/10.5235/152888712801753013

IFOAM EU Group (2015) IFOAM EU position paper on New Plant Breeding Techniques. http://www.ifoam-eu.org/sites/default/files/ifoameu_policy_npbts_position_final_20151210.pdf. Accessed 25 Aug 2018

IFOAM EU Group, Friends of the Earth Europe, Greenpeace EU, Slow Food, Réseau Semences Paysannes, Arche Noah, GeneWatch UK, GM Freeze, CEO, IG Saatgut, Friends of the Earth Germany, Test Biotech, GMWatch, Global 2000, EcoNexus, INSPRO, Arbeitsgemeinschaft bäuerliche Landwirtschaft e.V., Beyond GM, Compassion in world farming (2017) Joint Position Paper: New techniques of genetic engineering: Why EU GMO law must be fully applied to the so - called 'New Plant Breeding Techniques'. http://www.foeeurope.org/sites/default/files/gmos/2018/new_gm_techniques_joint_position_paper.pdf. Accessed 15 Feb 2018

IFOAM EU Group (2018) Press Release: IFOAM EU welcomes the European Court of Justice decision that new genetic engineering techniques will be regulated as GMOs. https://www.ifoam-eu.org/sites/default/files/ifoameu_policy_kgoo_pressreleasefinal.pdf. Accessed 4 Jan 2019

IFS Food (2017) Standard for auditing quality and food safety of food products: Version 6.1. https://www.ifs-certification.com/images/standards/ifs_food6_1/documents/standards/IFS_Food_V6_1_en.pdf. Accessed 12 Jan 2018

IG Saatgut (2016) Protecting GM-free seed in the EU: Preventing GMO thresholds and res pecting the zero tolerance policy: Background paper by IG Saatgut (Initiative for GM-free Seeds and Breeding). http://www.gentechnikfreie-saat.org/files/background_briefing_-_ig_saatgut_-_zero_tolerance_policy_and_labelling_obligation.pdf. Accessed 4 Mar 2018

Infantino M, Zervogianni E (2017) Causation in European tort law. The common core of European private law. Cambridge University Press, Cambridge

Institut d'études Opinion et Marketing (2016) Les Français et la technique du CRISPR-Cas9: Ifop pour Alliance Vita. https://www.ifop.com/wp-content/uploads/2018/03/3394-1-study_file.pdf. Accessed 24 Aug 2018

International Life Sciences Institute (2016) International Life Sciences Institute Crop Composition Database, Version 6. www.cropcomposition.org. Accessed 16 Aug 2018

International Organisation for Standardisation (2005a) International standard (ISO) 21569:2005/Amd 1:2013: Foodstuffs. Methods of analysis for the detection of genetically modified organisms and derived products – Qualitative nucleic acid based methods

International Organisation for Standardisation (2005b) International standard (ISO) 21570:2005/Amd 1:2013: Foodstuffs. Methods of analysis for the detection of genetically modified organisms and derived products – Quantitative nucleic acid based methods

International Organisation for Standardisation (2005c) International standard (ISO) 21571:2005/Amd 1:2013: Foodstuffs. Methods of analysis for the detection of genetically modified organisms and derived products – Nucleic acid extraction

International Organisation for Standardisation (2006) International standard (ISO) 24276:2006/Amd 1:2013: Foodstuffs. Methods of analysis for the detection of genetically modified organisms and derived products – General requirements and definitions

International Organisation for Standardisation (2015) International Standard (ISO) 9001:2015: Quality management systems. Requirements

International Service for the Acquisition of Agri-biotech Applications (2016) Global Status of Commercialized Biotech/GM Crops: 2016: ISAAA Brief No. 52

International Service for the Acquisition of Agri-biotech Applications (2017) Global Status of Commercialized Biotech/GM Crops in 2017: ISAAA Brief No. 53

International Union for the Protection of New Varieties of Plants (1991) Act of 1991: International Convention for the Protection of New Varieties of Plants. Of December 2, 1961, as Revised at Geneva on November 10, 1972, on October 23, 1978, and on March 19. http://www.upov.int/export/sites/upov/upovlex/en/conventions/1991/pdf/act1991.pdf. Accessed 31 Aug 2018

Ishii T, Araki M (2016) Consumer acceptance of food crops developed by genome editing. Plant Cell Rep 35:1507–1518. https://doi.org/10.1007/s00299-016-1974-2

Jack B (2009) Agriculture and EU environmental law. Ashgate Pub, Farnham

Jany K-D (2018a) Anträge auf Zulassung von gentechnisch veränderten Pflanzen zum Import als Lebens- und Futtermittel in die EU und Dauer der Zulassungsverfahren. https://www.biotech-gm-food.com/pflanzen/antraege-zulassung-gentech-pflanzen. Accessed 31 Aug 2018

Jany K-D (2018b) Zulassungen von gentechnisch veränderten Pflanzen 2018. https://www.biotech-gm-food.com/zulassungen-gv-Pflanzen-2018. Accessed 31 Aug 2018

Jiao Y, Peluso P, Shi J, Liang T, Stitzer MC, Wang B, Campbell MS, Stein JC, Wei X, Chin C-S, Guill K, Regulski M, Kumari S, Olson A, Gent J, Schneider KL, Wolfgruber TK, May MR, Springer NM, Antoniou E, McCombie WR, Presting GG, McMullen M, Ross-Ibarra J, Dawe RK, Hastie A, Rank DR, Ware D (2017) Improved maize reference genome with single-molecule technologies. Nature 546:524–527. https://doi.org/10.1038/nature22971

John Innes Centre, Rothamsted Research, NIAB, James Hutton Institute, Aberystwyth University, The Sainsbury Laboratory, Quadram Institute, Earlham Institute, Professor Denis Murphy, Professor Ian Crute, National Farmers Union, Tenant Farmers Association, Country Land & Business Association, Linking Environment And Farming, RASE and Innovation for Agriculture, British Society of Plant Breeders, DSV-UK, Germinal, Limagrain UK, RAGT UK, KWS UK, Elsoms Seeds, HL Hutchinson, Agrovista, Frontier Agriculture, Agrii, ProCam, Syngenta UK, Bayer, BASF Agricultural Solutions, Corteva Agriscience, Agricultural Biotechnology Council, Agricultural Industries Confederation (2018) Open letter to Defra Secretary of State Michael Gove. http://www.cpm-magazine.co.uk/wp-content/uploads/2018/09/180903-Michael-Gove-letter.FINAL_.pdf?mc_cid=2dddacaedb&mc_eid=b89825d9e5. Accessed 27 Jan 2019

Jones HD (2015) Regulatory uncertainty over genome editing. Nat Plants, 1. https://doi.org/10.1038/nplants.2014.11

Kahrmann J, Bömeke O, Leggewie G (2017) Aged GMO legislation meets new genome editing techniques. Zeitschrift für Europäisches Umwelt- und Planungsrecht 15:176–182

Kalaitzandonakes N, Alston JM, Bradford KJ (2007) Compliance costs for regulatory approval of new biotech crops. Nat Biotechnol 25:509–511. https://doi.org/10.1038/nbt0507-509

Kalaitzandonakes NG, Phillips PWB, Wesseler J, Smyth SJ (eds) (2016) The coexistence of genetically modified, organic and conventional foods: Government policies and market practices. Natural Resource Management and Policy, vol 49. Springer, New York

Kaminski R, Chen Y, Fischer T, Tedaldi E, Napoli A, Zhang Y, Karn J, Hu W, Khalili K (2016) Elimination of HIV-1 genomes from human T-lymphoid cells by CRISPR/Cas9 gene editing. Sci Rep 6:22555. https://doi.org/10.1038/srep22555

King's College London (2008) Do European consumers buy GM foods? Final report on "Consumer choice", a FP6 project. http://rahvatervis.ut.ee/bitstream/1/1969/1/Vokkjt2008.pdf. Accessed 26 October 2018

King A (2018) Crispr-edited crop research could be crippled by European court ruling. https://www.chemistryworld.com/news/crispr-edited-crop-research-could-be-crippled-by-european-court-ruling-/3009468.article. Accessed 5 Sept 2018

Koch BA (2007) Comparative report: damage caused by several successive, but independent events outside the victim's sphere. In: Winiger B, Koch BA, Koziol H, Zimmermann R (eds) Essential cases on natural causation. Springer, Vienna, pp 501–504

Koch BA (2008a) Comparative report. In: Koch BA (ed) Economic loss caused by genetically modified organisms: liability and redress for the adventitious presence of GMOs in Non-GM crops. Springer, Vienna, pp 585–651

Koch BA (2008b) Executive summary. In: Koch BA (ed) Economic loss caused by genetically modified organisms: liability and redress for the adventitious presence of GMOs in Non-GM crops. Springer, Vienna, pp 9–18

Koch BA (2010) Damage caused by GMOs: comparative analysis. In: Koch BA (ed) Damage caused by genetically modified organisms: comparative survey of redress options for harm to persons, property or the environment. de Gruyter, Berlin, pp 882–942

Koch BA (2013) Liability and redress options for damage caused by GMOs. In: Bertheau Y (ed) Genetically modified and non-genetically modified food supply chains: co-existence and traceability. Wiley-Blackwell, Chichester

Kohler J (2005) Schadensausgleich in Fällen des § 36a Gentechnikgesetz. Natur und Recht 27:566–575. https://doi.org/10.1007/s10357-005-0716-y

Koziol H (2007) Comparative report: damage caused by less than all possibly harmful events outside the victim's sphere. In: Winiger B, Koch BA, Koziol H, Zimmermann R (eds) Essential cases on natural causation. Springer, Vienna, pp 387–389

Kuntz M (2012) Destruction of public and governmental experiments of GMO in Europe. GM Crops Food 3:258–264. https://doi.org/10.4161/gmcr.21231

Laaninen T (2016) New plant-breeding techniques: applicability of GM rules. http://www.europarl.europa.eu/RegData/etudes/BRIE/2016/582018/EPRS_BRI(2016)582018_EN.pdf. Accessed 23 Feb 2018

Lappin J (2018a) Advisory Legal Opinion Expected for New Plant Breeding Techniques: USDA Foreign Agricultural Service - Gain Report. https://gain.fas.usda.gov/Recent%20GAIN%20Publications/Advisory%20Legal%20Opinion%20Expected%20for%20New%20Plant%20Breeding%20Techniques%20_Brussels%20USEU_EU-28_1-16-2018.pdf. Accessed 13 Feb 2018

Lappin J (2018b) EU Court Extends GMO Directive to New Plant Breeding Techniques: USDA Foreign Agricultural Service - Gain Report. https://gain.fas.usda.gov/Recent%20GAIN%20Publications/EU%20Court%20Extends%20GMO%20Directive%20to%20New%20Plant%20Breeding%20Techniques_Brussels%20USEU_Belgium%20EU-28_7-27-2018.pdf. Accessed 8 Aug 2018

Lawrence A (2016) Umeå researcher served a world first (?) CRISPR meal. https://www.umu.se/en/news/umea-researcher-served-a-world-first%2D%2Dcrispr-meal_5817178. Accessed 24 Aug 2018

Lawrenson T, Shorinola O, Stacey N, Li C, Østergaard L, Patron N, Uauy C, Harwood W (2015) Induction of targeted, heritable mutations in barley and Brassica oleracea using RNA-guided Cas9 nuclease. Genome Biol 16:258. https://doi.org/10.1186/s13059-015-0826-7

Le Déaut J-Y, Procaccia C (2017) Rapport au nom de L'Office Parlementaire d'Evaluation des Choix Scientifiques et Technologiques sur Les enjeux économiques, environnementaux, sanitaires et éthiques des biotechnologies à la lumière des nouvelles pistes de recherche: Tome I: Rapport. http://www.assemblee-nationale.fr/14/rap-off/i4618-tI.asp. Accessed 15 Feb 2018

Lee M (2008) EU regulation of GMOs: law and decision making for a new technology. Biotechnology regulation series. Edward Elgar, Cheltenham

Lee R (2017) Novel foods and risk assessment in Europe: separating science from society. In: Brownsword R, Scotford E, Yeung K (eds) The Oxford handbook of law, regulation, and technology, 1st edn. Oxford University Press, Oxford, pp 1209–1231

Leroux J (2016) New Breeding Techniques: New GMOs in a Legal Limbo. https://www.greeneuropeanjournal.eu/new-breeding-techniques-new-gmos-in-a-legal-limbo/. Accessed 1 Mar 2018

Lezaun J (2006) Creating a new object of government. Soc Stud Sci 36:499–531. https://doi.org/10.1177/0306312706059461

Lombardo L, Zelasco S (2016) Biotech approaches to overcome the limitations of using transgenic plants in organic farming. Sustainability 8:497. https://doi.org/10.3390/su8050497

Lombardo L, Coppola G, Zelasco S (2016) New technologies for insect-resistant and Herbicide-tolerant plants. Trends Biotechnol 34:49–57. https://doi.org/10.1016/j.tibtech.2015.10.006

Lucht JM (2015) Public acceptance of plant biotechnology and GM crops. Viruses 7:4254–4281. https://doi.org/10.3390/v7082819

Lusser M, Parisi C, Plan D, Rodriguez-Cerezo E (2011) New plant breeding techniques: state-of-the-art and prospects for commercial development. JRC Technical Report, EUR 24760. Publications Office of the European Union, Luxembourg

Madre Y, Agostino VD (2017) New plant-breeding techniques: what are we talking about? http://www.farm-europe.eu/travaux/new-plant-breeding-techniques-what-are-we-talking-about/. Accessed 15 Feb 2018

Malyska A, Bolla R, Twardowski T (2016) The role of public opinion in shaping trajectories of agricultural biotechnology. Trends Biotechnol 34:530–534. https://doi.org/10.1016/j.tibtech.2016.03.005

Mansour M, Key S (2004) From farm to fork: the impact on global commerce of the new European Union biotechnology regulatory scheme. Int Lawyer 38:55–69

Marchant GE, Stevens YA (2015) A new window of opportunity to reject process-based biotechnology regulation. GM Crops Food 6:233–242. https://doi.org/10.1080/21645698.2015.1134406

Martin de la Torre V (2016) S&Ds object to authorisation of GMOs and call for clear rules. http://www.socialistsanddemocrats.eu/newsroom/sds-object-authorisation-gmos-and-call-clear-rules. Accessed 31 July 2018

Masip G, Sabalza M, Pérez-Massot E, Banakar R, Cebrian D, Twyman RM, Capell T, Albajes R, Christou P (2013) Paradoxical EU agricultural policies on genetically engineered crops. Trends Plant Sci 18:312–324. https://doi.org/10.1016/j.tplants.2013.03.004

McDougall P (2011) Fact sheet: getting a biotech crop to the market. https://croplife-r9qnrxt3qxgjra4.netdna-ssl.com/wp-content/uploads/2014/04/Fact-Sheet-Getting-a-Biotech-Crop-to-Market.pdf. Accessed 25 Feb 2018

McEldowney J (2015) GMO cultivation in the EU: State of play. http://www.europarl.europa.eu/RegData/etudes/ATAG/2015/571330/EPRS_ATA(2015)571330_EN.pdf. Accessed 12 Mar 2018

McIntyre A (2016) Innovative farming could help solve global food crisis. https://www.theparliamentmagazine.eu/printpdf/3899. Accessed 9 July 2018

Meßerschmidt K (2011) Europäisches Umweltrecht: Ein Studienbuch. Juristische Kurz-Lehrbücher. Beck, München

Michalopoulos S (2018) Industry shocked by EU Court decision to put gene editing technique under GM law. https://www.euractiv.com/section/agriculture-food/news/industry-shocked-by-eu-court-decision-to-put-gene-editing-technique-under-gm-law/. Accessed 26 July 2018

Mielby H, Sandøe P, Lassen J (2013) Multiple aspects of unnaturalness: are cisgenic crops perceived as being more natural and more acceptable than transgenic crops? Agric Hum Values 30:471–480. https://doi.org/10.1007/s10460-013-9430-1

Modrzejewski D, Hartung F, Sprink T, Krause D, Kohl C, Wilhelm R (2018) 1. Aktualisierung der Übersicht über Nutz- und Zierpflanzen, die mittels neuer molekularbiologischer Techniken für die Bereiche Ernährung, Landwirtschaft und Gartenbau erzeugt wurden. https://www.bmel.de/SharedDocs/Downloads/Landwirtschaft/Pflanze/GrueneGentechnik/NMT_Stand-Regulierung_Anlage4-Aktualisierung.pdf;jsessionid=1AE43DEFCF9297059C192EB6C23F5E29.1_cid367?__blob=publicationFile. Accessed 11 Oct 2018

Monti A, Fusco F (2010) Damage caused by GMOs under Italian law. In: Koch BA (ed) Damage caused by genetically modified organisms: comparative survey of redress options for harm to persons, property or the environment. de Gruyter, Berlin, pp 313–333

Morgan T (2017) Breeding crops for the future. Food Sci Technol 31:14–16

Morineau C, Bellec Y, Tellier F, Gissot L, Kelemen Z, Nogué F, Faure J-D (2017) Selective gene dosage by CRISPR-Cas9 genome editing in hexaploid Camelina sativa. Plant Biotechnol J 15:729–739. https://doi.org/10.1111/pbi.12671

Mühlböck M, Tosun J (2015) Deciding over controversial issues: voting behavior in the Council and the European Parliament on genetically modified organisms. In: Seventh International Conference on Coexistence between Genetically Modified (GM) and non-GM based Agricultural Supply Chains (GMCC-15), Amsterdam

Mühlböck M, Tosun J (2017) How EU member states have tried (and failed) to reach agreement on GMOs – and what it could mean for EU decision-making. http://blogs.lse.ac.uk/europpblog/2017/11/13/how-eu-member-states-have-tried-and-failed-to-reach-agreement-on-gmos-and-what-it-could-mean-for-eu-decision-making/. Accessed 31 July 2018

Mühlböck M, Tosun J (2018) Responsiveness to different national interests: voting behaviour on genetically modified organisms in the council of the European Union. JCMS: J Common Mark Stud 56:385–402. https://doi.org/10.1111/jcms.12609

Nábrádi A, Popp J (2011) Economics of GM crop cultivation. Applied Studies in Agribusiness and Commerce (APSTRACT):7–14

Nap J-P, van Kessel G (2011) Epigenetics, an update: Commissioned by: The Netherlands Commission on Genetic Modification (COGEM). https://www.cogem.net/index.cfm/en/publications/publication/epigenetics-un-update. Accessed 23 Aug 2018
Nekrasov V, Wang C, Win J, Lanz C, Weigel D, Kamoun S (2017) Rapid generation of a transgene-free powdery mildew resistant tomato by genome deletion. Sci Rep 7:482. https://doi.org/10.1038/s41598-017-00578-x
New Breeding Techniques Platform (2015a). http://www.nbtplatform.org/. Accessed 13 July 2018
New Breeding Techniques Platform (2015b) Fact Sheet New Breeding Techniques: seizing the opportunity. http://www.nbtplatform.org/background-documents/factsheets/fact-sheet-on-nbts-in-general-2015.pdf. Accessed 11 Oct 2016
Nogué F, Vergne P, Chèvre A-M, Chauvin J-E, Bouchabké O, Déjardin A, Chevreau E, Hibrand-Saint Oyant L, Mazier M, Barret P, Guiderdoni E, Mathis L, Sallaud C, Matt M, Pierron J-P, Bonnel E, Foucrier S, Toppan A, Trannoy L, Rogowsky P (2015) GENIUS project - Genome ENgineering Improvement for Useful plants of a Sustainable agriculture: 15èmes Rencontres de Virologie Végétale, Aussois, France. Poster. https://agritrop.cirad.fr/575824/1/document_575824.pdf
Norer R, Preisig C (2016) Genetic technology in the light of food security and food safety - general report. In: Norer R (ed) Genetic technology and food safety. Springer, Heidelberg, pp 1–70
Nuffield Council on Bioethics (2016) Genome editing: an ethical review
OECD (1986) Recombinant DNA safety considerations: safety considerations for industrial, agricultural and environmental applications of organisms derived by recombinant DNA techniques, Paris
OECD (1993) Safety evaluation of foods derived by modern biotechnology: concepts and principles, Paris
OECD (2000a) Report of the task force for the safety of novel foods and feeds: C(2000)86/ADD1. https://www.oecd.org/chemicalsafety/biotrack/REPORT-OF-THE-TASK-FORCE-FOR-THE-SAFETY-OF-NOVEL.pdf. Accessed 19 Feb 2018
OECD (2000b) Report of the working group on harmonisation of regulatory oversight in biotechnology: C(2000)86/ADD2. https://www.oecd.org/chemicalsafety/biotrack/REPORT-OF-THE-WORKING-GROUP-ON-HARMONISATION-OF-REGULATORY.pdf. Accessed 19 Feb 2018
OECD (2018a) Abstracts OECD Conference on Genome Editing: Applications in Agriculture; Implications for Health, Environment and Regulation; 28–29 June 2018. http://www.oecd.org/environment/genome-editing-agriculture/oecd-conference-on-genome-editing-abstracts.pdf. Accessed 9 July 2018
OECD (2018b) Concentration in seed markets: potential effects and policy responses. OECD Publishing, Paris
OECD Working Group on the Safety of Novel Foods and Feeds (2018) Consensus documents: work on the safety of novel foods and feeds. http://www.oecd.org/science/biotrack/consensus-documents-safety-of-novel-foods-and-feeds.htm. Accessed 16 Aug 2018
O'Reilly E (2017) Decision in case 1285/2016/JAS on the failure by the European Commission to reply to a request for information concerning organisms developed using the new breeding technique CRISPR/Cas9. https://www.ombudsman.europa.eu/cases/decision.faces/en/75646/html.bookmark. Accessed 23 February 2018
Osborn MJ, Gabriel R, Webber BR, DeFeo AP, McElroy AN, Jarjour J, Starker CG, Wagner JE, Joung JK, Voytas DF, von Kalle C, Schmidt M, Blazar BR, Tolar J (2015) Fanconi anemia gene editing by the CRISPR/Cas9 system. Hum Gene Ther 26:114–126. https://doi.org/10.1089/hum.2014.111
Paarlberg RL, Hopkins RF, Ladewski L (2004) Regulation of GM crops: shaping an international regime. In: Evenson RE, Santaniello V (eds) The regulation of agricultural biotechnology. CABI Pub, Wallingford, pp 1–24
Palme C (2005) Das neue Gentechnik-Gesetz. Neue Zeitschrift für Verwaltungsrecht:253–257

Paoletti C, Flamm E, Yan W, Meek S, Renckens S, Fellous M, Kuiper H (2008) GMO risk assessment around the world: some examples. Trends Food Sci Technol 19:S70–S78. https://doi.org/10.1016/j.tifs.2008.07.007

Peters PG, Lambert TA (2007) Regulatory barriers to consumer information about genetically modified foods. In: Weirich P (ed) Labeling genetically modified food: the philosophical and legal debate. Oxford University Press, Oxford, pp 151–177

Phillips S, Flach B (2017) Dutch Proposal to Legislate NBTs: USDA Foreign Agricultural Service - Gain Report. https://gain.fas.usda.gov/Recent%20GAIN%20Publications/Dutch%20Proposal%20to%20Legislate%20NBTs_The%20Hague_Netherlands%20EU-27_9-29-2017.pdf. Accessed 4 Jan 2018

Plan D, van den Eede G (2010) The EU legislation on GMOs: an overview. EUR. Scientific and technical research series, vol 24279. Publications Office, Luxembourg

Plant Genome Editing Database (2018). http://plantcrispr.org/cgi-bin/crispr/index.cgi. Accessed 1 Dec 2018

Poli S (2006) The EU risk management of genetically modified organisms and the commission's defence strategy in the biotech dispute: are they inconsistent? In: Francioni F, Scovazzi T (eds) Biotechnology and international law. Hart, Oxford, pp 387–402

Pollack MA, Shaffer GC (2009) When cooperation fails: the international law and politics of genetically modified foods. Oxford University Press, Oxford

Potton E, Webb D (2017) Brexit: Agriculture and Trade: Briefing Paper Number 7974. http://researchbriefings.parliament.uk/ResearchBriefing/Summary/CBP-7974#fullreport. Accessed 11 July 2018

Prins TW, Kok EJ (2010) Food and feed safety aspects of cisgenic crop plant varieties: RIKILT report 2010.001. http://edepot.wur.nl/157733. Accessed 10 Sept 2017

Purnhagen KP, Kok E, Kleter G, Schebesta H, Visser RGF, Wesseler J (2018) The European Union Court's Advocate General's opinion and new plant breeding techniques. Nat Biotechnol 36:573–575. https://doi.org/10.1038/nbt.4174

Ribarits A, Brüller W, Hartmann J, Hochegger R, Mechtler K, Peterseil V, Söllinger J, Stepanek W, Widhalm I, Wögerbauer M, Leonhardt C (2014) Use of novel techniques in plant breeding and practical consequences concerning detection, traceability, labeling, and risk assessment. AgBioForum 17:183–190

Ricroch A, Clairand P, Harwood W (2017) Use of CRISPR systems in plant genome editing: toward new opportunities in agriculture. Emerg Top Life Sci 1:169–182. https://doi.org/10.1042/ETLS20170085

Ridley WP, Shillito RD, Coats I, Steiner H-Y, Shawgo M, Phillips A, Dussold P, Kurtyka L (2004) Development of the international life sciences institute crop composition database. J Food Compos Anal 17:423–438. https://doi.org/10.1016/j.jfca.2004.03.006

Riehm T (2017) §3 Gesetzliche Schuldverhältnisse. In: Langenbucher K (ed) Europäisches Privat- und Wirtschaftsrecht, 4. Auflage. Nomos Verlagsgesellschaft mbH & Co. KG, Baden-Baden, pp 207–237

Riggs AD, Martienssen RA, Russo VEA (1996) Introduction. In: Russo VEA (ed) Epigenetic mechanisms of gene regulation. Cold Spring Harbor Laboratory Press, Plainview, pp 1–4

Rodriguez-Cerezo E (2014) New plant breeding techniques. Conference "Scientific support to agriculture: Competitiveness, quality and sustainability", Athens

Rothamsted Research (2019) GM field trials approved. https://www.rothamsted.ac.uk/news/gm-field-trials-approved. Accessed 25 May 2019

Roïz J (2014) Limits of the current EU regulatory framework on GMOs: risk of not authorized GM event-traces in imports. OCL 21:D603. https://doi.org/10.1051/ocl/2014037

Rollin F, Kennedy J, Wills J (2011) Consumers and new food technologies. Trends Food Sci Technol 22:99–111. https://doi.org/10.1016/j.tifs.2010.09.001

Ronteltap A, van Trijp JCM, Renes RJ, Frewer LJ (2007) Consumer acceptance of technology-based food innovations: lessons for the future of nutrigenomics. Appetite 49:1–17. https://doi.org/10.1016/j.appet.2007.02.002

Sánchez-León S, Gil-Humanes J, Ozuna CV, Giménez MJ, Sousa C, Voytas DF, Barro F (2017) Low-gluten, nontransgenic wheat engineered with CRISPR/Cas9. Plant Biotechnol J. https://doi.org/10.1111/pbi.12837
Schaart JG, van de Wiel CCM, Lotz LAP, Smulders MJM (2016) Opportunities for products of new plant breeding techniques. Trends Plant Sci 21:438–449. https://doi.org/10.1016/j.tplants.2015.11.006
Schauzu M (2011) The European Union's regulatory framework: developments in legislation, safety assessment, and public perception. In: Baram M, Bourrier M (eds) Governing risk in GM agriculture. Cambridge University Press, Cambridge, pp 57–84
Schenkelaars P, Wesseler J (2016) Farm-level GM coexistence policies in the EU: context, concepts and developments. EuroChoices 15:5–11. https://doi.org/10.1111/1746-692X.12112
Schiemann J, Hartung F (2014) EU perspectives on new plant-breeding techniques. In: Eaglesham A, Hardy RWF (eds) New DNA-editing approaches: methods, applications and policy for agriculture, pp 201–210
Schiffers B (2017) Guest Editorial: New European Union plant health regime: a more stringent regulation that could impact trade from developing countries in the near future. Tunis J Plant Prot 12(1):1–3
Scholderer J (2004) Consumer attitudes towards genetically modified foods in Europe: structure and changeability. Universität Potsdam, Potsdam
Schwank G, Koo B-K, Sasselli V, Dekkers JF, Heo I, Demircan T, Sasaki N, Boymans S, Cuppen E, van der Ent CK, Nieuwenhuis EES, Beekman JM, Clevers H (2013) Functional repair of CFTR by CRISPR/Cas9 in intestinal stem cell organoids of cystic fibrosis patients. Cell Stem Cell 13:653–658. https://doi.org/10.1016/j.stem.2013.11.002
Science Media Centre (2018) expert reaction to Court of Justice of the European Union ruling that GMO rules should cover plant genome editing techniques. http://www.sciencemediacentre.org/expert-reaction-to-court-of-justice-of-the-european-union-ruling-that-gmo-rules-should-cover-plant-genome-editing-techniques/. Accessed 26 July 2018
Sciencewise, Nuffield Council on Bioethics (2016) Public dialogue on genome editing: Why? When? Who? http://nuffieldbioethics.org/wp-content/uploads/Public-Dialogue-on-Genome-Editing-workshop-report.pdf. Accessed 1 Mar 2018
Scott J (2005) European regulation of GMOs: thinking about judicial review in the WTO. In: Holder J, O'Cinneide C (eds) Current legal problems. Oxford University Press, Oxford, pp 117–147
Scott J, Vos E (2002) The juridification of uncertainty: observations on the ambivalence of the precautionary principle within the EU and the WTO. In: Joerges C (ed) Good governance in Europe's integrated market. Oxford University Press, Oxford, pp 253–286
Secretariat of the Convention on Biological Diversity (2016) Guidance on Risk Assessment of Living Modified Organisms and Monitoring in the Context of Risk Assessment: UNEP/CBD/BS/COP-MOP/8/8/Add.1. In: Commissioned by: Conference of the Parties to the Convention of Biological Diversity serving as the Meeting of the Parties to the Cartagena Protocol on Biosafety. Eighth meeting. Cancun, Mexico, 4–17 December 2016. https://www.cbd.int/doc/meetings/bs/mop-08/official/bs-mop-08-08-add1-en.pdf. Accessed 22 Feb 2018
Shew AM, Nalley LL, Snell HA, Nayga RM, Dixon BL (2018) CRISPR versus GMOs: public acceptance and valuation. Glob Food Secur 19:71–80. https://doi.org/10.1016/j.gfs.2018.10.005
Sikkema A (2018) 'Admission of GMOs up to Member States'. https://resource.wur.nl/en/show/Admission-of-GMOs-up-to-Member-States.htm. Accessed 23 Feb 2018
Sleenhoff S, Osseweijer P (2013) Consumer choice: linking consumer intentions to actual purchase of GM labeled food products. GM Crops Food 4:166–171. https://doi.org/10.4161/gmcr.26519
Smart RD, Blum M, Wesseler J (2017) Trends in approval times for genetically engineered crops in the United States and the European Union. J Agric Econ 68:182–198. https://doi.org/10.1111/1477-9552.12171
Snyder F (1993) The effectiveness of European community law: institutions, processes, tools and techniques. Mod Law Rev 56:19–54. https://doi.org/10.1111/j.1468-2230.1993.tb02852.x

Soyk S, Müller NA, Park SJ, Schmalenbach I, Jiang K, Hayama R, Zhang L, van Eck J, Jiménez-Gómez JM, Lippman ZB (2017) Variation in the flowering gene SELF PRUNING 5G promotes day-neutrality and early yield in tomato. Nat Genet 49:162–168. https://doi.org/10.1038/ng.3733

Sparrow P, Broer I, Hood EE, Eversole K, Hartung F, Schiemann J (2013) Risk assessment and regulation of molecular farming – a comparison between Europe and US. Curr Pharm Design 19:1–18

Spök A (2010) Assessing socio-economic impacts of GMOs: issues to consider for policy development; final report. Bundesministerium für Gesundheit, Wien

Sprink T, Eriksson D, Schiemann J, Hartung F (2016) Regulatory hurdles for genome editing: process- vs. product-based approaches in different regulatory contexts. Plant Cell Rep 35:1493–1506. https://doi.org/10.1007/s00299-016-1990-2

Stakeholder and Issue Mapping on New Breeding Techniques (2017) (Internal industry report, available as a leaked document). https://corporateeurope.org/sites/default/files/attachments/final_stakeholder_and_issue_mapping_on_nbts_04-2017.doc. Accessed 31 Aug 2018

Stein AJ, Rodríguez-Cerezo E (2009) The global pipeline of new GM crops: implications of asynchronous approval for international trade. JRC scientific and technical reports, JRC51799. Office for Official Publications of the European Communities, Luxembourg

Steinbrecher R (2015) Genetic Engineering in Plants and the "New Breeding Techniques (NBTs)": Inherent risks and the need to regulate. http://www.econexus.info/sites/econexus/files/NBT%20Briefing%20-%20EcoNexus%20December%202015.pdf. Accessed 24 Sept 2018

Stokstad E (2018) European court ruling raises hurdles for CRISPR crops. Science. https://doi.org/10.1126/science.aau8986

Sunstein CR (2017) On mandatory labeling, with special reference to genetically modified foods. Univ Pa Law Rev 165:1043–1095

Swedish Board of Agriculture (2015) CRISPR/Cas9 mutated Arabidopsis. https://www.upsc.se/documents/Information_on_interpretation_on_CRISPR_Cas9_mutated_plants_Final.pdf. Accessed 6 Sept 2017

Tagliabue G (2015) The nonsensical GMO pseudo-category and a precautionary rabbit hole. Nat Biotechnol 33:907–908. https://doi.org/10.1038/nbt.3333

Tagliabue G, Ammann K (2018) Some basis for a renewed regulation of agri-food biotechnology in the EU. J Agric Environ Ethics 31:39–53. https://doi.org/10.1007/s10806-018-9708-9

Tait J, Barker G (2011) Global food security and the governance of modern biotechnologies. EMBO Rep 12:763–768. https://doi.org/10.1038/embor.2011.135

Taschner HC, Frietsch E (1990) Produkthaftungsgesetz und EG-Produkthaftungsrichtlinie: Kommentar, 2nd edn. Beck, München

Taylor S (2008) Economic loss caused by GMOs in France. In: Koch BA (ed) Economic loss caused by genetically modified organisms: liability and redress for the adventitious presence of GMOs in Non-GM crops. Springer, Vienna, pp 203–212

The Greens/European Free Alliance in the European Parliament (2017) New GMOs: Not as precise as advertised… and still useless. https://www.greens-efa.eu/en/article/news/new-gmos-7284/. Accessed 23 Aug 2018

The Greens/European Free Alliance in the European Parliament (2018a) Gene editing techniques are GMOs says EU Court of Justice. https://www.greens-efa.eu/en/article/news/gene-editing-techniques-are-gmos-says-eu-court-of-justice/. Accessed 6 Aug 2018

The Greens/European Free Alliance in the European Parliament (2018b) New GMOs are not progress, but yet another tool of industrial farming. https://www.greens-efa.eu/en/article/news/new-gmos-are-not-progress-but-yet-another-tool-of-industrial-farming/. Accessed 20 July 2018

The Group of FAS Biotechnology Specialists in the European Union (2017) Agricultural Biotechnology Annual EU-28: USDA Foreign Agricultural Service - Gain Report. https://gain.fas.usda.gov/Recent%20GAIN%20Publications/Agricultural%20Biotechnology%20Annual_Paris_EU-28_12-14-2018.pdf. Accessed 1 May 2019

The Group of FAS Biotechnology Specialists in the European Union (2018) Agricultural biotechnology annual EU-28: USDA foreign agricultural service - gain report. https://gain.fas.usda.gov/Recent%20GAIN%20Publications/Agricultural%20Biotechnology%20Annual_Paris_EU-28_12-22-2017.pdf. Accessed 14 February 2018

The Netherlands Commission on Genetic Modification (2006) New techniques in plant biotechnology. COGEM report CGM/061024-02. http://www.cogem.net/index.cfm/en/publications/publication/new-techniques-in-plant-biotechnology. Accessed 23 Aug 2018

The Netherlands Commission on Genetic Modification (2009a) Zinc finger on the pulse: Developments and implications of zinc finger technology. COGEM Report CGM/090616-02. http://www.cogem.net/index.cfm/en/publications/publication/zinc-finger-on-the-pulse-developments-and-implications-of-zinc-finger-technology. Accessed 23 Aug 2018

The Netherlands Commission on Genetic Modification (2009b) Should EU Legislation Be Updated?: Scientific developments throw new light on the process and product approaches. COGEM Report CGM/090626-03. http://www.cogem.net/index.cfm/en/publications/publication/should-eu-legislation-be-updated-scientific-developments-throw-new-light-on-the-process-and-product-approaches?q=Should+EU+Legislation+Be+Updated%3F&category=&from=30-09-1998&to=01-09-2017&order=relevance. Accessed 23 Aug 2018

The Netherlands Commission on Genetic Modification (2014) CRISPR-Cas – Revolution from the lab: COGEM Report and advice CGM/141030-01. http://www.cogem.net/index.cfm/en/publications/publication/crispr-cas-revolution-from-the-lab?q=revolution+from+the+lab&category=&from=30-09-1998&to=01-09-2017&order=relevance. Accessed 23 Aug 2018

The Netherlands Commission on Genetic Modification, Health Council of the Netherlands (2016) Biotechnology Trend Analysis 2016: A Regulatory Disconnect. http://www.cogem.net/index.cfm/en/publications/publication/trend-analysis-biotechnology-2016?q=&category=all&from=30-09-1998&to=14-09-2016&order=date_desc. Accessed 8 Jan 2018

The Netherlands Government (2017) Proposal for discussion on actions to improve the exemption mechanism for genetically modified plants under Directive 2001/18/EC. https://webcache.googleusercontent.com/search?q=cache:RuFCginICAcJ:https://www.rijksoverheid.nl/binaries/rijksoverheid/documenten/kamerstukken/2017/09/13/proposal-for-discussion/proposal-for-discussion.pdf. Accessed 4 Jan 2018

TNS Opinion & Social (2010a) Biotechnology: Special Eurobarometer 341 / Wave 73.1. http://ec.europa.eu/commfrontoffice/publicopinion/archives/ebs/ebs_341_en.pdf. Accessed 15 Mar 2018

TNS Opinion & Social (2010b) Food-related risks: Special Eurobarometer 354 / Wave 73.5. https://www.efsa.europa.eu/sites/default/files/corporate_publications/files/reporten.pdf. Accessed 16 Feb 2018

TNS Opinion & Social (2014) Europeans, Agriculture and the Common Agricultural Policy (CAP): Special Eurobarometer 410 / Wave EB80.2. http://ec.europa.eu/commfrontoffice/publicopinion/archives/ebs/ebs_410_en.pdf. Accessed 7 Mar 2018

Trans Atlantic Consumer Dialogue (2016) Resolution on consumer concerns about new genetic engineering techniques. http://tacd.org/wp-content/uploads/2016/09/TACD-Resolution-new-genetic-engineering-techniques_with-appendix_7-September.pdf. Accessed 24 Oct 2017

Tzotzos GT, Hull R, Head GP (2009) Genetically modified plants: assessing safety and managing risk, 1st edn. Elsevier/Academic Press, Amsterdam/Boston

U.S. Department of Agriculture (2018) Secretary Perdue Statement on ECJ Ruling on Genome Editing: Release No. 0155.18. https://www.usda.gov/media/press-releases/2018/07/27/secretary-perdue-statement-ecj-ruling-genome-editing. Accessed 8 Aug 2018

U.S. Department of Agriculture (2019) Joint statement of Western Hemisphere agriculture leaders. https://www.usda.gov/media/press-releases/2019/05/13/joint-statement-western-hemisphere-agriculture-leaders. Accessed 15 May 2019

UK Advisory Committee on Releases to the Environment (2011) Advice on a Plant Breeding Technique Involving Oligo-Directed Mutagenesis: RTDS™. http://webarchive.

nationalarchives.gov.uk/20130702160853/http://www.defra.gov.uk/acre/files/20110319-Cibus-advice.pdf. Accessed 15 Feb 2018

UK Advisory Committee on Releases to the Environment (2013a) Report 2: Why a modern understanding of genomes demonstrates the need for a new regulatory system for GMOs. https://www.gov.uk/government/uploads/system/uploads/attachment_data/file/239852/genomes-and-gm-regulation.pdf. Accessed 6 Jan 2018

UK Advisory Committee on Releases to the Environment (2013b) ACRE advice: New techniques used in plant breeding. https://www.gov.uk/government/publications/genetically-modified-organisms-new-plant-growing-methods. Accessed 24 Sept 2018

UK Advisory Committee on Releases to the Environment (2018) Advice on genome-edited Camelina plants with increased levels of oleic acid. https://assets.publishing.service.gov.uk/government/uploads/system/uploads/attachment_data/file/708560/gmo-camelina-oleic-acre-advice.pdf.pdf. Accessed 22 July 2018

UK Biotechnology and Biological Sciences Research Council (2014) New Techniques for Genetic Crop Improvement: position statement. https://bbsrc.ukri.org/documents/genetic-crop-improvement-position-statement-pdf/. Accessed 24 July 2018

UK Department for Environment, Food and Rural Affairs (2019) Developing genetically modified organisms (GMOs) if there's no Brexit deal. https://www.gov.uk/guidance/developing-genetically-modified-organisms-gmos-if-theres-no-brexit-deal. Accessed 01 April 2019

UK Government (2016) Government statement: Evidence Check: GM and Gene Editing. https://www.parliament.uk/documents/commons-committees/science-technology/evidence-tests/GM-and-Gene-Editing.pdf. Accessed 21 July 2018

UK House of Commons Science and Technology Committee (2015) Advanced genetic techniques for crop improvement: regulation, risk and precaution: Fifth Report of Session 2014-15. HC 328. Tso, London

United Nations Environment Programme (1995) International technical guidelines for safety in biotechnology. https://unep.ch/biosafety/old_site/development/devdocuments/Techguidelines.pdf. Accessed 19 February 2018

Ulfbeck V (2008) Economic loss caused by GMOs in Denmark. In: Koch BA (ed) Economic loss caused by genetically modified organisms: liability and redress for the adventitious presence of GMOs in Non-GM crops. Springer, Vienna, pp 145–161

Valavanidis A (2016) New 'Plant Breeding' Techniques for Superior and Improved Cultivars in Agriculture: European Union Approval Challenges and Scientific Reports for Applications and Efficient Crop Developments. https://www.researchgate.net. Accessed 1 Mar 2018

van Dam C (2013) European tort law, 2nd edn. OUP, Oxford

van Diemen FR, Lebbink RJ (2017) CRISPR/Cas9, a powerful tool to target human herpesviruses. Cell Microbiol, 19. https://doi.org/10.1111/cmi.12694

van Heezik G, Tuinzing-Westerhuis F (2018) The consequences of the ECJ ruling for the European market. http://www.nbtplatform.org/stories/greetje-van-heezik-fleur-tuinzing-westerhuis. Accessed 5 Sept 2018

van Mil A, Hopkins H, Kinsella S (2017) Potential uses for genetic technologies: dialogue and engagement research conducted on behalf of the Royal Society: Findings Report. https://royalsociety.org/~/media/policy/projects/gene-tech/genetic-technologies-public-dialogue-hvm-full-report.pdf. Accessed 24 Aug 2018

VIB (2019) Permit for CRISPR maize field trial that aims to measure climate stress. http://www.vib.be/en/news/Pages/Permit-for-CRISPR-field-trial.aspx?mc_cid=4b21e7e5c7&mc_eid=b89825d9e5. Accessed 25 May 2019

Vogel B (2012) New Plant Breeding Techniques: Groundwork for the Clarification of Outstanding Questions on the legal Regulation of New Plant Breeding Techniques. Commissioned by the Swiss Federal Office for the Environment. https://awel.zh.ch/internet/baudirektion/awel/de/biosicherheit_neobiota/gvo/Neue_Pflanzenzuchtverfahren/_jcr_content/contentPar/downloadlist/downloaditems/542_1478606633506.spooler.download.1478606536961.pdf/NewPlantBreedingTechnologies__7Nov16_Endversion.pdf

Voigt B, Klima J (2017) CRISPR-Plants & Co. – the Quest for Adequate Risk Regulation: modern plant breeding techniques and the current legal framework for risk regulation in the European Union. Zeitschrift für Europäisches Umwelt- und Planungsrecht 15:319–338

von Kries C, Winter G (2011) Legal implications of the step-by-step principle. Environ Sci Eur 23:32. https://doi.org/10.1186/2190-4715-23-32

von Kries C, Winter G (2012) The structuring of GMO release and evaluation in EU law. Biotechnol J 7:569–581. https://doi.org/10.1002/biot.201100321

Wasmer M (2019) Roads forward for European GMO policy—Uncertainties in wake of ECJ judgment have to be mitigated by regulatory reform. Front Bioeng Biotechnol 7:1–12. https://doi.org/10.3389/fbioe.2019.00132

Weissenbacher M (2010) Damage caused by GMOs under Austrian law. In: Koch BA (ed) Damage caused by genetically modified organisms: comparative survey of redress options for harm to persons, property or the environment. de Gruyter, Berlin, pp 2–34

Winter G (2016a) Cultivation restrictions for genetically modified plants. Eur J Risk Regul 7:120–143. https://doi.org/10.1017/S1867299X00005444

Winter G (2016b) In search for a legal framework for synthetic biology. In: Engelhard M (ed) Synthetic biology analysed: tools for discussion and evaluation. Springer, Switzerland, pp 171–211

Wortmann L (2015) European political groups on genetically modified organisms: strong statements, weak programmes? Bachelor thesis. https://essay.utwente.nl/68117/1/wortmann_BA_BMS.pdf. Accessed 31 July 2018

Wolt JD, Wang K, Yang B (2016) The regulatory status of genome-edited crops. Plant Biotechnol J 14:510–518. https://doi.org/10.1111/pbi.12444

World Trade Organization Committee on Sanitary and Phytosanitary Measures (2018) International statement on agricultural applications of precision biotechnology - communication from Argentina, Australia, Brazil, Canada, the Dominican Republic, Guatemala, Honduras, Paraguay, the United States of America and Uruguay - Revision: G/SPS/GEN/1658/Rev.3

World Trade Organization Committee on Sanitary and Phytosanitary Measures (2019) Summary of the meeting of 1–2 November 2018: note by the secretariat. G/SPS/R/93

Yusuf N (2014) "One door, one key principle", does it really exist? Master Thesis. http://edepot.wur.nl/327217. Accessed 31 Aug 2018

Chapter 6
Regulation of Genome Editing in Plant Biotechnology: Japan

Tetsuya Ishii

Abstract To regulate the research and industrial uses of genetically modified organisms (GMOs), Japan enacted the Act on the Conservation and Sustainable Use of Biological Diversity through Regulations on the Use of Living Modified Organisms 2003. This law can be regarded as a product-based GMO regulation. To date, Japan has approved 133 GM crop varieties for cultivation, distribution, and import, thus becoming a major importer of GM crops in the world. However, no GM crops have been commercially cultivated in Japan, except one ornamental GM flower. A recent consumer survey showed that 40.7% of respondents expressed concern over the safety of GM food products. Meanwhile, some Japanese researchers have already used robust genome editing techniques, such as CRISPR-Cas9, and reported gene-disrupted apple, potato, soybean, tomato and rice. In 2017, a GM rice variety was approved as Japan's first field trial of a genome edited crop. In contrast, some citizen groups expressed opposition to the cultivation test and demanded the regulation of genome edited crops. However, relevant ministries have not considered the regulation of any uses of genome editing in earnest. The current state of Japan does not warrant a promising future of genome edited crops.

6.1 Introduction

Japan's Gross Domestic Product (GDP) value in 2016 represents 7.97% of the world economy that is the third GDP position followed by the U.S. and China.[1] With regard to farm products, Japan's trend in the import and export, however, has continued to record an import surplus trade balance for two decades. The import amount of farm products in 2016 was 5827.3 billion JPY, whereas the export amount

[1]IMF (2016).

T. Ishii (✉)
Office of Health and Safety, Hokkaido University, Sapporo, Japan
e-mail: tishii@general.hokudai.ac.jp

H.-G. Dederer, D. Hamburger (eds.), *Regulation of Genome Editing in Plant Biotechnology*, https://doi.org/10.1007/978-3-030-17119-3_6

was 459.3 billion JPY, thus resulting in the excess of imports of 5368 billion JPY (approx. US$47.5 billion).[2]

Of the imported farm products in 2016, major crops were maize (15,342 thousand tons), soy (3131 thousand tons), rapeseed (2366 thousand tons) and cotton (100 thousand tons).[3] In 2016, Japan imported the great part of maize from the U.S. (74.5%), most of soy from the U.S. (71.5%), the vast majority of rapeseed from Canada (95.1%) and about half of cotton from Australia (56%).[4] In the U.S., the cultivation area ratios of GM maize and soybean were approximately 92 and 94% in 2016, respectively. The cultivation area ratio of GM rapeseed in Canada and that ratio of GM cotton in Australia were about 93% and 98% in 2016, respectively.[5] Therefore, the estimated total amount of GM maize, soybean, rapeseed and cotton imported into Japan was at least 15,975 thousand tons in 2016. It is not an exaggeration to mention that Japan is one of the leading importers of GM crops in the world.

In 2003, Japan ratified the Cartagena Protocol on Biosafety to the Convention on Biological Diversity.[6] In 2004, the so-called Cartagena Law 2003 was enacted to domestically regulate the research use, cultivation, and industrial application of genetically modified organisms (GMOs). As of December 2017, competent ministers have approved 133 crop varieties for cultivation, distribution and import.[7] The commercial cultivation of GM crops in Japan began in 2009.[8] It was a GM flower for ornamental use. Namely, it was 'blue' rose developed through the introduction of two different transgenes (OECD UI: IFD-52401-4 and 9). However, it has been the sole example of commercial cultivation of GM crops in Japan thus far. The introduction of GM food crops into Japan, at present, completely depends on importation, not domestic cultivation. In 2017, the Consumer Affairs Agency issued a survey result of consumers' attitude towards GM food products: 40.7% of 10,648 people selected by stratified random sampling responded that they are anxious about the safety of GM food products.[9]

Recently, new plant breeding techniques (NPBTs), including genome editing, epigenome editing and oligonucleotide-directed mutagenesis (ODM), have been developed, which can produce plant varieties with no transgenes that are decisive indicators of GMOs in many countries.[10] Genome editing techniques, such as zinc-finger nucleases (ZFNs), transcription activator-like effector nucleases (TALENs), the clustered regularly interspaced short palindromic repeat (CRISPR)-Cas system,

---

[2]MAFF (2017d).

[3]MAFF (2017b).

[4]Ibid.

[5]Ibid.

[6]Biosafety Clearing House (2018).

[7]MAFF (2017a).

[8]Suntory (2018).

[9]CAA (2017b).

[10]Lusser et al. (2011).

are robust genetic engineering tools that employ artificially site-directed nucleases.[11] Directly introducing artificial, site-directed nucleases into plant cells can efficiently induce DNA double-strand breaks (DSBs) at target sites, and subsequently achieve non-homologous end-joining (NHEJ) as well as homology-directed repair (HR) in plant cells.[12] According to proposed categories of genetic modification by genome editing,[13] site-specific random mutagenesis achieved via NHEJ is called site-directed nucleases (SDN) Type 1 (SDN-1). SDN-2 generates site-specific desired point mutation via HR. In SDN-3, a larger stretch of DNA template is introduced at specific site in the plant genome. ODM employs approximately 20–100 nucleotides to attain mutagenesis similar to SDN-2, without using nucleases.[14]

Japanese researchers have used genome editing techniques and reported resultant crops, such as apple, grape, potato, soybean, tomato and rice that underwent a gene disruption due to the introduction of (insertions or deletions) indels.[15] In addition, Japan's first field trial of a genome edited crop took place in 2017. It used two different rice varieties, in which *TGW6* (IAA-Glucose hydrolase gene) and OsCKX2 (cytokinin oxidase/dehydrogenase) were disrupted using CRISPR-Cas9 to increase the grain size and number, respectively.[16] It is worth considering whether the technical advantage of genome editing can improve the social acceptance of crops developed by biotechnology in Japan where GM crops have been not well accepted.

The present report outlines the regulatory framework for GMOs, then shows the current state of the regulatory consideration of genome edited crops in Japan. After the further analysis of the regulatory position of genome edited crops, it considers public opinions towards genome edited crops. Taken together, this report presents the future of genome edited crops from Japan's perspective.

6.2 The Regulatory Framework for Genetically Modified Organisms (GMOs): An Overview

In 2004, Cartagena Law: The Act on the Conservation and Sustainable Use of Biological Diversity through Regulations on the Use of Living Modified Organisms 2003 was enacted. Its purpose is to ensure 'the precise and smooth implementation of the Cartagena Protocol on Biosafety to the Convention on Biological Diversity [. . .], thereby contributing to the welfare of humankind and helping to assure healthy cultural lives for the people now and in the future, by devising measures to regulate

[11]Gaj et al. (2013) and Jiang and Doudna (2017).

[12]Araki and Ishii (2015).

[13]EFSA (2012).

[14]Op. cit. 10.

[15]Nishitani et al. (2016), Nakajima et al. (2017), Sawai et al. (2014), Kanazashi et al. (2018), Ito et al. (2015), and Endo et al. (2016).

[16]MOE (2017).

the use of living modified organisms [LMOs, legal technical term of GMOs in Cartagena Law 2003] in order for the conservation and the sustainable use of biological diversity through international cooperation' (Article 1 of Chapter 1).[17]

Figure 6.1 structurally illustrates the governance of LMO uses by Cartagena Law 2003 and related regulations, including cabinet orders, ministerial ordinances and ministerial notifications. The enforcement of Cartagena Law 2003 has been carried out under the joint jurisdiction of six ministries: Ministry of Finance (MOF), Ministry of Education, Culture, Sports, Science and Technology (MEXT), Ministry of Health, Labour and Welfare (MHLW), Ministry of Agriculture, Forestry and Fisheries (MAFF), Ministry of Economy, Trade and Industry (METI) and Ministry of the Environment (MOE).

Cartagena Law 2003 defines 'organism' as: 'a single cell (excluding a single cell forming a cell colony) or a cell colony which is stipulated in the ordinance of the competent ministries as having the capacity to transfer or replicate nucleic acid, and viruses and viroids' [Article 2(1)]. A LMO shall mean 'an organism that possesses nucleic acid, or a replicated product thereof, obtained through use of the any of stipulated technologies' [Article 2 (2)].[18] Such technologies, which correspond to 'Modern Biotechnology' in the Cartagena Protocol,[19] are legally categorized into two types. The one type is 'technologies for processing nucleic acid extracellularly for the purpose of introducing the nucleic acid into cells, viruses or viroids to transfer or replicate the nucleic acid' [Article 2 (2) (i)] and the other type is 'technologies for fusing of the cells of organisms belonging to different taxonomical families' [Article 2 (2) (ii)].[20] This report focuses on the former technologies stipulated in Article 2 (2) (i). In the Regulations related to the Enforcement of the Cartagena Law (Fig. 6.1), some technologies are excluded from the technologies for processing nucleic acid extracellularly. Such excluded technologies stipulated in Article 2 of the Regulations are shown below.[21]

(i) Technology for processing by using, as nucleic acid to be introduced into cells, only the nucleic acid shown in the following

   A. The nucleic acid of living organism belonging to the same species as that of the living organism which the cells originate from.
   B. The nucleic acid of living organism belonging to the species that exchanges nucleic acid with the species of the living organism which the cells originate from in natural conditions

[17]Ministry of Justice (2018).

[18]Ibid.

[19]Biosafety Clearing House (2000).

[20]Op.cit 17.

[21]MEXT-MHLW-MAFF-METI-MOE (2003b).

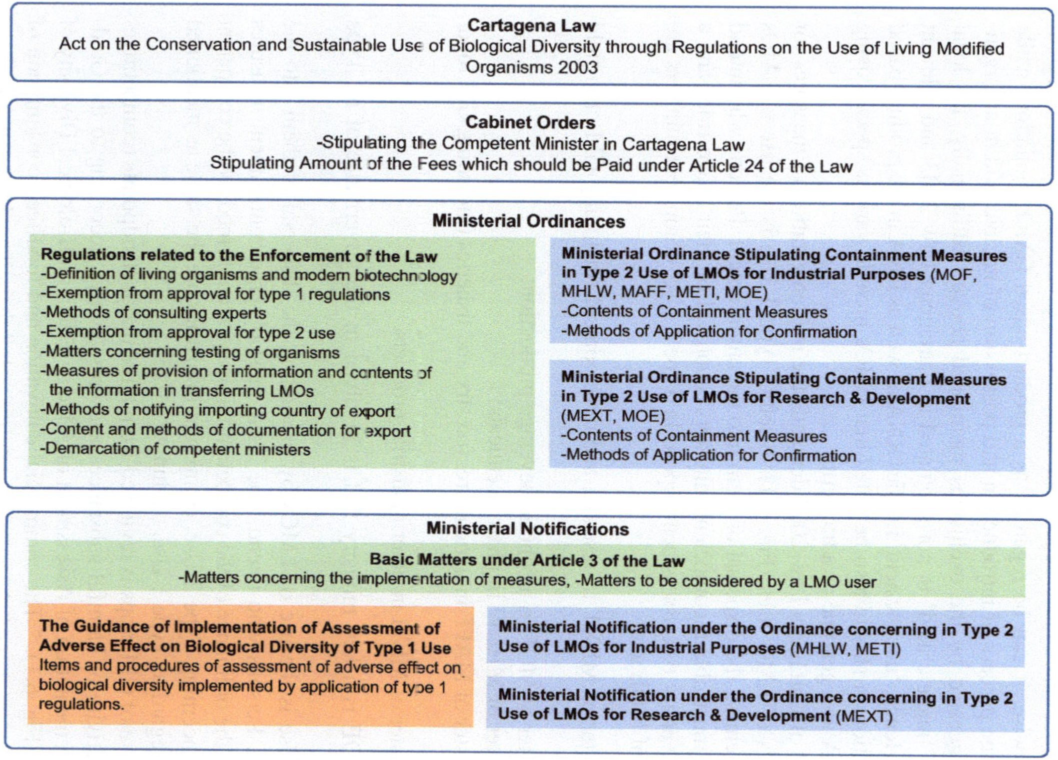

Fig. 6.1 The governance of uses of living modified organisms (LMOs) by Cartagena Law 2003 and relevant regulations. MOF Ministry of Finance, MEXT Ministry of Education, Culture, Sports, Science and Technology, MHLW Ministry of Health, Labour and Welfare, MAFF Ministry of Agriculture, Forestry and Fisheries, METI Ministry of Economy, Trade and Industry, MOE Ministry of the Environment. Green squares mean regulations related to the enforcement of, and basic matters under Cartagena Law 2003, which are relevant to Type 1 Use and Type 2 Use. Blue squares imply the regulation regarding Type 2 Use. The orange-coloured square is the regulation specific to Type 1 Use. Adapted from a figure shown on the website of Japan Biosafety Clearing House (http://www.biodic.go.jp/bch/english/law.html)

(ii) Technology for processing by using, as nucleic acid to be introduced into viruses or viroids, only the nucleic acid of viruses or viroids that exchanges nucleic acid with the viruses or viroids in natural conditions.

Such legal definitions in Cartagena Law 2003, in which nucleic acid or a replicated product possessed by an organism is emphasized, suggest that Japan adopts a product-based GMO regulation, rather than a process-based GMO regulation.[22] Otherwise, it can be regarded as a product-based GMO regulation, while considering the process of introduction and possession of nucleic acid in an organism,[23] as suggested by the exclusions mentioned above. Additionally, the legal definition of 'nucleic acid' is not stipulated in Cartagena Law 2003 and relevant regulations. Hence, 'nucleic acid' implies RNA as well as DNA in the product-based GMO regulation. It should also be noted that LMOs include various species using DNA or RNA as a genome or a transcript.

Chapter 2 of Cartagena Law 2003 demands the implementation of measures for the two types of LMO uses (Fig. 6.1). Briefly, Type 1 Use is the use of LMOs without preventing the dispersal of them into the environment. Persons who intend Type 1 Use must obtain competent ministers' approval of their usage rules and a biological diversity risk assessment report. In Type 1 Use, competent minsters are the Minster of MOE and:

- The Minster of MEXT in research and development (including GM crop field trial);
- The Minster of MOF in alcoholic beverage production;
- The Minster of MHLW in pharmaceuticals;
- The Minister of MAFF in agriculture, forestry and fisheries (including GM food/feed);
- The Minister of METI in other industrial products.

Thus, MOE is a key ministry that is involved in the governance of all Type 1 Uses.

Type 2 Use is the use of LMOs by preventing the dispersal of them into the environment. Persons who intend type 2 use must observe containment measures stipulated in the ordinance of the competent ministry or confirmed by the competent minister. Concerning Type 2 Use, competent ministries are the same as mentioned above with regard to Type 1 Use, excluding MOE.

Type 1 Use of GM crops is reviewed from three major standpoints (competitiveness, productivity of harmful substances and crossability), according to the Guidance of Implementation of Assessment of Adverse Effect on Biological Diversity of Type 1 Use (Fig. 6.1).[24] The commercial cultivation and food/feed consumption of GM crops is reviewed based on another administrative document, entitled

---

[22]Ishii and Araki (2017).

[23]SCJ (2014).

[24]MEXT-MHLW-MAFF-METI-MOE (2003a).

Table 6.1 Approved GM crop varieties in Japan, as of December 19, 2017

Crop	Import, distribution, or commercial cultivation as food/feed or an ornamental flower (cultivation also permitted)	Only for cultivation test in isolated fields	Major traits in approved varieties
Maize	83 (81[a])	11	Pest resistance, herbicide resistance
Cotton	33 (0[a])	3	Pest resistance, herbicide resistance
Soybean	28 (21[a])	3	Pest resistance, herbicide resistance, higher content of a specific ingredient
Rapeseed	16 (14[a])	2	Herbicide resistance
Alfalfa	5 (5)	0	Herbicide resistance
Papaya	1 (1)	0	Virus disease resistance
Sugar beat	1 (1)	0	Herbicide resistance
Carnation	8 (8)	1	Blue petals for ornamental use
Rose	2 (2[b])	0	Blue petals for ornamental use
Bent glass	0	1	–
Cyclamen	0	2	–
Rice	0	21	–
Total	177 (133)	44	

Adapted from Ministry of Agriculture, Forest, Fisheries website (http://www.maff.go.jp/j/syouan/nouan/carta/torikumi/index.html#1)
[a]Application for commercial cultivation was not submitted
[b]One of rose varieties is currently cultivated

'Concerning the application for approval of Type 1 Use Regulations with regard to the genetically modified plants, the production or circulation of which falls within the jurisdiction of the Minister of Agriculture, Forestry and Fisheries' (2007, last amended 2017, not shown in Fig. 6.1). MHLW receives and reviews the application for the safety review of GM foods. Then, MHLW requests the Food Safety Commission in the Cabinet Office to assess the effects of food on health of an applied GM food, in accordance with Food Safety Basic Act 2003 and Food Sanitation Law 1947. As of December 2017, 177 GM crop varieties, including maize, cotton, soybean and rapeseed, have been approved (Table 6.1).[25] Of them, 133 crop varieties are permitted for cultivation, distribution and import. As mentioned above, no GM crops have, however, been commercially cultivated in Japan thus far, except one GM rose variety. Of note, regarding rice that is a major food crop for Japanese people, 21 GM rice varieties were permitted for field cultivation test (Table 6.1).

[25]Op. cit.7.

Table 6.2 The review of GM plants and relevant regulars and legislation in Japan

Area of responsibility	Regulators	Relevant legislation
Type 2 Use: Research with measures for preventing the dispersal of GM plants	MEXT	Cartagena Law 2003
Type 1 Use: Cultivation test of GM crops in isolated fields, without measures for preventing the dispersal	MEXT and MOE	Cartagena Law 2003
Type 1 Use: Import, distribution or commercial cultivation of GM plants, without measures for preventing the dispersal	MAFF and MOE	Cartagena Law 2003
Safety of GM food/feed products	MHLW, MAFF and Food Safety Commission	Food Safety Basic Act 2003, Food Sanitation Law 1947 and Act on Safety Assurance and Quality Improvement of Feeds 1953
Inspection of the contamination of GM plants	MAFF	Cartagena Law 2003
Accreditation of inspection organizations on GM plants and food products	MAFF	Cartagena Law 2003
Labelling of GM food products	Consumer Affairs Agency	Food labelling Act 2013

MEXT Ministry of Education, Sports, Science and Technology, MAFF Ministry of Agriculture, Forest, Fisheries, MOE Ministry of the Environment, MHLW Ministry of Health, Labour and Welfare

However, no GM rice varieties were approved for cultivation, distribution nor import.

MAFF has other roles in the regulation of GM crops in Japan. This ministry is in charge of accreditation of inspection organizations that test the containment of GM crops and related foods. In addition, MAFF themselves also inspects whether the imported crop seeds and seedlings contain GM seeds or not (Article 31 of Cartagena Law 2003). In 2017, MAFF announced that its on-site inspection of GM seed contamination focuses on imported flax, cabbage, cauliflower, wheat, eggplant, green pepper, papaya and cotton.[26] However, a contamination of unapproved GM flower (petunia) seeds was found in the domestic market in May 2017. MAFF conducted an emergency survey on the contamination of GM petunia seeds. To date, they have found 60 unapproved GM petunia varieties among the 1359 petunia varieties commercially available in Japan.[27] Another grave incident or accident of GM crop contamination is so-called ‘overflowing rapeseed’ which resulted in the spontaneous growth of, and hybridization by GM rapeseed at the roadsides near ports and harbours, as reported by at least six papers, reviewed in.[28]

[26]MAFF (2017c).

[27]MAFF (2018b).

[28]Ryffel (2014).

In sum, the areas of review, relevant regulators and legislation regarding the research use, cultivation, distribution, import, safety and labelling of GM crops or food/feed products are shown in Table 6.2.

6.3 Regulatory Status of Genome Edited Plants

6.3.1 Applicability of the Regulatory Framework for GMOs

As discussed in the Sect. 6.2, it is suggested that Japan's Cartagena Law 2003 adopts a product-based GMO regulation, or a product-based GMO regulation considering the process of introduction and possession of nucleic acid in an organism. However, there have been no official decisions regarding the regulation of plants obtained through or plant breeding using NPBTs, such as genome editing.

Recently, MOE deliberated a periodic examination of the enforcement of Cartagena Law 2003 and called for public comments on the result. Such activities are mandatory by Article 34 (Measures for Progress of Scientific Knowledge) and 35 (Public Consultation) of Cartagena Law 2003. Its deliberation report and response to public comments in 2016 suggest that MOE recognizes that there is a pressing need for the regulatory consideration of crops developed using NPBTs.[29] However, it was also stated in the report that the regulation of such as genome edited crops should be meticulously considered, while collecting the latest scientific findings and relevant international movements. MOE further stated that relevant ministries should secure the system to provide guidance for developers who use genome editing for plant breeding. However, such statements by MOE are ambiguous for developers and consumers because they have, at present, no official guidance at all. According to the statement by MOE, MAFF recently showed the contact phone number for developers on its website, in order to respond to inquiries regarding the development of crops using NPBTs.[30]

At least, the genome edited crops that possess nucleic acid, or a replicated product thereof, obtained through use of technologies for processing nucleic acid extracellularly for the purpose of introducing the nucleic acid into plant cells are deemed to be the subject of Cartagena Law 2003, unless the possessed nucleic acid belongs to the same species as that of the plant from which plant cells originate, according to the stringent interpretation of definitions of LMOs and technologies for producing LMOs in Cartagena Law 2003.

More specifically, most plants obtained through SDN-3 would be subject to the product-based regulation if the integrated 'nucleic acid' is derived from another plant, animal, or microorganism species. In addition, plants derived from SDN-1 or SDN-2 may also fall under Cartagena Law 2003, if a *FokI* nuclease gene derived

[29] MOE (2016).

[30] MAFF (2018a).

from *Flavobacterium okeanokoites* in ZFN or *Xanthomonas oryzae pv. Oryzae* in TALEN,[31] *Cas9* endonuclease gene derived from *Streptococcus pyogenes* in CRISPR/Cas9,[32] or plasmid vector derived from such as *Agrobacterium tumefaciens* in *Agrobacterium*-mediated gene transfer is integrated in the plant genome and resultant plants possess such 'nucleic acid'. This indication is based on scientific evidence shown in a recent report; the integration of plasmid-derived DNA fragments in the plant genome was detected, when plasmid harbouring Cas9 gene were transfected into *Arabidopsis* protoplasts.[33]

Therefore, plants possessing other species' gene or 'nucleic acid' of any length via SDN-1, 2 and 3 are subject to Cartagena Law 2003.

6.3.2 Regulatory Classification of Genome Editing/Genome Edited Plants

Although Japan's regulators have not in detail considered the regulation of genome edited plants, presumed regulatory points to consider are discussed below from a process-based standpoint.

In Cartagena Law 2003, 'nucleic acid' implies DNA and RNA, as mentioned in Sect. 6.2. For this reason, single guide (sg) RNA, which is a guiding molecule of nuclease in CRISPR-Cas9, can legally be interpreted as 'nucleic acid'. In detail, the structure of a sgRNA is composed of crRNA and *trans*-activating crRNA (trancRNA) in addition to a sequence of 20 nucleotides in length that is complementary to a target sequence in the plant genome.[34] At least, crRNA and trancRNA are derived from *Streptococcus pyogenes*. Even if CRISPR-Cas9 in the form of ribonucleoprotein is introduced into plant cells for SDN-3, 2 or 1, Cartagena Law 2003 can regard crRNA and trancRNA as 'nucleic acid'. Similar points can be indicated regarding CRISPR-Cpf1 that is another CRISPR system derived from *Lachnospiraceae bacterium* or *Acidaminoncoccus sp.* and uses crRNA.[35] Although it is unlikely that such introduced RNA is reverse-transcribed and integrated into the plant genome (unless the plant is not infected with RNA virus), at least legal clarifications are required to exempt such methods from 'any of stipulated technologies' [Article 2 (2)]. [36]

In contrast, if SDN-1 is carried out using ZFN or TALEN in the form of protein, this case appears to fall out of Cartagena Law 2003. Likewise, ODM does not involve the introduction of nuclease gene or plasmid DNA at all, except

[31] Gaj et al. (2013).

[32] Jiang and Doudna (2017).

[33] Kim and Kim (2016).

[34] Op. Cit. 32.

[35] Tang et al. (2017).

[36] Op.cit.21.

oligonucleotide.[37] It appears that most of the crops which resulted from ODM using 20–100 nt are not subject to Cartagena Law 2003 if such a short DNA sequence can be found in the same plant species. Namely, such plants fall out of the GMO regulation if the gene modified by ODM is a genetic variant that exists in the plant species.

Return to a product-based standpoint. It is conceivable that some plant varieties with no transgenes, which can be efficiently generated via SDN-1 but are not subject to Cartagena Law 2003, may affect biological diversity by exerting competitiveness, production of harmful substances and crossability. For instance, *ALS*-disrupted maize and potato were generated via SDN-1,[38] imitating non-GM rice with a mutated *ALS* that confers herbicide resistance to the rice.[39] However, increasing crop varieties with such a disrupted *ALS* may worsen weedy plant issues, which have emerged through hybridization of a non-GM rice with a mutated *ALS* with wild rice varieties and became already worrisome situations in Italy and the USA.[40] In addition, SDN-1 produces various genetic variants with various length of indels. If CSR1 are mixed into the cultivars of plants with an ALS variant, the circumstances could further worsen because this variant enables outcrossing in at least a highly selfing species, *Arabidopsis thaliana*.[41] Such apprehensions might become real problems in the environment. Compared with ODM, genome editing can readily attain by multiplex editing that multiple genes in the genome are simultaneously modified using several types of artificial nucleases (in ZFN and TALEN) or sgRNA (in CRISPR/Cas9) in combination.[42] In addition, there is, at present, no consensus regarding the means of assessing off-target effects or mutations in genome edited plants.[43]

Although uncertainties will be likely high in the case of multiplex SDN-1 and no assessment of off-target mutations has been made, should the cultivation and food/feed consumption of such resultant crop products be deregulated? It seems vital to validate genetic modification and assess acquired traits in some cases of SDN-1.

[37]Op.cit.10.

[38]Svitashev et al. (2016) and Butler et al. (2015).

[39]Burgos et al. (2014).

[40]Busconi et al. (2012) and Burgos et al. (2014).

[41]Bergelson et al. (1998).

[42]Ishii and Araki (2016).

[43]Joung (2015).

6.4 Status Quo of Genome Edited Plants and Products Derived from Them

6.4.1 Type 2 Use of Plant Genome Editing

In Japan, the regulation of genome editing has not been discussed in earnest in relevant ministries, such as MOE, nor the National Diet. Meanwhile, basic research on genome editing (primarily CRISPR-Cas9) in several plant species has been reported by Japanese researchers (Table 6.3).[44] Such plant genome editing experiments were performed at laboratories, implementing containment measures (e.g. inactivation by autoclave, ethanol or hypochlorous acid; placing adhesive mats at exits; attach filters on exhaust vents) to prevent the dispersal of plants, seeds and pollen into the environment. This is because those experiments involving *Agrobacterium*-mediated transfer of plasmid DNA harbouring a genome editing construct (Table 6.3) must be in accordance with the ministerial ordinance of Type 2 Use.[45] An institutional genetic recombination committee in each research organization reviews such experiment protocols regarding whether appropriate containment measures can be carried out at laboratories. Likewise, the regulation on Type 2 Use is applied to plant experiments involving electroporation or ballistic transfer of genome editing system in DNA form.

Regarding plant genome editing experiments in which no nucleic acid nor transgenes are introduced into plant cells, as illustrated by a direct delivery of TALEN protein in *Nicotiana benthamiana* protoplasts,[46] such protocols are voluntarily reviewed by an institutional genetic recombination committee in Japan. Such voluntary reviews have been recommended by the Academic Association for Genetic Studies since 2014.[47]

6.4.2 Type 1 Use of Plant Genome Editing

Currently, there are no approved genome edited crops and their food products in Japan. However, in 2017, the National Agriculture and Food Research Organization (NARO) initiated Japan's first field trial of genome edited crop in Tsukuba city. It is a gene-disrupted rice produced using genome editing. Of note, the sequencing of rice genome (approximately 400 Mbp in 12 chromosomes) was finished in 2004, by the effort of a multinational project, in which Japan played a major role by sequencing 6 of total 12 chromosomes (55% of whole genome).[48]

---

[44] Op.cit.15.

[45] J-BCH (2018).

[46] Luo et al. (2015).

[47] AAGS (2014).

[48] Jackson (2016).

Table 6.3 Research examples of food crops modified using genome editing in Japan

Species	Target gene	Genome editing	Mutagenesis pathway	Gene transfer	Genotyped subject	Efficiency of gene disruption (allele)	Off-target mutation analysis	Confirmed trait	Reference
Potato	*SSR2*	TALEN	NHEJ (SDN-1)	Agrobacterium	T0	–	N.D.	Reduced levels of cholesterol and steroidal glycoalkaloids	Sawai et al. (2014)
Rice	*ALS*	CRISPR-Cas9	HR (SDN-2)	Agrobacterium	T0	0.24–0.36% (2)	N.D.	–	Endo et al. (2016)
Tomato	*RIN*	CRISPR-Cas9	NHEJ (SDN-1)	Agrobacterium	T1	0–100% (2)	N.D.	Changed fruit ripening	Ito et al. (2015)
Apple	*PDS*	CRISPR-Cas9	NHEJ (SDN-1)	Agrobacterium	T0	–	N.D.	Albinism	Nishitani et al. (2016)
Grape	*PDS*	CRISPR-Cas9	NHEJ (SDN-1)	Agrobacterium	T0	–	N.D.	Albinism	Nakajima et al. (2017)
Soybean	*PPD1*, *PPD2*	CRISPR-Cas9	NHEJ (simultaneous SDN-1 at two genes)	Agrobacterium	T2	33% or more (2)	Not found in 3 putative off-targeted loci	Gigantism of mature seeds	Kanazashi et al. (2018)

According to the approved Type 1 Use application, the field trial uses two rice varieties in which *TGW6* and OsCKX2 were specifically disrupted using CRISPR-Cas9 to increase grain size and number. In detail, it was suggested that the treatment of CRISPR-Cas9 in rice cells resulted in plants with several copies of Cas9 genes integrated per haploid due to *Agrobacterium*-mediated gene transfer.[49] The rice plants are cultivated and used in the field test to produce later progeny that lost integrated *Cas9* in the genome through genetic segregation. Given such a background, the applications of field trials using CRISPR-gene-disrupted rice varieties were filed as Type 1 Use of LMOs and reviewed and approved as such (MOE 2017). The permitted period of Type 1 Use is from 2017 to 2022. The isolated field used for this study is surrounded by a fence of 1.8 m in height, with signs of 'No Trespassing'. Moreover, the field has a catch-drain as well as a washing place for farm machines, tools and shoes, to prevent the dispersal of GM rice outside the isolated field. The isolated field is covered by a bird net in sprouting season, and by a windbreak net in stormy weather. Other than such facilities, work procedures were determined in the following.

(i) Weed control is conducted to avoid the growing of other plants within the field as much as possible.
(ii) Taking Type 2 Use is mandatory when the GM rice is transferred or stored outside the isolated field.
(iii) Drying and threshing are performed within the field after the cultivation of GM rice
(iv) Inactivation of rice seeds and straw by burning or autoclave. Burying or plowing-down of rice residuum and sucker
(v) Washing farm machines, tools and shoes within the isolated field.
(vi) Ensuring and managing all facilities to keep them functioning in the field.
(vii) Disseminating such procedures to all workers in charge of Type 1 Use.
(viii) Taking emergency measures if the GM rice may affect biological safety.

6.5 Reform Efforts

In 2012, two opinion articles were issued in two different newspapers, stating the same content that there was the pressing need for regulatory consideration regarding transgene-free organisms generated by genome editing in Japan.[50] In one of the articles, the officials of MOE just commented that they would gather relevant information.[51] On the other hand, a GM crop committee in the Science Council of Japan (SCJ) considered NPBTs and issued a report in 2014, entitled 'Current Status

[49]Op.cit.16.

[50]Segawa and Hatano (2012) and Nishiyama (2012).

[51]Segawa and Hatano (2012).

and Issues of New Plant Breeding Techniques'.[52] The report technically outlined five NPBTs, including genome editing and epigenome editing, and showed their great importance in future plant breeding. It also stressed the remaining issues regarding the detection of induced mutations, clarification of differences between induced mutations and naturally-occurring variants, and the assessment of unpredictable mutations and functional changes of endogenous genes. Furthermore, the report proposed that genome edited organisms should be, regardless of the existence of transgenes, regulated under Cartagena Law for the time being, while disclosing relevant information to the public and increasing international cooperation.

Following this, MAFF held a study group on new breeding techniques and issued a report with the subtitle 'Toward the Development and Application of Farm Products using New Plant Breeding Techniques such as Genome Editing' in 2015.[53] The report made two main proposals. One is funding to NPBT research by a national project 'Innovative Technologies for Next Generation Agriculture, Forest and Fisheries' in the Cabinet Office's Cross-ministerial Strategic Innovation Promotion Program (so-called 'SIP' program, 2014–2018, http://sip-nbt.agbi.tsukuba.ac.jp/). The other is a battery of efforts towards the social acceptance of farm products developed by NPBTs, which includes regulatory review of such NPBT products under Cartagena Law 2003 on case-by-case basis, information disclosure to and communication with the public, and international cooperation of regulatory framework of NPBTs. The report by MAFF's study group adopted some proposals made in the SCJ report. It appears that most actions proposed in the study group were carried out, except the regulatory consideration of NPBTs. Of note, it seems that the field trial of gene-disrupted rice complied with this proposal because the field trials were performed with approval of Type 1 Use.[54]

However, MOE has not taken the initiative in considering the regulation of NPBTs, as mentioned in Sect. 6.3.1. Although MAFF showed the phone number on its website to respond inquiries regarding NPBTs, can it effectively and concretely provide a regulatory guidance for developers, without any officially decided policies?

In my opinion, transgene-free, genome edited crops should also be subject to Cartagena Law 2003 or other regulation if the traits acquired in the crops are considered to potentially affect biodiversity through their competitiveness, productivity of harmful substances and/or crossability (Fig. 6.2).[55] In detail, it is important to carefully validate the target-specificity and off-target effects in plant cell cultures after designing sgRNA of the CRISPR/Cas9. Next, plant cells modified by validated CRISPR/Cas9 are subjected to an initial screen focused on on-target gene modifications. Regenerated plants with no significant off-target mutations are further

[52]Op.cit.23.

[53]MAFF (2015).

[54]Op.cit.16.

[55]Op.cit. 42.

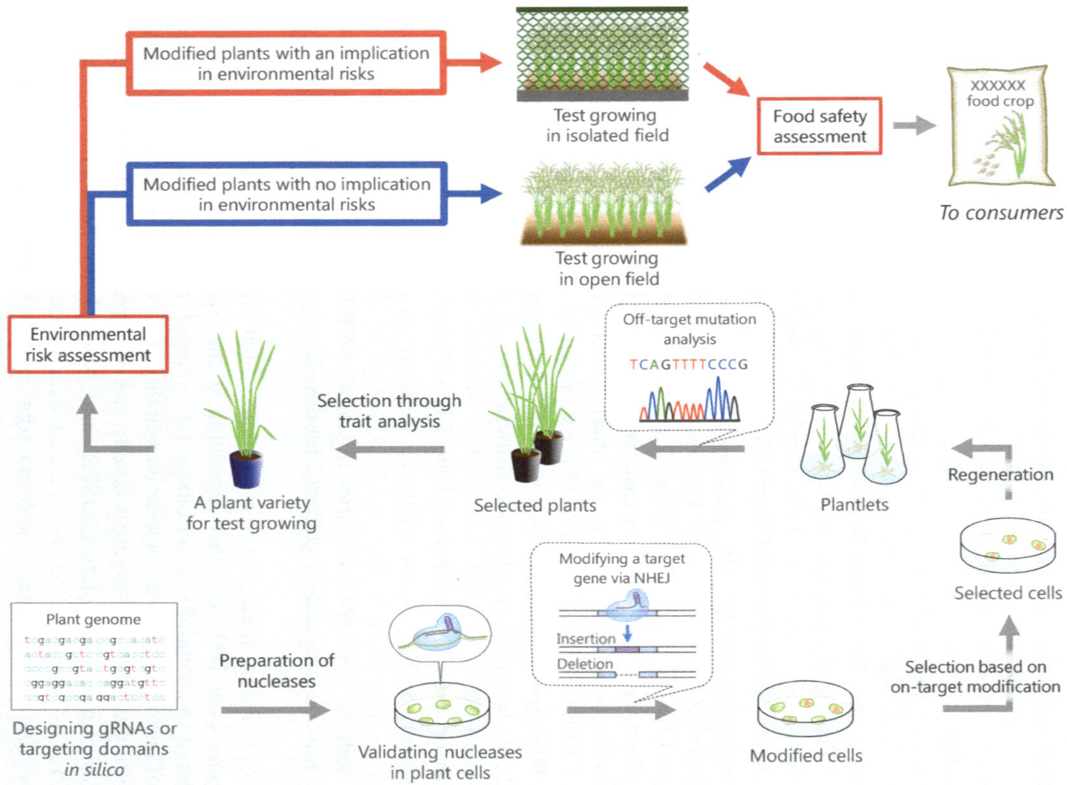

Fig. 6.2 A proposed scheme for the development of a transgene-free food crop using genome editing (Non-homologous end-joining). Note that this scheme considers only the case of the disruption of one gene. Multiple genome editing would require more stringent analysis prior to field test and food consumption. Adapted from: Ishii T, Araki M. Plant Cell Rep. 2016 Jul;35(7):1507–18

selected at laboratories. In addition to the acquired trait, the potential environmental impacts of the plants are evaluated. If the plants have an implication in environmental risks, such as the emergence of weeds by hybridization, test cultivation is carried out in an isolated field to evaluate their risks to the environment. Finally, the food product derived from such crops, regardless of cultivation in isolated or open fields, is subject to food safety assessment. Therefore, all food products derived from genome editing would require food safety assessment because no implication in environmental risks does not necessarily imply food safety.

6.6 GMO Labelling and Low Level Presence

If the safety of a GM food was confirmed, some of them are subject to mandatory GMO labelling under Food Labelling Act 2013 (Table 6.2). The policy of mandatory GMO labelling is currently applied to 8 GM crops (soybean, maize, potato, rapeseed, cotton, alfalfa, sugar beet and papaya) and their 33 processed food products in which the recombinant DNA or resultant protein remains detectable, in addition to GM soybean with high content of oleic acid and its processed food (soybean oil). Such food products are labelled as 'recombinant DNA technology applied'. By contrast, non-GM crops and related food products are allowed to be labelled as 'recombinant DNA technology not applied'. That is, the labelling of 'recombinant DNA technology not applied' is carried out on a voluntary basis.

However, the policy of mandatory GMO labelling in the 8 GM crops and their 33 processed foods can be exempted from two standpoints: in a final food product and during farm product handling.[56] Namely, mandatory GMO labelling is exempted from food products, such as soybean oil, soy sauce, corn oil, isomerized sugar, in which recombinant DNA or resultant proteins cannot be detected by using the latest analysis techniques. The exemption is also applied to a final food product in which GM ingredients (recombinant DNA and resultant proteins) do *not* present one of the three highest materials occupying more than 5% of the total weight. Such crops and food products can be labelled as 'recombinant DNA technology not applied'.

In the EU, the exemption of mandatory GM labelling is applicable to food products in which the level of introduced genetically engineered DNA or resultant proteins is less than 0.9% of the food/feed ingredients considered 'individually'.[57] On the other hand, 'non-GMO' labelling is not forbidden by the EU legislation and 'GM-free labels' are possible provided that they are not misleading for the consumer.[58] However, not only the threshold number but also the basis of calculating the containing ratio of GM ingredient are largely different between Japan's positive

[56] CAA (2017a).

[57] EC (2013).

[58] Ibid.

labelling policy and the EU's one. The standards of GMO labelling in Japan appear to be lax compared with that in the EU.

In Japan, the exemption can also be applied to Identity Preserved Handling (so-called IP Handling).[59] IP Handling means a management system where farm products to which GM technology has been applied and farm products to which GM technology has not been applied are separately managed in each process for production, distribution, and processing with the due care of a prudent manager and such fact is clearly certified with written documents. If IP Handling is confirmed to have been conducted in a farm product, the product is allowed to be labelled as 'recombinant DNA technology not applied'. However, unintentionally mixing of GM crops during farm product handling is also allowed up to total weight of 5%. In such farm products, the non-GMO labelling is possible.

The policy of GMO labelling has incurred tremendous debates in Japanese society. Currently, Consumer Affairs Agency in the Cabinet Office is considering whether the GMO labelling policy should be changed or not.[60] Some citizen groups asserted that the threshold of GMO ingredients should be more stringent, whereas some academic societies demanded that the irrational and controversial policy of non-GMO labelling should be abolished.[61] In the end of March 2018, the discussions surrounding GMO labelling in the Cabinet Office concluded that non-GMO labelling will be permitted only if GMO ingredients are not detectable.[62]

Japan's lax policy of mandatory GMO labelling and voluntary non-GMO labelling, which could be a product of compromise between GMO proponents and GMO opponents, probably supported the import of GM crops from abroad. Conversely, this policy has affected the acceptance of GMOs. According to a consumer survey in 2017 by the Consumer Affairs Agency, only 30.2% of respondents knew the crops and food products to which mandatory GMO labelling is applied, whereas as many as 60% of them knew non-GMO labelling.[63] It appears that biased perception of labelling of GMO and non-GMO jeopardizes the social discussion of GMOs in Japan.

In Sect. 6.5, I argued that transgene-free, genome edited crops should be subject to Cartagena Law 2003 or other regulation if the traits acquired are considered to potentially affect biodiversity. In this regard, I also asserted the use of DNA-tagging in transgene-free crops generated via NHEJ. Prior to this, it is necessary to address the existing issues of GMO labelling policy in Japan. The deliberation on GMO labelling in the Consumer Affairs Agency might provide the beneficial opportunity.

---

[59]Op.cit. 56.

[60]CAA (2018a).

[61]Hirasawa (2017).

[62]CAA (2018b).

[63]Op. Cit. 9.

6.7 Liability

With regard to food safety of LMOs and resultant products, penal provisions in Article 71–79 of Food Sanitation Law 1947 are applied. Penal provisions in Article 17–23 of Food Labelling Act 2013 are applied to the violation of GMO labelling.

Cartagena Law 2003 has penal provisions in Article 38–48 (Chapter V). The most severe punishment is show in Article 38, by which a person who violates orders under the provisions of Article 10 (Orders for Measures Concerning Type 1 Use), Article 11 (2) (Orders for Measures in the Event of Accidents Concerning Type 1 Use), Article 14 (Orders for Measures Concerning Type 2 Use), Article 15 (2) (Orders for Measures in the Event of Accidents Concerning Type 2 Use), Article 17 (5) (Orders for the Use of Organisms Subject to Testing), Article 26 (2) (Orders for Recalling Living Modified Organisms) or Article 29 (Orders Concerning Export) shall be punished by imprisonment with work of not more than 1 year or a fine of not more than one million yen, or a combination of these two.[64] Such penal provisions force persons who use or manage LMOs at laboratories, in the fields or in the market to take necessary measures to prevent or minimize the dispersal of LMOs, or inactivate or recall dispersed LMOs.

On December 5, 2017, Japan ratified the Nagoya–Kuala Lumpur Supplementary Protocol on Liability and Redress to the Cartagena Protocol on Biosafety, which will enter into force on March 5, 2018.[65] Therefore, Japan will legally establish response measures that will be taken in the event of damage resulting from LMOs, or where there is sufficient likelihood that damage will result if timely response measures are not taken, in addition to provisions in relation to civil liability.

The applicability of such penal provisions or liabilities to plants generated through NPBTs which are not subject to Cartagena Law 2003 and Food Labelling Act 2013 has not been determined yet. In contrast, the safety issues of imported food products that were manufactured from crops developed using NPBTs could be subject to penal provisions of Food Sanitation Law 1947. However, if any safety issues occur, is the national government responsible for the lack of regulatory responses to crops developed using NPBTs and derived food products?

6.8 Perception of Genome Editing

6.8.1 Position of Public Authorities

As mentioned in Sect. 6.3.1, the MOE deliberation report in 2016 and subsequent response by MAFF display that Japan's regulators are currently unwilling to consider the regulation of NPBTs.

---

[64]Op.cit. 17.

[65]Biosafety Clearing House (2017).

There has been another field trial regarding NPBTs in addition to the field trials of CRISPR-gene-disrupted rice. That is a field trial of epigenome editing (RNA-directed DNA methylation: RdDM) in potato.[66] Briefly, in a laboratory, transcriptional gene silencing (TGS) to two endogenous genes (*GBSSI* and *VInv*) is induced in potato rootstock by grafting of the scion that can produce two types of siRNA by *Agrobactrium*-mediated gene transfer. The field trial uses only the potato rootstock that has siRNA transmitted derived from the scion but has no transgenes according to the application. The application for a field trial was reviewed by applying the assessment method concerning the adverse effect on biological diversity of Type 1 Use of a LMO. However, no official approval of Type 1 Use was given by MEXT and MOE, because the potato rootstock is *not* an LMO under Cartagena Law 2003. The 3 year-field trial is underway in an isolated field established at the NARO in Tsukuba-city.

Such plant scientists may be satisfied with the tentative approval of field trial of epigenome-edited potato by regulators. However, it would be inappropriate to place, with no legal basis, the burden of Type 1 Use on all researchers who use NPBT-mediated crops in field trials if such crops have no 'nucleic acid' and no substantial impact on biodiversity. Moreover, such inappropriate handling by regulators might cause some people misunderstand that all crops produced using NPBTs are subject to Cartagena Law 2003. Furthermore, without clear regulations, NPBT-mediated crops which are approved abroad will likely be imported under existing GMO regulation, which does not reflect the technical issues such as off-target effects or multiplex editing in genome editing. It is crucial to draw a regulatory line among crop varieties that are generated using NPBTs.

6.8.2 Public Opinion

The two newspaper articles mentioned in Sect. 6.3.1 did not effectively raise the regulatory issues of genome edited organisms in 2012. However, the situation changed in 2017.

Some citizen groups expressed their opposition to the field trial of CRISPR-gene-disrupted rice when public consultation was announced regarding its Type 1 Use application. Such opposing groups include one farm-consumer group[67] and one consumer-related group.[68] Their main reason of opposition to the field trial is based on the fact that rice, a widely cultivated and major food crop in Japan, is being developed using genome editing, which regulation is not considered at all in Japan. Subsequently, the farmer-consumer group sent a written request to the minsters of MEXT, MAFF, MOE, MHLW, and Consumer Affairs Agency that a

[66]Hirosaki University and NARO (2017).

[67]Tane to Syoku to Hito @Forum (2017a).

[68]Seikatsu Club (2017).

regulatory framework is required for NPBTs such as genome editing and epigenome editing.[69]

In Japan, no GM rice varieties have been approved for cultivation, distribution nor food consumption. Only 21 field trials of GM rice have been approved thus far (Table 6.1). The field trial of CRISPR-gene-disrupted rice is just one of them. However, the citizen groups' indication seems appropriate in that rice plants generated using CRISPR-Cas9 was not reviewed based on proper regulations that demand cautious reviews regarding environmental risks in genome edited crops.

6.9 Conclusion

Previous sociological studies suggested that people's negative attitude toward GMOs is associated with the lack of trust in developers and/or relevant regulations, insufficient knowledge of GMOs, poor risk–benefit communication, and ethical values.[70]

Likewise, such psychological factors should be considered in the social introduction of food products developed using NPBTs such as genome editing. People's ethical values regarding crops produced using genome editing must be respected by appropriately labelling genome edited crops and resultant food products, because there are currently no compelling reasons to force people to have such food products without knowing them. Poor risk-benefit communication can be improved though careful dialogues with the provision of well-balanced information on plant genome editing. However, the terms, 'genome' and 'editing' will likely make it more difficult for the public to understand technical differences between GM crops and genome edited crops. Most importantly, regulations on genetic engineering must not only take into account the scientifically appropriate use of biotechnology in agriculture but also social norms which offer a basis for promoting public dialogues surrounding resultant crops and food products. However, Japan's ministries do not consider the regulation of any uses of NPBTs such as genome editing in earnest. Will Japan become a leading importer of genome edited crops in addition to GM crops? Without efforts to enact clear regulation, the situations surrounding NPBTs will likely become more controversial in Japan.

In sum, the current state of Japan does not warrant a promising future of genome editing and other NPBTs.

---

[69]Tane to Syoku to Hito @Forum (2017b).

[70]Op.cit. 42.

References

Academic Association for Genetic Studies (AAGS) (2014) Statement, opinion and policy on the management of organisms generated using genome editing. http://www.idenshikyo.jp/genome-editing/genome-editing_1.html. Accessed 26 Jan 2018

Araki M, Ishii T (2015) Towards social acceptance of plant breeding by genome editing. Trends Plant Sci 20:145–149

Bergelson J, Purrington CB, Wichmann G (1998) Promiscuity in transgenic plants. Nature 395:25

Biosafety Clearing House (2000) Article 3 in Cartagena Protocol on Biosafety. https://bch.cbd.int/protocol/text/. Accessed 25 Jan 2018

Biosafety Clearing House (2017) Nagoya – Kuala Lumpur supplementary protocol on liability and redress to the Cartagena Protocol on Biosafety. http://bch.cbd.int/protocol/supplementary/. Accessed 26 Jan 2018

Biosafety Clearing House (2018) Cartagena Protocol Country Profile: Japan. https://bch.cbd.int/about/countryprofile.shtml?country=jp. Accessed 25 Jan 2018

Burgos NR, Singh V, Tseng TM, Black H et al (2014) The impact of herbicide-resistant rice technology on phenotypic diversity and population structure of United States weedy rice. Plant Physiol 166:1208–1220

Busconi M, Rossi D, Lorenzoni C et al (2012) Spread of herbicide-resistant weedy rice (red rice, Oryza sativa L.) after 5 years of Clearfield rice cultivation in Italy. Plant Biol 14:751–759

Butler NM, Atkins PA, Voytas DF et al (2015) Generation and inheritance of targeted mutations in Potato (Solanum tuberosum L.) using the CRISPR/Cas system. PLoS One 10:e0144591

Consumer Affairs Agency (CAA) (2017a) The situations on labelling system of food products applied genetic recombination technology. http://www.caa.go.jp/policies/policy/food_labeling/other/pdf/genetically_modified_food_170426_0003.pdf. Accessed 26 Jan 2018

Consumer Affairs Agency (CAA) (2017b) Report on consumer's attitude towards food labelling 2016 (April 26, 2018). http://www.caa.go.jp/policies/policy/food_labeling/information/research/2016/. Accessed 25 Jan 2018

Consumer Affairs Agency (CAA) (2018a) Study group on the labelling of genetic recombination technology. http://www.caa.go.jp/policies/policy/food_labeling/other/genetically_modified_food.html. Accessed 26 Jan 2018

Consumer Affairs Agency (CAA) (2018b) Report on 28 March 2018: study group on the labelling of genetic recombination technology. http://www.caa.go.jp/policies/policy/food_labeling/other/review_meeting_010/pdf/review_meeting_010_180419_0003.pdf. Accessed 5 Jun 2018

Endo M, Mikami M, Toki S (2016) Biallelic gene targeting in Rice. Plant Physiol 170:667–677

European Commission (EC) (2013) Traceability and labelling of GMOs and GM food/feed products. https://ec.europa.eu/food/plant/gmo/traceability_labelling_en. Accessed 26 Jan 2018

European Food Safety Authority (EFSA) (2012) Scientific opinion addressing the safety assessment of plants developed using Zinc Finger Nuclease 3 and other site-directed nucleases with similar function (18 October 2012). http://onlinelibrary.wiley.com/doi/10.2903/j.efsa.2012.2943/epdf. Accessed 25 Jan 2018

Gaj T, Gersbach CA, Barbas CF 3rd (2013) ZFN, TALEN, and CRISPR/Cas-based methods for genome engineering. Trends Biotechnol 31:397–405

Hirasawa Y (2017) Call for reconsideration: labelling of 'genetic recombination technology not applied' may misleading. Sankei Shimbun, 6 November 2017

Hirosaki University and NARO (2017) Press release by Hirosaki University and National Agriculture and Food Research Organization (NARO): starting field trial of a new potato variety produced by a new breeding technique. http://www.hirosaki-u.ac.jp/wordpress2014/wp-content/uploads/2017/04/20170420.pdf. Accessed 26 Jan 2018

International Monetary Fund (IMF) (2016) World Economic Outlook Database, October 2016. http://www.imf.org/external/pubs/ft/weo/2016/02/weodata/weoselgr.aspx. Accessed 25 Jan 2018

Ishii T, Araki M (2016) Consumer acceptance of food crops developed by genome editing. Plant Cell Rep 35:1507–1518
Ishii T, Araki M (2017) A future scenario of the global regulatory landscape regarding genome-edited crops. GM Crops Food 8:44–56
Ito Y, Nishizawa-Yokoi A, Endo M et al (2015) CRISPR/Cas9-mediated mutagenesis of the RIN locus that regulates tomato fruit ripening. Biochem Biophys Res Commun 467:76–82
Jackson SA (2016) Rice: the first crop genome. Rice 9:14
Japan Biosafety Clearing House (J-BCH) (2018) Containment measures to be taken in Type 2 use of living modified organisms for research and development. http://www.biodic.go.jp/bch/english/law.html. Accessed 25 Jan 2018
Jiang F, Doudna JA (2017) CRISPR-Cas9 structures and mechanisms. Annu Rev Biophys 46:505–529
Joung JK (2015) Unwanted mutations: standards needed for gene-editing errors. Nature 523:158
Kanazashi Y, Hirose A, Takahashi I et al (2018) Simultaneous site-directed mutagenesis of duplicated loci in soybean using a single guide RNA. Plant Cell Rep 37:553–563
Kim J, Kim J-S (2016) Bypassing GMO regulations with CRISPR gene editing. Nat Biotechnol 34:1014
Luo S, Li J, Stoddard TJ et al (2015) Non-transgenic plant genome editing using purified sequence-specific nucleases. Mol Plant 8:1425–1427
Lusser M, Parisi C, Pan D et al (2011) New plant breeding techniques – state-of-the-art and prospects for commercial development. http://ftp.jrc.es/EURdoc/JRC63971.pdf. Accessed 25 Jan 2018
MEXT-MHLW-MAFF-METI-MOE (2003a) Guidance of implementation of assessment of adverse effect on biological diversity of Type 1 use 2003. http://www.biodic.go.jp/bch/english/law.html. Accessed 25 Jan 2018
MEXT-MHLW-MAFF-METI-MOE (2003b) Regulations related to the enforcement of the law concerning the conservation and sustainable use of biological diversity through regulations on the use of living modified organisms. http://www.biodic.go.jp/bch/english/law.html. Accessed 25 Jan 2018
Ministry of Agriculture, Forestry and Fisheries (MAFF) (2015) Report by the study group on new breeding techniques in Ministry of Agriculture, Forestry, and Fisheries: 'Toward the Development and Application of Farm Products using New Plant Breeding Techniques such as Genome Editing'. http://www.affrc.maff.go.jp/docs/commitee/nbt/top.htm. Accessed 26 Jan 2018
Ministry of Agriculture, Forestry and Fisheries (MAFF) (2017a) Approved living modified crops (as of Dec 19, 2017). http://www.maff.go.jp/j/syouan/nouan/carta/torikumi/index.html#1. Accessed 25 Jan 2018
Ministry of Agriculture, Forestry and Fisheries (MAFF) (2017b) Major crop exporters to Japan and their status of cultivation 2016. http://www.maff.go.jp/j/syouan/nouan/carta/zyoukyou/attach/pdf/index-13.pdf. Accessed 25 Jan 2018
Ministry of Agriculture, Forestry and Fisheries (MAFF) (2017c) Notification of inspection of living modified organisms in 2017. http://www.maff.go.jp/j/syouan/nouan/carta/torikumi/attach/pdf/index-67.pdf. Accessed 25 Jan 2018
Ministry of Agriculture, Forestry and Fisheries (MAFF) (2017d) Outline of Japan's export and import of agricultural forestry and fishery products. http://www.maff.go.jp/j/tokei/kouhyou/kokusai/attach/pdf/houkoku_gaikyou-1.pdf. Accessed 25 Jan 2018
Ministry of Agriculture, Forestry and Fisheries (MAFF) (2018a) On the organisms produced by new plant breeding techniques. http://www.maff.go.jp/j/syouan/nouan/carta/tetuduki/index.html#3-2. Accessed 25 Jan 2018
Ministry of Agriculture, Forestry and Fisheries (MAFF) (2018b) On unapproved living modified petunia. http://www.maff.go.jp/j/syouan/nouan/carta/torikumi/petunia.html. Accessed 25 Jan 2018

Ministry of Justice (2018) Act on the conservation and sustainable use of biological diversity through regulations on the use of living modified organisms 2003. http://www.japaneselawtranslation.go.jp/law/detail/?id=132&vm=04&re=01. Accessed 25 Jan 2018

Ministry of the Environment (MOE) (2016) Regarding the results of examination of Cartagena Law enforcement and its public consultation. http://www.env.go.jp/press/102964.html. Accessed 25 Jan 2018

Ministry of the Environment (MOE) (2017) Public consultation on type 1 use of living modified rice (No. 2 FY2016). http://www.env.go.jp/press/103721.html. Accessed 25 Jan 2018

Nakajima I, Ban Y, Azuma A et al (2017) CRISPR/Cas9-mediated targeted mutagenesis in grape. PLoS One 12:e0177966

Nishitani C, Hirai N, Komori S et al (2016) Efficient genome editing in Apple using a CRISPR/Cas9 system. Sci Rep 6:31481

Nishiyama A (2012) More complex techniques emerged in crop genetic recombination: where can a line of genetic recombination be drawn? Response by Japan government delayed in global discussions. Nikkei Shimbun on 28 August 2018

Ryffel GU (2014) Transgene flow: facts, speculations and possible countermeasures. GM Crops Food 5:249–258

Sawai S, Ohyama K, Yasumoto S et al (2014) Sterol side chain reductase 2 is a key enzyme in the biosynthesis of cholesterol, the common precursor of toxic steroidal glycoalkaloids in potato. Plant Cell 26:3763–3774

Science Council of Japan (SCJ) (2014) Report: current status and issues of new plant breeding techniques. http://www.scj.go.jp/ja/info/kohyo/pdf/kohyo-22-h140826.pdf. Accessed 25 Jan 2018

Segawa S, Hatano Y (2012) No vestiges after genetic modification: new techniques applied in crops by Tokushima University and Hiroshima University. Asahi Shimbun on 22 August 2018

Seikatsu Club (2017) Opposition to the approval of type 1 use of genetic recombinant rice. http://tokyo.seikatsuclub.coop/news/2017/04/post_87.html. Accessed 26 Jan 2018

Suntory (2018) History of Blue Rose. https://www.suntory.co.jp/sic/research/s_bluerose/story/. Accessed 25 Jan 2018

Svitashev S, Schwartz C, Lenderts B et al (2016) Genome editing in maize directed by CRISPR-Cas9 ribonucleoprotein complexes. Nat Commun 7:13274

Tane To Syoku To Hito @FORUM (2017a) Opposition to the cultivation test of GM rise produced using genome editing in isolated fields. http://nongmseed.jp/?p=1923. Accessed 26 Jan 2018

Tane To Syoku To Hito @FORUM (2017b) Request of a regulatory frame work for new plant breeding techniques on 8 June 2017. http://nongmseed.jp/?p=1994. Accessed 26 Jan 2018

Tang X, Lowder LG, Zhang T et al (2017) A CRISPR-Cpf1 system for efficient genome editing and transcriptional repression in plants. Nat Plants 3:17018

Chapter 7
Genetic Engineering in the United States: Regulation of Crops and Their Food Products

Margaret Rosso Grossman

Abstract In the United States, federal administrative agencies, including the US Department of Agriculture (USDA), implement statutes and regulations that govern genetically engineered plants and their products. US regulatory measures were developed in light of genetic modification using rDNA, before the advent of new technologies, including gene editing, that offer simple, elegant paths to genetic improvement in plants. After a brief review of global and US production of GE crops, this Chapter analyzes US policy and regulation of these crops and their food products, with emphasis on the role of USDA. The Chapter addresses significant issues raised by GE crops: coexistence of GE and non-GE crops, low-level presence, tort liability, and the new US labeling law for bioengineered foods. Recent policy statements encourage modernized science-based regulation, but some uncertainty applies to regulation of crops developed with new genetic technologies. The Chapter analyzes USDA's governance of new crop varieties with a focus on the "Am I Regulated?" process, which determines whether new organisms are subject to USDA regulation. The USDA has declined regulatory jurisdiction for a number of products of gene editing and has indicated its intention not to regulate crops developed with certain new technologies. Finally, the Chapter outlines some regulatory challenges posed by advances in biotechnology.

This Chapter is based on work supported by USDA, National Institute of Food and Agriculture, Hatch Project No. ILLU-470-348. The final version of this Chapter was submitted on 29 June 2018, with selected updates in January 2019.

M. R. Grossman (✉)
Department of Agricultural and Consumer Economics, University of Illinois, Urbana, IL, USA
e-mail: peggyg@illinois.edu

H.-G. Dederer, D. Hamburger (eds.), *Regulation of Genome Editing in Plant Biotechnology*, https://doi.org/10.1007/978-3-030-17119-3_7

7.1 Introduction

Genetic engineering of plants using recombinant DNA technology, beginning in the 1980s, has led to commercialized varieties of insect-resistant and herbicide-tolerant crops (predominantly soy, maize, cotton), grown globally on millions of hectares. GE technology has been used to develop plants with other traits (for example, resistance to browning in apples and potatoes), grown only on a small number of hectares. Most genetically engineered traits and varieties are not yet sold and grown commercially, however, for a variety of reasons, including regulatory concerns.[1] In the past decade, scientists have developed a number of innovative techniques, many characterized as gene editing, to engineer plants. GE technology is developing rapidly, and the US National Academies of Sciences, Engineering, and Medicine concluded in a 2017 report that the "scale, scope, complexity, and tempo of biotechnology are likely to increase in the next 5–10 years."[2]

Gene editing, an important new technology, offers a precise way to "insert, delete, or replace DNA at a specific location within a crop's DNA. The crop's DNA is cut at the specific location using different biological molecules, such as Zinc Finger Nucleases or CRISPR/Cas 9."[3] Although scientists have developed a number of crops with gene editing, farmers are not yet growing most of these new varieties.[4]

Among gene editing techniques, CRISPR/Cas9 has assumed particular importance. CRISPR (clustered regularly interspaced short palindromic repeats) are "segments of bacterial DNA that, when paired with a specific guide protein, such as Cas9 . . ., can be used to make targeted cuts in an organism's genome."[5] CRISPR offers more precision and lower cost than other technologies for gene editing (for example, TALEN and ZFN),[6] making CRISPR accessible to a wide range of plant developers, including small companies, universities, and other public institutions. Moreover, CRISPR is versatile. For example, CRISPR/Cas9 has facilitated recent

---

[1]NASEM (2016a), pp. 5–6. These crops are referred to as genetically engineered (GE) or, in the European Union, genetically modified (GM) or genetically modified organisms (GMOs). USDA documents use the term genetic engineering, so this Chapter often refers to GE (rather than GM) crops. The term gene editing applies to technologies such as CRISPR.

[2]NASEM (2017), p. 172. Scope refers to new types biotechnology products not yet seen by regulators; scale, to the number of products and variants of products; complexity, to the number of traits in a single product and interactions between elements in a product. NASEM (2017), p. 139.

[3]Jaffe (2017). See Enríquez (2017a) for a summary of types of genome editing and their applications in agricultural and other fields. See also Schaart et al. (2015), pp. 3–17 (explaining new breeding techniques).

[4]Jaffe (2017). Jaffe noted that farmers grow a gene-edited canola variety developed by Cibus.

[5]NASEM (2016b), p. 1, n.1. NASEM (2017), p. 187 defines CRISPR: "A naturally occurring mechanism of immunity to viruses found in bacteria that involves identification and degradation of foreign DNA. This natural mechanism has been manipulated by researchers to develop genome-editing techniques." Cas9 refers to CRISPR associated protein 9. Another definition from Jaffe (2017): CRISPR is "a molecule that identifies the location where the DNA should be cut, and Cas 9 is the enzyme that cuts the DNA at the identified location."

[6]Baumann (2016), p. 139.

developments in gene drive research, first explored in the 1960s. Organisms modified with gene drives are designed to spread in the environment,[7] and preliminary laboratory studies indicate that gene drives developed with CRISPR/Cas9 "could spread a targeted gene through nearly 100% of a given population of yeast, fruit flies, or mosquitoes."[8]

Reflecting scientific enthusiasm for CRISPR technology, a 2014 symposium in the influential journal *Science* focused on CRISPR/Cas9, asserting that "the CRISPR-Cas9 system is revolutionizing genomic engineering and equipping scientists with the ability to precisely modify the DNA of essentially any organism."[9] Moreover, "[t]he elegance and simplicity of Cas9 have sparked the imagination of scientists across many scientific disciplines."[10] Indeed, the National Academy of Sciences awarded its 2018 NAS Prize in Food and Agriculture Sciences to a professor whose "groundbreaking research established CRISPR as the adaptive immune system of bacteria, a discovery which promoted the practical use of CRISPR-Cas systems for genome editing . . . work [that] has tremendous worldwide applications in food and agriculture."[11]

Despite scientific advances, however, "CRISPR continues to be an arcane secret in the legal realm."[12] Therefore, as one researcher notes, "[t]he neglect of CRISPR in legal scholarship poses grave uncertainty regarding how the law will treat this emerging technology going forward. Legal scholars have either largely ignored this field or kept a distance from it, presumably due, in part, to the challenges that complex scientific principles often pose to non-scientists in the legal and legislative arenas."[13]

Regulatory uncertainty is not unique to CRISPR, but also affects other innovative genetic technologies. The US Department of Agriculture's Advisory Committee on Biotechnology and 21st Century Agriculture noted in 2016 that "the regulatory status of products developed using new precise breeding technologies sometimes referred to as 'gene editing techniques' (which include the use of tools such as CRISPR-Cas9, TALENs, meganucleases, and others) is not clear."[14] Although some of these new products may be similar to comparable existing products, others may

---

[7]NASEM (2016b), pp. 1, 5, 12–13 (noting that these organisms may pose environmental risks).

[8]NASEM (2016b), p. 3. Gene drives are beyond the scope of this Chapter.

[9]Hicklin (2014), p. 2. CRISPR has been used to modify human embryos, raising numerous ethical and other issues beyond the scope of this Chapter. Baumann (2016).

[10]Zhang (2014), p. 3.

[11]NAS (2018) ($100,000 prize to Rodolphe Barrangou, North Carolina State University). Other scientists, including Jennifer Doudna and Emmanuelle Charpentier, have received awards that recognize significant contributions to CRISPR research.

[12]Enríquez (2017a), p. 608.

[13]Enríquez (2017a), pp. 608–609.

[14]USDA, AC21 (2016), p. 32. US regulatory agencies and others refer to the "product" of biotechnology. In many instances the term "resulting organism" would be more appropriate. Professor Pieter van der Meer, Ghent University and Free University Brussels, Belgium, provided this helpful distinction.

differ. As new processes are introduced, and products that differ from existing crop varieties are developed, the lack of comparators (similar organisms) will require new approaches to risk analysis.[15] Without a clear regulatory process and reliable risk analysis, cultivation of crops produced with these important plant breeding techniques may raise issues for both coexistence and international trade.[16]

In the United States, regulation of crops produced by genetic technology is based on decades-old policy, governed by a number of statutes and regulations, and implemented by several federal administrative agencies. Policy discussions in recent years point to possible regulatory amendments. Even without regulatory changes, however, a January 2018 policy document from the interagency Task Force on Agriculture and Rural Prosperity indicated that "better coordination of the Department of Agriculture, Environmental Protection Agency, and Food and Drug Administration regulations on genetic modification of crops and livestock is needed to reduce barriers to commercialization of safe, beneficial and improved genetically engineered entities. Our regulatory system must protect public health, welfare, safety, and our environment while promoting economic growth, innovation, competitiveness, and job creation."[17]

In light of recent developments in genetic technology and questions about its regulation in the United States, this Chapter focuses first on production of genetically engineered crops in the United States and globally. It then turns to an analysis of GE regulation, with a discussion of federal policy and the responsibilities and requirements of the US Department of Agriculture (USDA), the Environmental Protection Agency (EPA), and the Food and Drug Administration (FDA) for authorization of GE crop varieties. The Chapter then outlines recent developments in federal policy and some agency proposals for regulatory change. It focuses on US measures for the protection of non-GE crops (coexistence and low-level presence), followed by an outline of recently-enacted requirements for labels for GE food and a brief discussion of tort liability. The Chapter then turns to a discussion of the uncertainty of regulation of innovative GE technology in the United States and USDA decisions that decline to regulate some products of gene editing. Finally, the Chapter considers recommendations for future development of the US regulatory system.

[15]NASEM (2017), p. 172. A comparator is a "known nonbiotechnology organism that is similar to the engineered organism except for the engineered trait" (p. 187).

[16]USDA, AC21 (2016), p. 32.

[17]Task Force (2018), p. 33. Genetic engineering of livestock is beyond the scope of this Chapter. For a discussion of GE animal and cloning, see Grossman (2018), pp. 348–357.

7.2 Production of GE Crops in the United States

The US government supports and promotes the application of new technologies in agriculture "as a matter of policy" and also as a means to enhance US competitiveness.[18] Thus, GE crops have played an increasingly important role in US agriculture. In 2017, producers in the United States planted 75 million hectares of GE crops, 40% of the global total of 189.8 million hectares. Global production in 2017 was 3% higher than in 2016. Farmers in 24 countries planted GE varieties. Five industrial countries accounted for 47% of the world's GE crops; 19 developing countries, 53%. Since 1996, when farmers produced only 1.7 million hectares, GE crop production increased about 112-fold.[19] Researchers have noted, however, that "countries that have adopted GM crops are planting them on increasingly larger acreages year after year as opposed to broader diffusion of the technology to previously non-adopting countries."[20]

Global production of GE crops included soybeans (94.1 million hectares, 50% of global GE hectares), maize (59.7 million), cotton (24.1 million), and canola (10.2 million), with fewer hectares of alfalfa, sugar beet, papaya, and other crops.[21] These crops have significant economic value. Global value of biotech seed (seed price plus technology fees) in 2017 was estimated at US $17.2 billion – $8.72 billion for maize, $6.33 billion for soy, $1.42 billion for cotton, and $0.46 billion for canola.[22] The value of harvested products, including grain, is significantly higher than seed value.

Between 1992 and 2017, 40 countries issued 1995 food approvals, 1338 feed approvals, and 800 cultivation approvals. These involved 29 crops and 498 events.[23] Since 2017, other varieties have been approved. Countries are often reluctant to approve crops for cultivation.[24] Indeed, as the US National Academies of Science, Engineering, and Medicine noted in 2016, "[f]or a variety of scientific, economic, social, and regulatory reasons, most genetically engineered (GE) traits and crop varieties that have been developed are not in commercial production."[25]

The United States is a leader in approvals, as well as cultivation, of GE varieties. As the International Service for the Acquisition of Agri-Biotech Applications

---

[18]Bergeson (2017), p. 33.

[19]ISAAA (2018c), pp. 3–5. Brazil, Argentina, Canada, and India planted more than 10 million hectares; other countries planted fewer hectares. Eighteen countries, including 14 developing countries in Latin America, Asia and Africa, grew 50,000 ha or more.

[20]Gleim et al. (2016), p. 112.

[21]ISAAA (2018c), p. 101. Other GE crops were squash, potato, eggplant, and apple.

[22]ISAAA (2018c), pp. 122–123. GE seeds cost more than conventional seeds, but "farmers realize economic benefits from growing GE crops through higher crop yields, and/or lower pesticide costs, and management time savings." Fernandez-Cornejo et al. (2014), p. 47.

[23]ISAAA (2018c), p. 108 (excluding ornamentals). See ISAAA (2018a), a global database of approvals.

[24]Gleim et al. (2016), p. 102. A study of corn found more approvals for import, often for feed, than for cultivation (pp. 108–109).

[25]NASEM (2016a), p. 5.

(ISAAA) reported in 2018, "Since 1996, USA has approved 197 single trait events in 19 crop species: alfalfa (3 events), apple (3), Argentine canola (21), chicory (3), cotton (28), creeping bentgrass (1), flax (1), maize (43), melon (2), papaya (3), plum (1), potato (43), rice (3), soybeans (25), squash (2), sugar beets (3), tobacco (2), tomato (8), and wheat (1)."[26] In 2017, USDA approved varieties of canola and creeping bentgrass for cultivation.[27]

Development of new crops is expensive, with significant costs for meeting regulatory requirements. Moreover, the regulatory process is time-consuming. One study found that assessment of scientific evidence in the United States took 686 days; another study of 95 applications calculated the mean time for the whole approval process (field trials, petition process) as 1321 days between 1988 and 1997 and 2467 days between 1998 and 2016.[28]

The USDA issues an annual crop acreage report, which includes estimated production of major GE varieties. In 2017, biotech varieties were 92% of corn, 94% of soy, and 96% of upland cotton; in 2018, the percentage of biotech corn and soy varieties remained stable, with a slight reduction to 94% of cotton.[29] Most GE corn and cotton varieties have stacked traits that provide both insect resistance and herbicide tolerance; GE soy offers herbicide tolerance. In 2017, 100% of US sugar beets and canola were GE herbicide tolerant varieties, but GE alfalfa made up only 14.4% of harvested alfalfa acres.[30] Alfalfa is planted every 6–7 years, and (after litigation beginning in 2007) deregulation occurred only in 2011.[31] Only a small proportion of producers grow GE sweet corn, squash, and papaya, as well as non-browning apples and potatoes.[32]

For producers, GE varieties with insect resistance and herbicide resistance offer good economic outcomes and management flexibility.[33] The United States has shared significantly in income benefits from GE crops. Researchers calculated global farm income benefits, related both to higher production and lower costs, from GE crops. For 2016, that amount was $18.2 billion, with a total increase in farm income since 1996 of $186.1 billion.[34] Farm income benefits result from yield gains from insect-resistant maize, lower costs and higher yields from insect-resistant cotton, and increased income from cost savings for herbicide-tolerant soybeans and yield gains

[26]ISAAA (2018c), p. 9 (using 2017 data).

[27]APHIS, USDA (2018d). In 2016, "food, feed and cultivation approvals were made for apple (1 event), maize (2) and potato (3)." ISAAA (2017), p. 8.

[28]Smart et al. (2017), pp. 183, 187, 192. In the EU, the mean approval time for authorized organisms from 1995 to 2015 was 1758 days.

[29]NASS (2017), pp. 29–31; NASS (2018), pp. 31–33.

[30]ISAAA (2018c), pp. 11–12.

[31]Fernandez-Cornejo et al. (2016), pp. 4–6. See Monsanto Co. (2010).

[32]Greene et al. (2016), p. 12; ISAAA (2018c), pp. 12–14.

[33]NASEM (2016a), p. 21. See also Smyth (2017), pp. 79–81.

[34]Brookes and Barfoot (2018), p. 71.

for canola.[35] The US captured the largest share of these farm income benefits, gaining $80.3 billion since 1996.[36] This large benefit for US farming "is not surprising given that US farmers were first to make widespread use of GM crop technology and for several years the GM adoption levels in all 4 US crops have been in excess of 80%."[37]

Similarly, the United States has shared in some positive environmental effects of GE technology. GE traits have led to "a significant reduction in the environmental impact associated with insecticide and herbicide use" on global GE crop areas.[38] Since 1996, global pesticide use declined by 8.1%, and environmental impacts of herbicides and insecticides declined by 18.6%. Varieties with insect-resistant traits achieved the most environmental gain.[39] Between 1990 and 2015, herbicide use increased more in non-GE crops—maize, rice, wheat, and cotton—than in GE crops.[40] Overuse of glyphosate, however, has fostered weed resistance in some areas.[41] GE crops can also help to mitigate climate change. Emissions of greenhouse gases are reduced when farmers use less fuel for chemical applications and soil cultivation and sequester carbon by applying no-till and reduced till cultivation.[42]

US regulation of GE crops and food, described below, is intended to ensure the safety of those products. A number of US and international professional organizations have attested to the safety of biotech crops.[43] In the US, these include the American Association for the Advancement of Science,[44] the National Academy of

[35]Brookes and Barfoot (2017a), pp. 9–10.

[36]Developing country farmers gained $96 billion through 2016. Brookes and Barfoot (2018), p. 71. The US level of benefits was followed by benefits to Argentina, India, China, and Brazil. The EU gained relatively little: $251.3 million in Spain; $23.6 million in a few other EU countries. Brookes and Barfoot (2017a), pp. 10–11.

[37]Brookes and Barfoot (2017c), p. 163.

[38]Brookes and Barfoot (2017a), p. 13.

[39]Brookes and Barfoot (2017a), pp. 13–15. See Brookes and Barfoot (2017b), for data on environmental benefits in individual countries, including the United States.

[40]Kniss (2017), p. 1. No GE wheat is approved for cultivation.

[41]Resistance occurs where producers do not follow management strategies, including refuges of non-GM crops and high-dose applications to kill partially-resistant insects. NASEM (2016a), p. 13.

[42]Brookes and Barfoot (2017a), pp. 16–17. Although a global ban on GE crops is unlikely, it could affect land use and increase greenhouse gas emissions, adding more than a billion tons of carbon dioxide to the atmosphere. A ban would also increase food prices and impose other significant welfare losses. Mahaffey et al. (2016).

[43]ISAAA (2017), pp. 104–105. In addition to the American Association for the Advancement of Science, the National Academy of Science, and the American Medical Association, others include the World Health Organization, European Commission, UK Royal Society of Medicine, French Academy of Science, Food Standards Australia and New Zealand, Union of German Academics of Sciences and Humanities, and 7 other world academies of sciences. In 2010, the European Commission concluded that "biotechnology, and in particular GMOs, are not per se more risky" than conventional crops. European Commission (2010), p. 16.

[44]AAAS (2012): "[T]he science is quite clear: crop improvement by the modern molecular techniques of biotechnology is safe," and GM varieties are the "most extensively tested crops ever added to our food supply," and pose no special risk.

Sciences, and the American Medical Association.[45] Indeed, GE crops and their food and feed products have caused no known health effects, nor have adverse effects on livestock been detected. No substantial evidence shows that GE foods are less safe or pose higher health risks than foods from conventional counterparts.[46] Nonetheless, an important research report cautioned that "any new food–GE or non-GE–may have some subtle favorable or adverse health effects that are not detected even with careful scrutiny and that health effects can develop over time."[47] Even so, commentators have emphasized the safety of GE crops and their products: "To our knowledge, no GM product has been explicitly rejected for health or environmental safety reasons by any regulator anywhere."[48] Indeed, "[b]iotech crops have now had an unblemished record of safe use and consumption for over 20 years."[49]

7.3 Regulation of GE Crops and Food Products in the United States

7.3.1 Policy

US statutes and agency regulations that govern biotechnology are designed to ensure that GE crops and their food products are safe for agriculture, the environment, and health.[50] The Coordinated Framework for Regulation of Biotechnology, published in 1986 and updated in 1992,[51] established US policy. Drafted in cooperation with administrative agencies, the Coordinated Framework recognized the commercial importance of biotechnology, but acknowledged concerns about regulation. Its primary focus was the then-new technology using recombinant DNA (rDNA). The United States did not immediately enact new statutes to govern biotechnology; instead, the Coordinated Framework relied on existing federal law, later supplemented by new laws and regulations. The document drew three conclusions that continue to influence US policy for biotechnology: products of biotechnology are not fundamentally different from conventional products; regulations should focus on the product, rather than on the process; and regulatory jurisdiction for biotechnology products should be based on their use.[52]

---

45 AMA (2012).

46 NASEM (2016a), pp. 2, 19.

47 NASEM (2016a), p. 19.

48 Gleim et al. (2016), p. 99.

49 ISAAA (2017), p. 1.

50 This section relies in part on Grossman (2018, 2016b), pp. 306–314, and Grossman (2012).

51 OSTP (1986, 1992). A 2002 policy document led to stronger regulation and enhanced oversight. OSTP (2002). The 2017 Update to the Coordinated Framework is discussed below.

52 OSTP (1986), pp. 23,303–23,304.

The Coordinated Framework was accompanied by proposed USDA guidelines for biotechnology research.[53] Shortly thereafter, however, USDA decided not to finalize those guidelines, but to contribute agricultural provisions to the National Institutes of Health (NIH) guidelines; the agency proposed provisions for agricultural research to be included in the NIH guidelines.[54] The NIH guidelines provide federal oversight of rDNA research.[55] These guidelines prescribe safety practices and containment procedures designed to protect researchers, the public, and the environment. Appendix P describes physical and biological containment for recombinant or synthetic nucleic acid molecule research that involves plants.[56] Researchers at institutions that receive NIH funding for rDNA research must comply with NIH guidelines, and other federal agencies and research sponsors may require compliance. Institutional Biosafety Committees provide oversight, helping researchers to assess risks and implement good laboratory practices.[57] USDA refers to these guidelines, as well as good scientific practices, for laboratory research.

The United States relies on a "risk-based, scientifically sound approach" to regulatory approval of genetically engineered crops and their products.[58] Regulation depends on characteristics of products or their intended use.[59] Accordingly, major responsibility for GE crops and their products is divided among three agencies. The US Department of Agriculture (USDA) ensures that GE crops are safe to grow; the US Environmental Protection Agency (EPA) ensures that they are safe for the environment; the US Food and Drug Administration (FDA, along with the EPA) ensures that they are safe to eat. US agency decisions about GE crops and their products are generally technical decisions that focus on whether products meet statutory requirements. US law allows little consideration of socio-economic issues.[60]

[53]USDA (1986).

[54]USDA (1991), p. 4134.

[55]NIH (2016). An earlier version of NIH guidelines predated the Coordinated Framework.

[56]NIH (2016), pp. 105–116. Appendix Q pertains to animals.

[57]NASEM (2017), pp. 69–70; NASEM (2016b), pp. 153, 170.

[58]OSTP (1992), p. 6753. See generally Biology Fortified (2018), a database of peer-reviewed research on the risks of GE crops.

[59]NASEM (2016a), p. 467. Although US regulation focuses on the product, whether USDA and EPA regulate is based in part on process (p. 25). The US approach has been characterized as "category-based," somewhere between product-based and process-based. For USDA, regulation focuses on varieties that contain a plant pest; for EPA, insecticidal properties; for FDA, food additives. Peck (2017), p. 332.

[60]NASEM (2016a), pp. 472–473.

7.3.2 *Authorization*

7.3.2.1 US Department of Agriculture

The USDA's Animal and Plant Health Inspection Service (APHIS) and its Biotechnology Regulatory Services govern environmental release (field testing), interstate movement, and import of GE plants that may pose a plant pest risk. Under the Coordinated Framework, APHIS coordinates its regulatory activities with the EPA and FDA.

The Plant Protection Act of 2000[61] gives USDA authority to regulate plant pests, broadly defined.[62] APHIS regulations enacted in 1987, with subsequent amendments, govern authorization;[63] these continue to apply, after APHIS proposed, but then withdrew, amended regulations. Under current regulations, most GE varieties created by rDNA technology could harbor a plant pest and are therefore considered "regulated" articles, defined in part as an "organism which has been altered or produced through genetic engineering, if the donor organism, recipient organism, or vector or vector agent belongs to any genera or taxa" designated or designed as a plant pest.[64] Regulated articles cannot be sold commercially until they are evaluated and authorized—that is, classified as "nonregulated."

7.3.2.1.1 Field Trials

Under the APHIS science-based regulatory framework, developers of GE plants carry out field trials to assess safety. Field trials occur while the new variety is still a regulated article. They require a permit or notification to APHIS.[65] Biotechnology

---

[61]7 United States Code [USC] §§ 7701–7772 (replacing the Plant Pest and Plant Quarantine Acts).

[62]7 USC § 7711. Pests are defined at 7 Code of Federal Regulations [CFR] § 340.1 and listed in § 340.2.

[63]7 CFR part 340.

[64]7 CFR § 340.1:

> *Regulated Article.* Any organism which has been altered or produced through genetic engineering, if the donor organism, recipient organism, or vector or vector agent belongs to any genera or taxa designated in §340.2 [listing organisms that are plant pests] and meets the definition of plant pest, or is an unclassified organism and/or an organism whose classification is unknown, or any product which contains such an organism, or any other organism or product altered or produced through genetic engineering which the Administrator, determines is a plant pest or has reason to believe is a plant pest. Excluded are recipient microorganisms which are not plant pests and which have resulted from the addition of genetic material from a donor organism where the material is well characterized and contains only non-coding regulatory regions.

[65]A policy document from 2002 focused on principles for field testing. The level of confinement should be consistent with the level of risk to health and environment. Strict confinement measures should apply to GE traits that carry unknown or unacceptable risks. Though out-crossing and

Regulatory Services provides extensive resources to help applicants submit required documents and comply with regulatory requirements. In addition, a voluntary biotechnology quality management support program helps organizations to comply with regulatory requirements for GE plants.

Notifications, available since 1993, can be used for plants that do not present novel plant risks. Notifications cannot be used for perennials, noxious weeds, or products for pharmaceutical or industrial use. Regulations define the notification procedure and require information about the location and size of the trial and extensive technical data about the regulated article.[66] Performance standards help to avoid environmental or economic harm, and the developer must allow inspection of facilities and records and report results to APHIS. The majority of field trials do not require permits, but occur after notification to APHIS.

Field trials of GE varieties that may pose risks to health or the environment require a permit.[67] This includes plants for pharmaceutical or industrial use, plants modified with animal or human genetic material, and other organisms precluded from the notification procedure. The permit application requires detailed information about geographic location, technical data about the GE regulated article, experimental design, plans to prevent escape, and final disposal of the trial plants. Permits may trigger obligations for APHIS under the National Environmental Policy Act (NEPA),[68] which requires federal agencies, including USDA, to prepare an environmental impact statement (EIS) for "major federal actions significantly affecting the quality of the human environment."[69] As a first step, APHIS prepares an environmental assessment. If the GMO may have a significant environmental impact, APHIS prepares an EIS to disclose environmental effects from, and alternatives to, the proposed federal action.[70] An EIS is a disclosure document, but does not impose duties on the agency.

APHIS will grant or deny a permit for field trials after reviewing the application, including the data submitted. If granted, the permit will impose regulatory requirements and plant-specific conditions—for example minimum separation distances between regulated crops and their sexually compatible non-regulated relatives. APHIS requires the permit holder to allow inspections, provide notice of unauthorized or accidental releases of the trial plants, and report results of field trials, including any unexpected characteristics of the organisms. A Permit User's

commingling should be minimized, low levels of GE gene presence from field tests might be acceptable. OSTP (2002). See also APHIS (2007) on low-level presence.

[66]7 CFR § 340.3. Other limitations on use of notification apply. USDA proposed regulations, withdrawn in November 2017, would have ended the notification procedure.

[67]7 CFR § 340.4. Interstate movement and import also require permits.

[68]42 USC §§ 4321–4370f.

[69]42 USC § 4332(2)(C). Not all experts agree that GE approvals should be subject to NEPA. Conko et al. (2016), p. 497.

[70]If the environmental assessment finds no significant impact, the agency does not prepare an EIS. APHIS's failure to prepare an EIS for a GM alfalfa resulted in extensive litigation that reached the US Supreme Court in *Monsanto Co.* (2010). For details see Grossman (2010).

Guide from Biotechnology Regulatory Services provides information to assist developers in the permit process.[71] Between June 1987 and December 2018, APHIS had received more than 21,400 notifications and permits; notifications were a majority. APHIS acknowledged notifications and issued permits for thousands of field trials or for interstate movement.[72]

7.3.2.1.2 Nonregulated Status

If field trials indicate that a new GE variety is not a plant pest or a threat to agriculture or the environment, the developer can petition APHIS for a determination of nonregulated status so that the variety can be sold commercially.[73] Regulations prescribe the format and content of the petition, which must state the factual grounds for nonregulation, including published and unpublished scientific studies and data from tests on the GE organism, as well as information unfavorable to the petition. The petition must include detailed scientific descriptions of the organism specified in the regulations, as well as field test reports for all trials completed under a notification or permit. Data must indicate "that the regulated article is unlikely to pose a greater plant pest risk than the unmodified organism from which it was derived."[74] APHIS must publish notice of petitions for nonregulated status in the Federal Register and accept comments from the public. Again, APHIS must comply with NEPA by preparing an environmental assessment and, if required, an EIS. After reviewing the risks of the organism, APHIS will make a decision to approve or deny the petition. If the GE variety poses no environmental or agricultural risk, APHIS will grant nonregulated status. By March 2018, APHIS had issued 123 determinations of nonregulated status. Recent determinations included several Simplot potatoes, an apple that resists browning, and glyphosate-resistant bentgrass.[75] If the nonregulated variety also meets the requirements of the FDA and the EPA, it can move freely in commerce. APHIS has no further jurisdiction over organisms that are nonregulated[76] and does not impose post-regulation monitoring requirements on those organisms.

Developers can also request that APHIS extend a determination of nonregulated status for an organism on the basis of its similarity to an organism with nonregulated

---

[71]BRS (2017c).

[72]APHIS (2018a).

[73]7 CFR § 340.6.

[74]7 CFR § 340.6(c)(4).

[75]APHIS (2018d). Other petitions were pending (6), withdrawn (32), or incomplete (1). The number on the APHIS website differs from the information from ISAAA (2018c) cited in note 26 above. Since March 2018, APHIS has made several more determinations of nonregulated status.

[76]NASEM (2017), p. 15 (stating that USDA did not have authority to reassess products with nonregulated status); NASEM (2016a), p. 471. The National Academies recommended that regulatory agencies have authority to require monitoring of GE crops after approval. NASEM (2016a), p. 507.

status.[77] APHIS must find that the new organism is similar to the antecedent organism used for comparison, which does not pose plant pest risks and is therefore no longer regulated. The extension procedure is not an exemption, but instead requires significant information from the developer to ensure that the organism raises no new safety issues. Like petitions for nonregulated status, extensions trigger NEPA and may require a new environmental assessment; if possible APHIS will re-use or update earlier environmental assessments. As APHIS advised in its guidance for developers, "to be considered for the extension process, (1) APHIS must have made a prior determination of nonregulated status of the mechanism-of-action for the trait of interest in any crop, and (2) APHIS must have made a determination of nonregulated status for the phenotype category (e.g. herbicide tolerance, insect resistance) in the subject crop."[78] Only a small number of recent determinations of nonregulated status rely on the extension procedure. Companies that plan to sell GE organisms in foreign markets may need to comply with normal authorization requirements.[79]

7.3.2.2 Environmental Protection Agency

The EPA regulates GE plants with pesticidal substances under authority of the Federal Insecticide, Fungicide, and Rodenticide Act (FIFRA),[80] and the agency governs pesticide residues in GE foods under the Food, Drug and Cosmetic Act (FDCA).[81] Although the EPA addresses food safety issues associated with pesticides, the FDA governs other food safety issues in products from GE plants.

GE plants that express pesticidal substances (for example, the insecticide *bacillus thuringiensis* (*Bt*)) are called plant incorporated protectants (PIPs).[82] PIPs differ from traditional chemical pesticides because they are produced and used in plants, so EPA regulatory requirements also differ somewhat from those for chemical

[77]7 CFR § 340.6(e).

[78]BRS (2016), p. 4.

[79]Conko et al. (2016), p. 497 (suggesting that APHIS decisions are designed in part "to prepare a paper trail to safeguard against abusive, harassing procedural lawsuits under NEPA").

[80]7 USC §§ 136-136y.

[81]21 USC §§ 301–399f; 21 USC §§ 2201–2252 (Food Safety Modernization Act of 2011).

[82]PIPs are considered pesticides because they are introduced in plants as way of "preventing, destroying, repelling, or mitigating any pest." 7 USC § 136(u). EPA regulations define PIP as "a pesticidal substance that is intended to be produced and used in a living plant, or in the produce thereof, and the genetic material necessary for [its] production." 40 CFR § 174.3.

pesticides.[83] EPA's rules for PIPs, which govern PIPs under FIFRA and the FDA, date from 2001, with some later amendments.[84]

FIFRA provides that no pesticide can be sold or used in the United States until it is registered, a process that requires the developer to demonstrate that its composition warrants the registrant's claims, its label complies with FIFRA, and the pesticide will not "cause unreasonable adverse effects on the environment."[85] The EPA can consider economic costs and benefits in determining whether a pesticide has an unreasonable adverse environmental effect.[86]

Under the FIFRA registration system, the EPA collects data on the efficacy, safety, and environmental effects of PIPs.[87] To provide data for registration, developers conduct field tests, authorized by experimental use permits. Some field tests do not require permits (e.g., laboratory or greenhouse tests). In addition, small-scale tests (a maximum of 10 acres per target pest) do not require permits, but crops must be destroyed or eaten by experimental animals unless EPA has granted a tolerance or exemption from a tolerance for pesticide residues.[88] EPA encourages developers to consult with the agency about physical and biological controls, even for small-scale field tests; compliance with APHIS requirements will often satisfy EPA requirements.[89] Moreover, unlike USDA, EPA can require post-market monitoring of approved PIPs. The agency may impose requirements—for example, planting non-GE refuges to manage insect resistance—for their use.[90]

The EPA governs pesticide residues in GE foods. Under the FDCA, raw or processed food with a pesticide residue (including PIPs) is considered adulterated; it cannot be sold in interstate commerce unless the residue meets an established tolerance (a legal limit on the maximum amount of a substance in or on food) or is exempt from the tolerance requirement.[91] The EPA may establish a pesticide

[83]40 CFR part 154 (pesticide registration), part 172 (experimental use permits), and part 174 (PIPs). Under 40 CFR part 155 (registration review process), the EPA is expected to re-evaluate registered pesticides every 15 years. NASEM (2017), p. 153.

[84]EPA (2001). In May 2018, EPA withdrew 1994 and 2001 regulatory proposals affecting PIPs, noting that possible new proposals would take account of more current science and recent developments in biotechnology. EPA (2018), pp. 20,006–20,008.

[85]7 USC § 136a(c)(5). Unreasonable adverse effects are defined in § 136(bb).

[86]7 USC § 136(bb). See NASEM (2016a), pp. 474–477 (details of risk assessment).

[87]See Conko et al. (2016), p. 501 (asserting that the PIP category is "too artificial and contrived" and that no data or experience suggests that genetic modifications should be regulated "as if they were pesticides sprayed on crops").

[88]40 CFR § 172.3.

[89]EPA (2007). EPA may require an experimental use permit or a temporary food tolerance.

[90]NASEM (2016a), pp. 471–472. The EPA requires warning labels on PIP crops. CAST (2018), p. 13.

[91]21 USC § 346a. Until 1996, pesticide residues were considered food additives and were subject to the Delaney clause that prohibits approval of substances that contain carcinogens. 21 USC § 348(c)(3)(a).

tolerance for food only if scientific data establishes that the tolerance is safe.[92] But the EPA may also grant temporary or permanent exemptions from the tolerance requirement if the pesticidal protein is not toxic or allergenic and data shows a reasonable certainty that aggregate exposure will not cause harm to human health. If a field test for a PIP may result in a food residue, the agency may establish a temporary food tolerance. For most PIPs in GE crops (e.g., *Bt* protein in corn, cotton, and soy), however, EPA regulations grant temporary and permanent exemptions from the tolerance requirement.[93] The FDA enforces pesticide tolerances.

7.3.2.3 Food and Drug Administration

The FDA governs food safety under the FDCA. Developers of new foods have legal responsibility to evaluate the safety of a new food and to comply with the statute and FDA regulations. The agency does not conduct pre-market safety review of new plant foods. To ensure food safety, the FDA relies primarily on statutory provisions that prohibit adulteration and misbranding of food and govern food additives.[94]

In 1992, the agency issued a Policy Statement on GE foods. The FDA indicated that its focus was the food product, rather than the process: "The regulatory status of a food, irrespective of the method by which it is developed, is dependent upon objective characteristics of the food and the intended use of the food .... [K]ey factors in reviewing safety concerns should be the characteristics of the food product, rather than the fact that the new methods are used."[95] Thus, the FDA did not plan to review new GE foods unless changes in structure, function, or composition made it necessary to review and approve a new GE food to protect public health. Moreover, the FDA relied on its existing regulatory system to govern the safety of GE foods.

Under the 1992 Policy, the FDA does not require premarket approval of all GE foods, but relies on the concept of substantial equivalence, a comparative approach rather than a safety assessment. Substantial equivalence ascertains whether a GE food shares characteristics—nutritional components—with its conventional counterpart.[96] If nutritional components of the GE food and its conventional counterpart do not differ, the GE food is substantially equivalent. The agency requires premarket

---

[92]Safe means "a reasonable certainty that no harm will result from aggregate exposure to the pesticide chemical residue, including all anticipated dietary exposures and all other exposures for which there is reliable information." 21 USC § 346a(b)(2).

[93]40 CFR §§ 174.501–174.538 (listing permanent and temporary exemptions from the requirement of a tolerance).

[94]21 USC §§ 342, 343, 348.

[95]FDA (1992), pp. 22,984–22,985.

[96]FDA (1992), p. 22,992 (citing to documents on substantial equivalence from the Organisation for Economic Co-operation and Development, the UN Food and Agriculture Organization, and the World Health Organization). On the substantial equivalence of GE and non-GE crops, see ISAAA (2018b).

review if foods lack substantial equivalence because their nutritional components (for example, allergens, toxins) differ or pose more risk.

The FDA has authority to regulate GE foods under the adulteration and food additive provisions of FDCA. A food is adulterated if it bears or contains, among other things, a "poisonous or deleterious" added substance harmful to health or an unsafe additive.[97] Food additives are considered unsafe (and therefore adulterated) until they have been granted premarket approval or are exempt from approval. FDA regulations prescribe an approval process that requires data to demonstrate a reasonable certainty that the additive is safe. If the FDA finds the additive safe, it will issue a regulation to approve the food additive.[98]

The FDCA definition of food additive excludes substances that are "generally recognized as safe" (GRAS).[99] Because most new plant foods have been accepted as safe, the FDA decided that most GM foods would be considered GRAS.[100] Therefore, most non-pesticidal GE foods have escaped premarket review as food additives. Although a GRAS determination requires the same level of scientific evidence required for approval of a food additive, manufacturers themselves make the GRAS determination and may, but need not, notify FDA of their decision.

In 2016 the FDA issued new GRAS regulations.[101] GRAS notifications to FDA remain voluntary—a decision that remains controversial. The regulations establish a seven-part format for the GRAS notice, which requires comprehensive information about the new substance as well as data about its safety and dietary exposure. The regulations emphasize the stringent scientific requirements for a GRAS determination. The FDA will evaluate voluntary GRAS notices and inform the notifier of the results of the evaluation.[102]

Although the FDA does not mandate review of new GE plant foods or GRAS substances, the agency issues guidance documents, which are less formal than regulations and do not bind the agency or the public. In its 1992 Policy Statement, the FDA noted that "prudent" developers of food using new technologies would consult with the agency before commercial distribution of those foods.[103] Guidance documents facilitate that consultation, which should occur before new foods are

[97] 21 USC § 342(a), defining as adulterated any food that contains an additive unsafe under § 348.

[98] 21 USC § 348; 21 CFR part 170.

[99] 21 USC § 321(s): "The term 'food additive' means any substance the intended use of which results or may reasonably be expected to result, directly or indirectly, in its becoming a component or otherwise affecting the characteristics of any food ... if such substance is not generally recognized, among experts qualified by scientific training and experience to evaluate its safety, as having been adequately shown through scientific procedures ... to be safe under the conditions of its intended use." Pesticide chemical residues in raw or processed agricultural foods are not defined as food additives.

[100] FDA (1992), p. 22,990. See Alliance for Bio-Integrity (2000), p. 177, which upheld FDA's presumption of GRAS status for GM foods.

[101] FDA (2016) and Grossman (2017).

[102] 21 CFR §§ 170.203–170.285 (GRAS regulations).

[103] FDA (1992), p. 22,991.

commercialized. The FDA's 1997 guidance document on consultation procedures for industry suggests that developers consult with the agency early in the development process and in a final consultation when data—including nutritional and safety assessments—indicate that the new product is safe. The FDA reviews industry data to identify unresolved scientific and regulatory issues.[104]

Early food safety evaluations, recommended in a 2006 guidance document, are carried out during field tests.[105] These evaluations are intended to ensure that new proteins are safe before low levels could enter the food supply via inadvertent cross-pollination in field tests or commingling of seeds. Early food safety evaluations, which can identify allergens or toxins, precede biotechnology consultations, but developers can use data from the earlier evaluations in later biotechnology consultations that precede commercialization of new foods.

Although these biotechnology consultations are voluntary, developers of GE foods do seek consultations, perhaps to avoid liability triggered by unsafe foods. By March 2018, the FDA had completed 16 early food safety evaluations (with 2 others withdrawn), and by December 2018, the agency had completed 183 biotechnology consultations on food from GE plants.[106]

7.4 Recent Regulatory Initiatives

Innovations in biotechnology may eventually lead to amendments in the regulatory system. Policy documents have been published, but little regulatory change has occurred.

7.4.1 Policy Developments

President Obama initiated a focus on improved regulation with a January 2011 Executive Order that instructed regulatory agencies to strive for scientific integrity and to use the "most innovative and least burdensome tools" for regulation.[107] The White House Office of Science and Technology Policy (OSTP) then issued a March 2011 memorandum that set out principles for regulating emerging technologies, including genetic engineering. Agencies "share a fundamental desire for regulation and oversight that ensure the fulfillment of legitimate objectives such as the protection of safety, health, and the environment. Regulation and oversight should avoid unjustifiably inhibiting innovation, stigmatizing new technologies, or creating trade

[104]FDA (1997).

[105]FDA (2006).

[106]FDA (2018c) (evaluations); FDA (2018a) (consultations).

[107]Obama (2011), pp. 1–2.

barriers."[108] The OSTP memorandum focused in part on risk assessment and risk management and noted that "[f]ederally mandated risk management actions should be appropriate to, and commensurate with, the degree of risk identified in the assessment."[109]

In 2015, the Obama administration issued a memorandum on modernizing the biotechnology regulatory system, which established a Biotechnology Working Group, with representatives of the President's executive office, USDA, EPA, and FDA. Noting again that regulation of biotechnology "must protect public health, welfare, safety, and our environment while promoting economic growth, innovation, competitiveness, and job creation," the memorandum set up a process intended both to modernize the current regulatory system and to ensure that periodic updates would occur.[110] Its goals were "to ensure public confidence in the regulatory system and to prevent unnecessary barriers to future innovation and competitiveness by improving the transparency, coordination, predictability, and efficiency of the regulation of biotechnology products while continuing to protect health and the environment."[111] To achieve these goals, the working group was directed to update the Coordinated Framework and to develop a long-term strategy to ensure that the regulatory system can efficiently use the best science available to evaluate risks of new biotechnology products.[112] In addition, the working group was to commission an independent analysis to identify risks (or lack of risks), both well understood and unfamiliar, from future biotechnology products.[113]

In September 2016, the Biotechnology Working Group issued its National Strategy for Modernizing the Regulatory System for Biotechnology Products, which reflects "a vision for ensuring that the Federal regulatory system is prepared to efficiently assess the risks, if any, of the future products of biotechnology,"[114] while at the same time "supporting innovation, protecting health and the environment, promoting public confidence in the regulatory process, increasing transparency and predictability, and reducing unnecessary costs and burdens."[115] The National Strategy describes existing mechanisms and activities, as well as future plans, of the major regulatory agencies for increasing transparency, predictability and efficiency, and supporting the science of biotechnology. Plans include more interagency communication to ensure timely decisions about regulatory jurisdiction, collaboration in scientific oversight, and review of regulatory structures and

[108] Holdren et al. (2011), p. 1.

[109] Holdren et al. (2011), p. 2.

[110] Holdren et al. (2015), p. 1.

[111] Holdren et al. (2015), p. 1.

[112] Holdren et al. (2015), pp. 4–5.

[113] The analysis was intended to survey the "future landscape" of biotechnology products. A report from the National Academies of Science, Engineering, and Medicine fulfilled this requirement. NASEM (2017).

[114] US White House (2016), p. 13.

[115] US White House (2016), p. 8.

requirements. Individual agencies also plan to review and update various policies and guidance for regulating products of biotechnology.[116] The National Strategy, however, failed to describe plans for assessing the risks of future products of biotechnology.[117]

The Biotechnology Working Group finalized an Update to the Coordinated Framework for the Regulation of Biotechnology in January 2017.[118] Among the principles set out in the Update is that regulation will continue to focus on the intended uses of products rather than the process of development. Characteristics and uses of the product and the environment should determine risk. Under the risk-based approach, "the regulatory system should distinguish between those biotechnology products that require a certain level of Federal oversight and those that do not."[119] Scientific advances may require regulatory amendments to address perceived risks, but the Update did not focus on those possible requirements.

The 2017 Coordinated Framework sets out the current responsibilities of the USDA, EPA, and FDA in regulation of products of biotechnology, identifying each agency's statutory authority and responsibilities for assessment of GE foods and other products.[120] A detailed table identifies numerous biotechnology product areas (not only agricultural products) and the responsibility and coordination among regulatory agencies for oversight of organisms derived from GE plants, animals, microbes and cells.[121] The 2017 Coordinated Framework also describes agency cooperation in the regulation of biotechnology products. Eight hypothetical case studies provide practical examples for GE developers of the process of regulation and agency roles.[122] The Update, however, provides no regulatory guidance for new technologies like CRISPR and other forms of genome editing.[123]

Despite the 2015 charge to the Biotechnology Working Group, which focused in part on a strategy for regulating innovative technologies, the National Strategy and the 2017 Update represent a "missed opportunity," and the "process has prevented clarification and restructuring of the regulatory system, [and has missed] opportunities to increase the legitimacy of that system in the eyes of the public and stakeholders."[124] At the outset the OSTP had indicated that the working group would "clarify existing authorities" rather than propose new regulatory options, and the process included no public consideration of new federal legislation on

[116]US White House (2016), pp. 16–18.

[117]See Peck (2017), p. 324.

[118]US White House (2017). In 2015, the Executive Office of the President had directed the agencies to update the Coordinated Framework (p. 5).

[119]US White House (2017), p. 8.

[120]US White House (2017), Table 1, pp. 9–10.

[121]US White House (2017), Table 2, pp. 28–35.

[122]US White House (2017), pp. 39–51.

[123]Peck (2017), p. 324.

[124]Kuzma (2016), p. 1211.

biotechnology.[125] Nevertheless, the 2017 Coordinated Framework did not meet the goals of the 2015 memorandum on modernizing the US biotechnology regulatory system.[126]

7.4.2 *Agency Activities*

7.4.2.1 USDA

USDA's APHIS proposed amendments to its GE regulations in January 2017, but in less than a year, USDA withdrew the proposal. In February 2016, APHIS had published notice of a proposed Environmental Impact Statement (to satisfy the requirements of NEPA) for revisions to its biotechnology regulations.[127] In that notice, APHIS published possible new definitions of critical terms—biotechnology, product of biotechnology, and regulated organism. Its proposed definition of biotechnology was not limited to rDNA technology, but included other forms of genome editing.[128] The proposed definition of regulated organism was an "organism developed using biotechnology that poses plant pest or noxious weed risks as documented in an APHIS risk analysis that APHIS has determined to regulate."[129] APHIS asked for public input on these definitions. In the notice, APHIS suggested four possible regulatory alternatives. One approach would revise current regulations with a process "to review and regulate certain products of biotechnology to protect plant health; analyze potential plant pest and/or noxious weed risks first; and

[125]Kuzma (2016), pp. 1212–1213.

[126]CAST (2018), p. 7. See the quotation from Holdren et al. (2015), p. 1, cited at note 111 above.

[127]APHIS (2016) (referring to regulations at 7 CFR part 340). See APHIS (2017a), pp. 2–36, which presented three alternatives, including regulations to facilitate coexistence through its authority over noxious weeds: "incorporate the noxious weed authority under the PPA as inclusive of GE plants that cause economic harms due to the mere presence of GE plant material in non-GE crops or crop products, regardless of whether this occurs as a result of cross-pollination, or commingling of GE plant material with non-GE crops or their products during harvest, post-harvest shipping and processing, or other means." That is, economic harms from mere presence would trigger regulation. The Draft EIS alternatives differed from the APHIS proposed regulations that were withdrawn.

[128]APHIS (2016), p. 6227:

> *Biotechnology*. Laboratory-based techniques to create or modify a genome that result in a viable organism with intended altered phenotypes. Such techniques include, but are not limited to, deleting specific segments of the genome, adding segments to the genome, directed altering of the genome, creating additional genomes, or direct injection and cell fusion beyond the taxonomic family that overcomes natural physiological reproductive or recombination barriers. This definition does not include and is intended not to include traditional breeding, marker assisted breeding, or chemical or radiation-based mutagenesis.

[129]APHIS (2016), p. 6227. "Product of biotechnology. An organism developed using biotechnology." APHIS requested public comments on these definitions, in light of "the potential, or lack of potential" of products of biotechnology to pose risks as plant pests or noxious weeds.

thereafter regulate only when appropriate and necessary."[130] Other alternatives would take no action, repeal the regulations and implement a voluntary consultation process, or increase the scope of regulation of potential plant pests or noxious weeds.

In January 2017, APHIS published proposed amendments to its 1987 rule governing the release and commercialization of GE organisms.[131] Although this proposal was withdrawn in November 2017,[132] a brief consideration is instructive. In its proposal, APHIS noted that advances in understanding of GE organisms indicate that most GE technologies do not pose plant pest risks. Some technologies, not defined as regulated articles under the 1987 rule, use no plant pests as donor or recipient organisms, but could nonetheless involve risks. GE plants should also be evaluated as noxious weeds, because some GE plants (for example, switchgrass used to produce ethanol) may become weedy.[133] GE biological control agents and GE plants that produce plant-made industrials and pharmaceuticals require regulation to protect the food supply.

As APHIS noted when it withdrew the proposal, the rule would have been a major change from its "regulate first/analyze later" approach because it would first assess new GE organisms to determine if they posed risks and then regulate only those organisms that did present risks.[134] In the proposal, APHIS identified two situations that require regulation of GE plants: "when APHIS has reached a determination that the plant and trait combination associated with the GE plant causes it to act as a plant pest or noxious weed"[135] and "when APHIS is presented with a GE plant with a novel plant and trait combination, and has not yet evaluated this plant and trait combination for its plant pest and noxious weed risk."[136] If a plant falls within the regulatory definition of GE organism, a developer who wants to release the organism must request an APHIS evaluation of its potential to be a plant pest or noxious weed. If APHIS concludes that the organism is unlikely to pose a risk, it would no longer be regulated. If a risk may exist, the developer must obtain a permit to release, import, or move the organism interstate and comply with regulatory requirements under the permit.[137] In March 2017, Biotechnology Regulatory Services published a provisional list of the regulatory status of numerous crops under the proposed rule. Only a

[130]APHIS (2016), p. 6227 (italics omitted). This alternative would eliminate notification, because many products would not be regulated, and also eliminate the petition for nonregulated status.

[131]APHIS (2017b), proposing to amend 7 CFR part 340 (Docket APHIS-2015-0057). The 2017 proposed rule responds to a request from the USDA Office of the Inspector General and a directive from Congress to amend some aspects of APHIS regulations.

[132]APHIS (2017c).

[133]APHIS (2017b), pp. 7009–7011.

[134]APHIS (2017c), p. 51,582.

[135]APHIS (2017b), p. 7018.

[136]APHIS (2017b), p. 7018.

[137]APHIS (2017b), pp. 7014–7015.

few varieties were regulated or "proposed regulated"; most were currently not regulated or "proposed not regulated."[138]

Public comments on the proposed rule revealed substantial disagreement. Some comments noted that the proposal would increase the regulatory burden by expanding regulation of GE organisms; others noted that exemptions and exclusions would increase risk by narrowing the scope of regulation. Some comments asserted that the proposed risk assessment would hinder innovation; others claimed that it lacked rigor.[139] In light of this disagreement, and perhaps reflecting the changed political climate in the United States, APHIS withdrew the proposal, effective 7 November 2017. Before publishing a new proposal, the agency plans to evaluate new scientific knowledge about GE organisms and to involve stakeholders in development of alternative policy approaches.[140]

In January 2018, an interagency Task Force on Rural Agriculture and Prosperity focused on biotechnology in its call to action for harnessing technological innovation. Improvements in the biotechnology regulatory system should "(1) maintain high standards that are based on the best available science and that deliver appropriate health and environmental protection; (2) establish transparent, coordinated, predictable, and efficient regulatory practices across agencies with overlapping jurisdiction; and (3) promote public confidence in the oversight of the products of biotechnology through clear and transparent public and diplomatic engagement."[141] To achieve these improvements, the Task Force recommended that the federal Administration take three actions:

(a) "Coordinate federal regulation of biotechnology products" by supporting the Updated Coordinated Framework and the National Strategy.
(b) "Coordinate interagency action through the Office of Science and Technology Policy" by empowering the OSTP Biotechnology Working Group to encourage inter-agency cooperation, improve science-based regulation, help biotechnology innovators, and promote consumer understanding of risk-based regulation.
(c) "Expedite commercialization of biotechnology products" by connecting regulators with funding and research and development agencies.[142]

These recommendations focus on policy, however, and do not suggest regulatory amendments.

[138]BRS (2017a).

[139]One comment included a petition signed by 10,261 members of the Organic Consumers Association, with 2741 unique comments. See documents collected at www.regulations.gov under Docket APHIS-2015-0057.

[140]APHIS (2017c). APHIS provided no time frame for developing a new regulatory proposal.

[141]Task Force (2018), p. 34. The Task Force, established in April 2017 by Executive Order 13790, Trump (2017), was asked "to identify key legislative, regulatory, and policy changes to achieve rural prosperity in seven areas," including technological innovation.

[142]Task Force (2018), p. 34 (capital letters removed from quotations).

7.4.2.2 FDA

The FDA is also focusing on innovative technologies. In January 2017, the agency requested comments about genome editing in new plant varieties used for food.[143] The National Strategy mentioned the FDA's intention to "clarify its policy for the regulation of products derived from genome editing techniques," that is, modification by "targeted DNA sequence alterations" including CRISPR, ZFN, and other methods.[144] The FDA invited comments from various stakeholders and the public to inform its thinking about food from these new plant varieties. The FDA sought information about food safety risks from new technologies, data relevant to safety assessment and the regulatory status of these new organisms, and advice about voluntary consultations or other methods of informing the FDA about new products. FDA's request does not propose new regulatory measures, and the agency plans to continue to offer voluntary consultation with developers of new food varieties.

Among the many comments submitted to the FDA, the American Soybean Association noted that it supported USDA's now-withdrawn proposal to exclude some varieties created by genome-editing techniques from regulation. Focusing on genome editing within a plant species, the Association indicated that varieties created by many of these techniques (citing "insertion, deletions, substitutions, including mutagenesis, and changes to endogenous DNA sequence and function"[145]) result in products no different from those created by cross-breeding and other traditional breeding processes. In most cases, food safety risks are no different. The Association insisted that no reasonable scientific basis existed to include varieties indistinguishable from those created by traditional breeding in the FDA's voluntary consultation process.[146]

7.5 Protecting Non-GE Crops

7.5.1 Coexistence

In the US, GE varieties make up over 90% of total production acreage of maize, soy, cotton, canola, and sugar beets, and nearly half of US cropland is planted with GE varieties. Recent approval of GE varieties of apples and potatoes is likely to add to the small amount of GE fruit and vegetable acreage.[147] Because GE varieties

[143]FDA (2017).

[144]FDA (2017), p. 6565.

[145]American Soybean Association (2017).

[146]American Soybean Association (2017). The Association acknowledged that "gene editing techniques used for transgenic varieties could result in novel functional gene(s)" that may pose food safety risks; FDA's voluntary consultation process has mitigated these risks.

[147]Greene et al. (2016), pp. 11–13, 29.

predominate for major crops, the possibility of cross-pollination or commingling of GE and other crops (traditional, organic, or identity preserved) raises concerns that may become more critical as producers cultivate more GE varieties. But as a recent report of the National Academies of Sciences, Engineering, and Medicine noted, "neither EPA nor APHIS addresses the economic conflicts that arise from the coexistence of commercial GE and non-GE crops as part of the regulatory-approval process. Neither agency requires post-approval monitoring nor management plans to prevent the low-level presence of GE traits in non-GE crops or foods."[148] Indeed, the US Government Accountability Office recently indicated that USDA should work toward enhanced oversight and a better understanding of the impacts of the commingling of crops.[149] Both government and the private sector make some efforts to address coexistence, as producer organizations encourage farmers who cultivate GE crops to follow good stewardship practices.

The USDA Advisory Committee on Biotechnology and 21st Century Agriculture (AC21), established in 2003, developed recommendations to strengthen coexistence in an increasingly complex agricultural landscape. In 2012, the Advisory Committee recommended that USDA take specific measures to facilitate co-existence. USDA responded to some of the recommendations, but lacked authority to carry out others (for example, providing compensation for losses from unintended GE presence and incentives for neighbors to develop cooperative coexistence plans). In 2016, the Advisory Committee published a report that included two documents on coexistence.[150] One document was intended to encourage discussions among neighboring farmers about production challenges surrounding coexistence; the other outlined factors for farmers to consider when a nearby farmer grows an identity-preserved crop. These documents encouraged voluntary cooperation.

The risk of unintended presence of GE material in non-GE crops exists despite efforts by producers and supply chains to avoid pollen drift and commingling.[151] With GE varieties planted on almost half of US cropland, adventitious presence has economic consequences for producers of non-GE and organic varieties. Practices required to avoid cross-pollination (for example, buffer strips, delayed planting to stagger pollination) reduce land in production and may lead to lower yields. Segregation and testing of non-GE and organic varieties impose costs on producers and the grain industry. Pollen drift and commingling have imposed other economic losses on producers of non-GE varieties. Some organic farmers have reported

[148]NASEM (2016a) p. 474. EPA has authority to require monitoring.

[149]US GAO (2016).

[150]USDA, AC21 (2016). Coexistence is "the concurrent cultivation of conventional, organic, IP, and GE crops consistent with underlying consumer preferences and farmer choices" (p. 13).

[151]NASEM (2016a), p. 501. Indeed, adventitious presence has been considered normal. Farming "is practiced in the open air" subject to the vagaries of nature, and "100% purity is impossible." Grossman (2007), p. 329.

economic losses ($6.1 million between 2011 and 2014) from commingling with GE varieties.[152]

7.5.2 *Low-Level Presence*

Closely related to the issue of coexistence is the problem of low-level presence of GE material in non-GE crops. The regulatory process, with USDA and EPA requirements for field trials and FDA early food safety assessments, minimizes the likelihood of unauthorized GE varieties in other crops or food.[153] Nonetheless, low-level presence can result from natural processes (pollen movement), failure of coexistence, or escape of GE material during field tests. Asynchronous approvals of GE varieties may cause problems when "small amounts of grain with biotech events approved in an exporting country but not in an importing one, known as low-level presence (LLP), may be contained in an international shipment of agricultural commodities."[154] Low-level presence of approved GE varieties does not raise a food safety issue, but instead is often "an issue of noncompliance with the importing country's regulations for a product with at least one existing completed full safety assessment and a history of safe use."[155]

To address the issue, APHIS developed a policy for responding to low-level presence of regulated GE plant materials in commercial seeds or grain.[156] APHIS responds to occurrences of low-level presence with remedies appropriate to the risk in each case, coordinating with EPA and FDA to investigate the occurrence, evaluate risk, and determine what remediation measures are required. When an incident would result in introduction or dissemination of a plant pest or threaten plant health or the environment, APHIS will require remedial measures authorized by the Plant Protection Act.[157] APHIS will generally not require remediation if the regulated GM material poses no significant safety risk to plant health or the environment. Two situations are likely to require no remediation: if the GE material comes from plants that qualify for the APHIS notification process (that is, the majority of GE plants) or

[152]Greene et al. (2016), pp. 27–28. Even if organic producers meet USDA organic process standards for certification, crops with GE material may fail to meet stricter private industry standards for purity.

[153]For example, EPA (2007) (guidance for avoiding low-level presence from field tests).

[154]CAST (2016), pp. 3, 5. Greater asynchronies between pairs of countries result in less trade.

[155]GAABT (2015), p. 6.

[156]APHIS (2007), p. 14,469 (referring to existence of "low-level mixing of genes and gene products from unintended plant sources. This is true for both conventionally bred plants as well as biotechnology-derived plants. These occurrences can result from natural processes such as the movement of seeds or pollen, or human-mediated processes associated with field testing, plant breeding, or seed production").

[157]7 USC § 7714 (authorizing remedial measures for plant pests and weeds, including seizure or destruction of regulated materials).

if the GE plant is similar to another GE plant that has been deregulated and therefore poses no safety risk.[158]

Recognizing the importance of biotechnology in satisfying international food demand and the risk of trade disruptions from low-level presence, 13 importing and exporting countries agreed to cooperate to develop practical approaches for managing low-level presence.[159] The 2012 International Statement on Low Level Presence focused on low levels of "recombinant DNA plant materials."[160] Cooperating countries, led by Canada, participate in a global low-level presence initiative to develop a policy model for grain, food and feed. That policy is designed to minimize trade disruptions, protect health and safety, help regulators develop a risk-based approach to low-level presence, and facilitate transparency and predictability for both importers and exporters. It recommends a compliance threshold of 3% for low-level presence involving a commercialized GE crop not yet approved in the importing country.[161] An industry coalition, the Global Alliance for Ag Biotech Trade, recommended a more lenient threshold of 5% as "technically feasible, cost effective, and practical" for GE food, feed, and grain for processing, if the GE variety has been tested in accordance with recommendations of the Codex Alimentarius Commission in the exporting country.[162]

Unless workable thresholds apply, the unintended presence of GE varieties causes significant trade risks. Industry stewardship is critical. Both the American Soybean Association and the National Corn Growers Association developed stewardship programs.[163] Moreover, seed firms may delay commercial planting of new GE crops until major import markets have granted approval, a strategy that deprives producers and consumers of the benefits of the new crop but may prevent trade incidents.[164]

Although commercial grain contracts normally allow some impurities,[165] a zero or extremely low tolerance may apply to unapproved GE varieties, even when the presence is adventitious. Grain shipments that exceed the importing country's tolerance can be—and have been—rejected. Low-level presence "can cause abrupt,

---

[158]APHIS (2007), p. 14,651.

[159]FAS (2012). Countries are Australia, Argentina, Brazil, Canada, Chile, Costa Rica, Mexico, Paraguay, Philippines, Russia, United States, Uruguay, and Vietnam.

[160]FAS (2012). The statement defined low-level presence as "low levels of recombinant DNA plant materials that have passed a food safety assessment according to Codex Guideline for the Conduct of Food Safety Assessment of Foods Derived from Recombinant-DNA plants (CAC/GL 45-2003) in one or more countries but may on occasion be present in food in importing countries in which the safety of the relevant recombinant-DNA plants has not yet been determined."

[161]Agriculture and Agri-Food Canada (2017). The policy model refers to organisms modified by rDNA technology.

[162]GAABT (2015), p. 4. See CAC (2003a, b).

[163]Grossman (2016b), p. 328.

[164]CAST (2016), p. 5.

[165]For certified seed, the normal purity level is 99.5% (allowing 0.25% weed seed and 0.25% other crop varieties). Grain contracts often allow 3–5% impurity. Smyth (2017), pp. 82–83.

large-scale trade disruptions, sustained changes in international trade patterns, and significant economic losses that are borne by both importers and exporters."[166] Recent trade incidents have involved GE crops produced in the United States. Some resulted from adventitious presence of still-regulated varieties; others, from asynchronous approvals.

A recent incident affected trade and cost the developer billions of US dollars. Syngenta's Viptera® (MIR 162) was commercialized in the United States in 2011. China had not approved the crop and in 2013 rejected shipments, resulting in lower corn prices and significant losses. Thousands of producers and grain handlers filed individual and class action lawsuits against Syngenta.[167] In June 2017, a jury found Syngenta negligent and awarded $217.7 million to a class of 7000 Kansas farmers. Cases in other states were scheduled for trial, but in September 2017, Syngenta agreed to a global settlement of cases involving more than 100,000 farmers in Minnesota and elsewhere.[168] The class has been expanded to include more than 600,000 producers and others and would end most claims against Syngenta; the settlement, with preliminary court approval, will cost $1.51 billion.[169] In another incident, traces of Herculex™ corn (DAS 59122-7), developed by Dow AgroSciences and Pioneer Hi-Bred, appeared in shipments to the EU before the crop was approved for import in the EU in 2007. This violated the EU's zero tolerance policy and disrupted the corn market for 3 years, significantly decreasing sale of US corn gluten feed to the EU.[170] Another example involved adventitious presence of regulated crops. In 2006, traces of Bayer's LibertyLink® rice (LL601), not yet approved in the US, were discovered in samples of commercial long-grain rice including exports. LibertyLink® posed no threats to health or environment and was soon authorized, but the incident affected US-EU trade in rice as late as 2012. Producers and others sued Bayer, and the company paid over $2 billion in settlements.[171]

7.6 Labels for GE Food

Labeling of food in the United States is governed mainly by the FDCA and FDA regulations, with USDA requirements for some meat products. These measures (for example, nutrition labels) apply to all foods, including GE foods. FDA policy,

[166]CAST (2016), p. 8.

[167]For details on the Syngenta cases, see Redick (2017), pp. 45–54. China approved Viptera® in 2014, but not for cultivation.

[168]Feeley and Fisk (2017).

[169]Tidgren (2018).

[170]Smyth (2017), p. 83.

[171]Smyth (2017), p. 83. See Grossman (2012), pp. 93–95 for a discussion of LibertyLink® (LL601) rice and StarLink™ corn, an earlier incident.

however, did not require special labels to inform consumers of GE content in foods.[172] Under the FDCA, a food is "misbranded" if its label is "false or misleading,"[173] which is determined in part by "the extent to which the labeling . . . fails to reveal [material] facts."[174] The FDA did not consider the process used to develop new plant varieties to be "material" information and therefore concluded that it had no legal or scientific basis to require labels for GE food unless the food itself differed materially (for example, by presence of an allergen) from similar foods. A federal court decision upheld FDA's conclusion,[175] and several respected scientific organizations, including the American Medical Association, supported FDA's decision not to require labels.[176] The FDA allowed voluntary labeling, however, if labels were truthful and not misleading. To assist industry, the FDA published a guidance document for voluntary labeling of GE food.[177]

Consumers and non-governmental organizations demanded labels for GE foods, and a few states enacted labeling laws. In 2016, Congress enacted the National Bioengineered Food Disclosure Standard,[178] which preempts state and local labeling laws and requires labeling of statutorily-defined bioengineered food and food with bioengineered ingredients. The GE disclosure law states that the term "bioengineering" and similar terms refer to food "(A) that contains genetic material that has been modified through in vitro recombinant deoxyribonucleic acid (DNA) techniques; and (B) for which the modification could not otherwise be obtained through conventional breeding or found in nature."[179] The law applies to food intended for human consumption, as defined in the FDCA.[180] The law does not apply to all food, but requires GE labeling primarily for food subject to FDA labeling requirements and some food subject to USDA labeling requirements.

USDA, acting through its Agricultural Marketing Service (AMS), is responsible for implementing the GE Disclosure Standard.[181] USDA must determine when a food is bioengineered, including the GE threshold for labeling, as well as the form of disclosure. The law permits several forms of labeling—text on the package, a symbol, a QR code, and a digital link. USDA was directed to conduct a study to

[172]For more details on labeling see Grossman (2016a) and references therein.

[173]21 USC § 343(a).

[174]21 USC § 321(n).

[175]Alliance for Bio-Integrity (2000).

[176]Grossman (2016b), pp. 317–318.

[177]FDA (2015).

[178]7 USC §§ 1639-1639c, 1639i-1639j.

[179]7 USC § 1639(1).

[180]21 USC § 321(f): "The term 'food' means (1) articles used for food or drink for man or other animals, (2) chewing gum, and (3) articles used for components of any such article." Small manufacturers face less burdensome obligations and a later compliance date, and very small manufacturers and restaurant foods are exempt. 7 USC § 1639b.

[181]For an analysis of mandatory labeling, with special reference to GM foods, and the conclusion that "it will not be easy for the USDA to show that the benefits of the [labeling] mandate justify the costs," see Sunstein (2017), p. 1069.

identify technological challenges that could limit consumer access to label information. The study, published in September 2017, identified obstacles for retailers and consumers.[182]

USDA considered comments from consumers and industry stakeholders in response to questions that focused on terminology and definitions, types of disclosure, application of exemptions, and compliance.[183] Although AMS had planned to publish draft regulations in Fall 2017, the agency finally published draft regulations in the Federal Register in early May 2018, with a comment period ending 3 July 2018.[184] Draft regulations focused on applicability of the disclosure requirement), the form of disclosure (text on labels, symbol, electronic or digital link, or text message), recordkeeping, and enforcement. They did not resolve all issues raised by the Disclosure Standard, including a threshold for labeling[185] and whether highly refined ingredients with undetectable rDNA should be defined as bioengineered food, nor did they address foods produced by gene editing.

In December 2018 AMS published final regulations[186] to establish disclosure requirements for bioengineered food or food that contains a bioengineered food ingredient. The AMS definition of bioengineered food mirrors the statutory definition quoted above, but food in which modified genetic material cannot be detected is not considered bioengineered.[187] AMS did not adopt a more expansive definition of bioengineering that would include gene editing technologies (for example, CRISPR) and future biotechnology developments.[188] Products developed without rDNA technology, or with no detectable modified genetic material, will normally not be considered bioengineered and will therefore not bear BE labels.

Some exemptions to the disclosure requirement apply. Although consistency with the EU labeling threshold of 0.9% would have facilitated trade, AMS adopted a higher threshold. The final rule exempts from BE labeling "food in which no ingredient intentionally contains a bioengineered (BE) substance, with an allowance for inadvertent or technically unavoidable BE presence of up to five percent (5%) for

[182]Deloitte (2017). Although many Americans own a smartphone and nearly all national and regional grocery chains provide Wi-Fi in stores, not all small and rural retailers provide Wi-Fi, which would impose costs. Moreover, scanning QR codes or digital links is difficult for many consumers. Results of the study may have influenced AMS's addition of text message as an alternative method of disclosure.

[183]AMS (n.d.).

[184]AMS (2018a).

[185]AMS requested comments on three possible thresholds. Two alternatives would have exempted "[f]ood in which an ingredient contains a bioengineered substance that is inadvertent or technically unavoidable," with thresholds of no more than 5%, or alternatively 0.9%, by weight of that specific ingredient. The third alternative would exempt food from labeling if ingredients containing a bioengineered substance are no more than 5% "total weight of the food in final form." AMS (2018a), p. 19,886.

[186]AMS (2018b). AMS failed to meet the 29 July 2018 statutory deadline for final regulations.

[187]66 CFR § 66.1, with analysis and recordkeeping requirements for detectability at 66 CFR § 66.9.

[188]AMS (2018b), p. 65,835.

each ingredient."[189] Other exemptions include restaurant food, food produced by very small food manufacturers (annual receipts of less than $2.5 million), food from animals that consumed BE food, and certified organic food.

The final regulations prescribe standards, including the required location, for four types of disclosure: text, symbol, electronic or digital link, and text message. USDA's List of Bioengineered Foods prescribes the foods that must bear a disclosure.[190] Raw agricultural commodities and processed food with only BE ingredients must be labeled "bioengineered food." Multi-ingredient foods must be labeled "contains a bioengineered food ingredient."[191] Entities that identity BE food by symbol must use the symbol approved in the regulations.[192] Entities that disclosure bioengineered food via an electronic or digital link (for example, a QR code) or text message must comply with regulatory requirements that ensure information is available at all times.[193] Importers are subject to the disclosure requirements. AMS may eventually establish mutual recognition arrangements to recognize BE labeling standards from other governments.[194]

The regulations impose implementation dates, a compliance period, and a compliance date. Food manufacturers must implement the Disclosure Standard by 1 January 2020; small food manufacturers must implement by 1 January 2021.[195] During the compliance period, however, entities may comply voluntarily, using existing labels with stickers or ink stamps to identify BE food. By 1 January 2022, all regulated entities must be in full compliance.[196]

Although the food industry seems to have accepted the law, advocacy groups have been critical.[197] For example, many foods without detectable BE content, especially highly-refined products like cooking oils, will escape labeling. Consumers without access to cell phones or other electronic devices may be unable to access electronic disclosure. Moreover, the Disclosure Standard negates the FDA's decision that genetic modified foods do not require special labels, and AMS regulations have potential to conflict with some FDA requirements, especially the principle that

[189]66 CFR § 66.5.

[190]66 CFR § 66.6. The List, to be updated annually, currently includes the following foods: "Alfalfa, apple (Arctic™ varieties), canola, corn, cotton, eggplant (BARI Bt Begun varieties), papaya (ringspot virus-resistant varieties), pineapple (pink flesh varieties), potato, salmon (AquAdvantage®), soybean, squash (summer), and sugarbeet." Non-BE varieties of these foods (and foods with nondetectable modified genetic material) are not subject to disclosure. AMS (2018b), p. 65,826. Foods not on the list are subject to disclosure if the manufacturer has actual knowledge that the food or an ingredient is bioengineered. 66 CFR § 66.109.

[191]66 CFR § 66.102.

[192]66 CFR § 66.104.

[193]66 CFR §§ 66.106, 66.108.

[194]AMS (2018b), p. 65,825. US exports must meet requirements of import countries.

[195]66 CFR § 66.13. Small food manufacturers have annual receipts of at least $2.5 million but less than $10 million. Provisions for voluntary disclosure may apply. 66 CFR § 66.116.

[196]66 CFR § 66.13; AMS (2018b), pp. 65,861–65,862.

[197]Grossman (2016a), pp. 504–507.

disclosures must appear on food labels. Moreover, although knowing failure to meet BE disclosure requirements violates the Disclosure Standard law, neither the statute nor the final regulations impose stringent measures for enforcement.[198]

7.7 Liability

Developers of GE crops and their products are obligated to comply with federal laws and regulations that govern authorization. Failure to comply may result in penalties established by statute or regulation. Actions that violate FDCA provisions (for example, adulteration, misbranding) may be enjoined, and the FDCA authorizes civil and criminal penalties.[199] Between GE developers and producers, breach of contract obligations (often imposed by developers to protect intellectual property) may lead to liability for producers, and even unintentional presence of protected varieties in a producer's crops may trigger litigation.

The US regulatory system, however, does not assign liability for harm from sale, cultivation, and use of GE varieties and their products. Instead, tort liability, governed by state common-law principles or statutes, may redress those harms; other state statutes (for example, consumer fraud) may also apply. Developers face extensive liability if a regulated GE variety, or a variety not approved by trading partners, enters the stream of commerce, with significant economic consequences. Producers of non-GE crops may incur costs and suffer losses caused by gene transmission from pollen drift or commingling, but organic and non-GE producers usually bear those costs and losses.[200] Significant incidents that resulted in liability have involved GE crops, rather than GE food products.

Tort causes of action to redress damage from GE crops include nuisance, trespass, negligence, strict liability, or a combination of these claims. Proof of causation—the source of harm from the GE crop—is a critical element for liability in tort, but each cause of action also requires specific elements of proof.[201] The most common tort remedy is an award of monetary damages; less often, an injunction will prohibit defendant's harmful activities.

Liability has followed discovery of regulated crops in food or the grain supply, often causing costly disruptions of trade. An early case involved StarLink™ corn, found in food products when it had been approved only for feed and industrial uses, but not yet for food. A court decision in 2000 held that the class action for damages could be tried in negligence and private and public nuisance. Because the parties

[198] 66 CFR §§ 66.400–66.406. After an investigation and a hearing (if requested by the food producer), the results of an investigation of possible violations will be made public.

[199] 21 USC §§ 331–337a.

[200] Greene et al. (2016), pp. 25–28.

[201] For details on liability, see Grossman (2016b), pp. 326–330.

settled before trial, those claims were not litigated.[202] Similarly, litigation involving Bayer CropScience's LibertyLink® rice, detected in the US rice supply before approval, involved claims in negligence and private nuisance. Farmers and other plaintiffs won a large jury verdict, and Bayer CropScience settled other pending cases. A decade later, an unapproved Monsanto wheat variety, evidently escaped from field tests, was discovered in an isolated field, triggering lawsuits alleging nuisance, negligence and strict liability.

Other incidents resulted from asynchronous approvals that caused market disruptions. A recent example is the Syngenta litigation mentioned above, related to the commercialization of Viptera® corn not yet approved by China, an important trading partner. Individual and class action lawsuits raised claims in negligence, nuisance, trespass, consumer fraud, and other causes of action. Syngenta has negotiated a costly settlement with class action plaintiffs.

7.8 US Regulation of Innovative GE Technology

7.8.1 Regulatory Uncertainty

In the United States, the regulatory treatment of plants developed by gene editing and other innovative technologies remains uncertain.[203] Indeed, new "genetic-engineering technologies challenge most existing regulatory systems by blurring the distinction between genetic engineering and conventional plant breeding while enabling increasingly profound alterations of plant metabolism, composition, and ecology."[204] Although these technologies offer precision and diversity, they raise difficult regulatory issues. Some scientists and lawyers believe that these new crops should not be subject to regulatory strictures applied to rDNA crops.[205] Others, including green groups, may disagree.[206]

Regulation of GE technology requires an applicable statute or regulation—a "regulatory trigger." As the discussion of the US regulatory system indicates, GE crops and their products are governed by statutes that authorize USDA, FDA, and EPA to promulgate and implement regulations. USDA governs GE crops under the

---

[202]StarLink (2000). The court decision in *StarLink* indicated that the economic loss doctrine often precludes recovery for damages unrelated to injury to property or physical loss (called pure economic loss).

[203]European Union law also poses uncertainties. See Kahrmann et al. (2017).

[204]NASEM (2016a), p. 26.

[205]See, for example Kerr (2017), p. 69: "Genomics and its associated techniques should not raise the issues pertaining to risks to the environment that arose in the case of transgenic crops. . . . There is no need for a special regulatory regime, nor for special international trade rules such as those associated with the creation of plants with transgenic gene combinations." Moreover, "food safety implications should be no different than with food products arising from conventional breeding."

[206]Wolt et al. (2016), p. 514.

Plant Protection Act, which gives the agency authority over plant pests. APHIS regulations define plant pests broadly, but define genetic engineering narrowly as "genetic modification of organisms by recombinant DNA techniques."[207] Regulated articles are genetically engineered organisms produced with a plant pest or constituting a plant pest. USDA also has statutory authority over noxious weeds, and can regulate a new product defined as a noxious weed under that authority even if the product is not a regulated article under APHIS GE crop regulations.[208]

The EPA's authority over GE varieties under FIFRA is limited to varieties with pesticidal properties and under the FDCA, to pesticidal residues in food. Thus, few products of new technologies, particularly those in which knockouts result in plants that resist pests or viruses, are likely to be subject to EPA regulation because they contain "no new genetic material from a nonsexually compatible source."[209] The FDA, with FDCA authority for food safety, regulates to prevent misbranding and adulteration and to authorize food additives. FDA's definition of genetic engineering is broader than USDA's definition.[210] Nonetheless, unless a new GE variety is allergenic, toxic, or not substantially equivalent to a comparator, FDA relies on voluntary consultation. Because USDA's authority (or lack of authority) to regulate innovative products of biotechnology is critical, the focus in the following discussion is USDA jurisdiction and recent treatment of innovative GE varieties.[211]

7.8.2 USDA Regulation

An important issue is whether crops produced with innovative technologies "will fall within the definition of GE crops used by various regulatory agencies as a regulatory trigger and therefore be subject to premarket safety reviews."[212] Some statutes and regulations, like the USDA regulation quoted above, define the regulated article or product in terms of genetic modification with rDNA techniques. Genome editing, however, does not transfer genetic material, but can "make direct modifications or deletions of traits within the organism's genome without inserting new material at all."[213] Thus USDA's definition is likely to exclude many products of innovative biotechnology. Similarly, the definition of biotechnology in the GE labeling law (the GE Disclosure Standard discussed above) refers to food modified by "in vitro recombinant deoxyribonucleic acid (DNA) techniques" with modifications not

[207] 7 CFR § 340.1. At the time of the 1986 Coordinated Framework, GE plants were produced with rDNA technology using *agrobacterium tumefaciens*, a plant pathogen. NASEM (2016a), p. 498.

[208] 7 CFR part 360.

[209] NASEM (2016a), p. 498.

[210] FDA (1992).

[211] See Enríquez (2017b), p. 538.

[212] NASEM (2016a), p. 493.

[213] Peck (2017), p. 321.

possible through "conventional breeding or found in nature."[214] This narrow definition makes it unlikely that the Disclosure Standard will apply to most foods from varieties developed with innovative technologies.

The definitional issue may be critical for regulating new crops, especially if the regulatory trigger is based on the process of development, rather than the product.[215] That is, process-based regulations are "always a step behind the introduction of new techniques."[216] Another critical question is whether traditional types of risk assessment, which assume that a GE crop was created with rDNA technology, address risk characteristics of crops produced with new technologies.[217] Moreover, some plant breeding techniques "provide no identifiable markers that can be used to verify the technology utilized to create the new variety . . . . Examples of these technologies are targeted mutagenesis techniques such as oligonucleotide-directed mutagenesis (ODM), zinc finger nuclease (ZFN), meganuclease technique, and transcriptional activator-like effector–nuclease (TALEN), and gene silencing techniques such as clustered regularly interspaced short palindromic repeats (CRISPR)."[218] The lack of markers to verify technology is likely to raise issues connected with coexistence of these new crops with traditional varieties or organic production.

7.8.2.1 Am I Regulated?

As the discussion above indicates, an individual or company with plans to release a regulated GE organism into the environment, move it interstate, or import it must obtain authorization from USDA (currently a notification or a permit). A developer who is unsure whether the organism is a regulated article can request a determination from APHIS Biotechnology Regulatory Services using "Am I Regulated?".[219] This procedure informs developers about their regulatory obligations and alerts APHIS to new developments. The procedure requires a signed letter of inquiry accompanied by data about the GE organism. In addition to information about the developer and the intended activity (release or movement), the letter must include descriptions of the intended phenotype, genetic change (insertion, deletion, substitution), vector or vector agent (biolistic, nuclease, *Agrobacterium*), and scientific information about the construct. Letters with confidential business information require both a full version and a version with confidential information deleted, with justification for confidentiality. BRS provides a user's guide to help in submitting documentation.[220]

[214] 7 USC § 1639(1).
[215] NASEM (2016a), p. 493.
[216] Conko et al. (2016), p. 502.
[217] NASEM (2016a), p. 494.
[218] Smyth (2017), p. 81.
[219] BRS (2017b).
[220] BRS (2011).

A database of regulated article letters of inquiry and USDA's responses indicates that between January 2011 and November 2018, BRS had received and responded to inquiries about 74 products. BRS concluded that many, but not all, were not regulated because they included no plant pest or could be created by conventional breeding.[221] Some of these conclusions are described below.

7.8.2.2 USDA Decisions

USDA, acting through APHIS and its Biotechnology Regulatory Services, has declined regulatory jurisdiction over numerous varieties developed with new genetic engineering technologies. Beginning in 2004, USDA indicated that specific crops developed by OMM (oligonucleotide mediated mutagenesis), ZFN (zinc finger nuclease), EMN (engineered mega nuclease), and TALEN (transcriptional activator-like effector nuclease) were not regulated articles under its biotechnology regulations.[222] Since about 2010, USDA declined to regulate numerous products developed with innovative technologies that have no plant-pest components and therefore do not pose a risk of plant pests.[223]

For example, in 2010, the agency determined that Dow AgroScience's ZFN-12 maize, engineered with zinc finger nuclease technology to reduce production of phytate (an anti-nutritional component of feed grain) was not regulated. The agency stated, "[N]o plant pest was used to create the ZFN-12 maize plants, which contain deletions at the IPK1 gene. There is no reason to believe that *Zea mays* containing an IPK1 deletion is a plant pest or is likely to pose a plant pest risk. Therefore, the ZFN-12 maize plants with induced deletions due to the use of zinc finger nuclease technology are not considered regulated articles."[224] In 2012, APHIS responded to Dow AgroScience with a more general statement: "GE plants containing targeted deletions, caused by naturally-occurring DNA repair after the targeted break is made by zinc-finger nucleases, and in which no genetic material is inserted into the plant genome, are not regulated articles .... The nucleases used are not from a plant pest and no plant pest sequences are inserted into the plant genome using this technology."[225]

---

[221]APHIS (2018b).

[222]7 CFR part 340; Wolt et al. (2016), pp. 511, 515, Tables 1, 3.

[223]Waltz (2016), p. 293 (mentioning 30 products); NASEM (2016a), p. 495. Other techniques include cisgenesis and intragenesis, developed in part because of "legislative, regulatory, marketing, and public-perception concerns." NASEM (2016a), p. 357. Cisgenesis involves modification of an organism with a gene from a different variety of a crop or a sexually compatible species. Intragenesis recombines plant DNA from the crop itself or its sexually compatible relatives into a genetic construct introduced into the new plant. Simplot's Innate potato, approved for cultivation in the US, is intragenic.

[224]Gregoire (2010).

[225]Gregoire (2012). ZFN products involving plant pests will be regulated, and insertions of genetic material must be evaluated on a case-by-case basis.

Similarly, in 2016 APHIS declined to regulate the Simplot low-PPO5 potato, engineered with TALEN technology to knock out the PPO5 gene and resist black spot. It contains no foreign genetic material and is neither a plant pest nor a noxious weed.[226] Therefore it is not regulated under APHIS's regulations for GE organisms, but import would be subject to plant protection and quarantine requirements and to other regulatory authorities. Also in 2016, APHIS decided not to regulate a Calyxt potato engineered with TALEN technology[227] and in 2018, a Calyxt nutritionally-enhanced wheat engineered with TALEN.[228] APHIS has declined to regulate varieties, including biolistically-derived maize, produced by other techniques.[229] Am-I-Regulated? inquiries indicate that APHIS carries out careful review; recently APHIS decided that a new variety, a plant pathogen made less virulent by gene deletion, remains regulated as a plant pest.[230]

APHIS has considered varieties created with CRISPR/Cas9 technology, which also lack the plant pest regulatory trigger. In 2016, the agency evaluated the status of the first two varieties developed with CRISPR/Cas9 technology to reach the regulatory system.[231] A researcher at the University of Pennsylvania developed a white button mushroom with an anti-browning phenotype, which improves appearance, lengthens shelf life, and facilitates mechanical harvesting. The developer had used CRISPR/Cas9 to delete genes related to an enzyme that caused browning. In a confirmation letter to the researcher, APHIS reviewed its provisions for determining whether a GE organism is a regulated article and the characteristics of the mushroom with gene deletions but no foreign DNA. Because APHIS concluded that the mushroom does not include "introduced genetic material" and is not a plant pest, the mushroom is not a regulated article nor is it regulated as a noxious weed.[232] Importation of the mushroom, however, is subject to APHIS plant protection and

---

[226] APHIS (2018b). Simplot had followed USDA's petition process for other potatoes and received nonregulated status. The USDA APHIS database includes Am-I-Regulated? inquiries and Biotechnology Regulatory Services responses since 2011. Citations to APHIS (2018b) link to these documents for each variety.

[227] APHIS (2018b).

[228] APHIS (2018b) (also considering the potential for weediness).

[229] In both 2016 and 2018, BRS confirmed that biolistically derived maize, created by direct physical gene transfer, included no plant pests and posed no threat of weediness. Therefore, the varieties are not regulated. APHIS (2018b). In March 2018, APHIS declined to regulate a salinity-tolerant cisgenic rice, developed with biolistics, as a plant pest, but planned to consider weediness in more detail. The agency also did not regulate a fragrant moss created with PEG-mediated transformation and in May 2018 declined to regulate two biolistic soybeans. APHIS (2018b).

[230] APHIS (2018b). *Erwinia amylovora*, modified for application to apple trees, remained a plant pathogen, which is a plant pest.

[231] Waltz (2016), p. 582. The corn may be commercialized within 5 years; plans for commercialization of the mushroom are uncertain.

[232] 7 CFR part 340; 7 CFR part 360.

quarantine requirements, and the mushroom may be subject to FDA and EPA requirements.[233]

Shortly thereafter, APHIS made a similar determination about a waxy corn developed by DuPont Pioneer, also developed with CRISPR/Cas9 technology. APHIS concluded that the waxy corn, which may achieve higher yields, is neither a plant pest nor a noxious weed.[234] Like the mushroom, the importation of corn is subject to APHIS plant protection and quarantine requirements, and the waxy corn may be subject to FDA and EPA regulation.

In 2017, APHIS determined that a variety of *Setaria viridis* (green foxtail or bristlegrass) engineered with CRISPR/Cas9 was not a regulated article, but had potential to be a problematic weed in agricultural environments. Therefore APHIS recommended maintenance of isolation distances to avoid possible crosses with *Setaria italica* (foxtail millet), used for hay.[235] APHIS also declined to regulate a false flax (*Camelina sativa*), used as a biofuel and to replace fish oil for aquaculture. Developers of the *Camelina* variety, Yield10 Bioscience, indicated that their testing and data collection took 2 years, and the USDA decision, 2 months. They saved time and spent far less than the $30–$50 million that would have been required to comply with more comprehensive GE regulation. In October 2017, APHIS declined to regulate a soybean (*Glycine max*), which is salt-tolerant and resistant to drought. The soybean developers, USDA's Agricultural Research Service, plan to comply with FDA's voluntary consultation process.[236] In January 2018, APHIS declined to regulate a DuPont Pioneer corn, developed with CRISPR/Cas9 technology, with improved resistance to northern leaf blight. Similarly, in May 2018, the agency declined to regulate a tomato variety developed with CRISPR/Cas9 at the University of Florida.[237]

USDA's decisions not to regulate crops developed with CRISPR and other innovative technologies indicate, as a 2016 National Academies report insisted, that "any attempt by regulators to define the scope of a regulatory system through the definition of specific technologies will be rapidly outmoded by new approaches."[238]

In March 2018, USDA issued a formal statement about its regulation of innovative plant breeding. The agency noted that plants developed with innovative

[233]APHIS (2018b). Okanagan Specialty Fruits' Arctic apples, which do not brown when cut, were also developed by gene silencing, which shut down genes for the browning enzyme. USDA granted nonregulated status to Arctic Golden and Granny apples in 2015, and later extended that status to Arctic Fuji apples. The nonbrowning trait can be characterized as a "loss of function" trait, in contrast to other GE crops with "gain of function" traits (for example, herbicide tolerance or insecticide resistance). NASEM (2016a), p. 272.

[234]APHIS (2018b).

[235]APHIS (2018b).

[236]APHIS (2018b), Waltz (2018), p. 6. The USDA-ARS Plant Science Research Unit that developed the crop is located in St Paul, Minnesota.

[237]APHIS (2018b).

[238]NASEM (2016a), p. 509.

techniques are often indistinguishable from plants developed by traditional breeding and that innovative techniques, including genome editing, offer benefits for both farmers and consumers. USDA stated: "Under its biotechnology regulations, USDA does not currently regulate, or have any plans to regulate plants that could otherwise have been developed through traditional breeding techniques as long as they are developed without the use of a plant pest as the donor or vector and they are not themselves plant pests."[239] Examples of genetic changes that USDA will not regulate are deletions of any size, single base pair substitutions, insertions from compatible plant relatives,[240] and complete null segregants.[241] USDA promised to advance a "science-based and practical approach that protects plant health while allowing for technical advancements."[242]

In a press release that publicized USDA's statement, the US Secretary of Agriculture noted that new plant breeding tools can "introduce new plant traits more quickly and precisely, potentially saving years or even decades in bringing needed new varieties to farmers."[243] USDA's approach to genetic technologies will continue to exercise regulatory oversight when necessary to protect plant health, but will encourage innovation that does not pose risk, allowing USDA to "do right and feed everyone."[244]

One informed commentator called USDA's statement "a big deal and a very good thing" that "doesn't go far enough, [but] promises to help protect the powerful new tool of 'gene editing,' from being smothered in its cradle by the irrational fears that have hobbled traditional 'GMOs' over the past three decades."[245] This commentator, like others, emphasized the importance of regulating innovation according to the properties of the resulting organism, rather than the method of development.

7.8.3 FDA Consultations

APHIS letters declining to regulate products of new technology indicate to developers that those products may be subject to EPA and FDA requirements. Unlike USDA, FDA definitions do not exclude new technology. In its 1992 policy, the FDA defined genetic modification as "the alteration of the genotype of a plant using any

[239] APHIS (2018c).

[240] APHIS (2018c) explains: "the change to the plant solely introduces nucleic acid sequences from a compatible relative that could otherwise cross with the recipient organism and produce viable progeny through traditional breeding."

[241] APHIS (2018c) explains: "off-spring of a genetically engineered plant that does not retain the change of its parent."

[242] APHIS (2018c), referring to recommendations in Task Force (2018), discussed above, Sect. 7.4.2.1.

[243] USDA (2018).

[244] USDA (2018).

[245] Giddings (2018).

technique, new or traditional"; modification is broadly defined as "the alteration in the composition of food that results from adding, deleting, or changing hereditary traits, irrespective of the method."[246] In practice, FDA regulation, when it has applied to GE food products, focuses on food additives. Most GE foods, considered generally regarded as safe (GRAS), are not defined as food additives and are therefore excluded from regulation as food additives. New technologies often do not add genetic material to the modified food, so scholars have suggested that the FDA may lack regulatory jurisdiction in the absence of food additives.[247]

Developers of new GE foods normally participate in FDA's voluntary biotechnology consultations, but the agency's database of consultations does not yet include the innovative varieties discussed above.[248] The database does, however, include numerous other varieties to which USDA had granted nonregulated status—for example, Okanagan Specialty Fruits' Arctic apples genetically engineered to resist browning. FDA's response letter, sent at the end of the consultation process, notes that it relied on information supplied by Okanagan about the apples and Okanagan's steps to ensure that the apples complied with FDA legal requirements. FDA reminded the developer about its responsibility to obtain clearances from USDA and (if necessary) EPA and to comply with applicable label requirements. FDA indicated that it had no further questions, but emphasized that it is Okanagan's "continuing responsibility to ensure that foods marketed by the firm are safe, wholesome, and otherwise in compliance with all applicable legal and regulatory requirements."[249]

The FDA is likely to provide similar responses to products of new GE technology, relying on an agency review of the information provided by the developer.

7.9 US Regulation of Innovative GE Technologies: What Direction?

As the discussion of recent documents from the Biotechnology Working Group indicated, neither the 2016 National Strategy for regulation of biotechnology nor the 2017 Updated Coordinated Framework provided a roadmap for regulation of new GE crops. The National Strategy recommended that regulation focus on products that require federal oversight, and the Coordinated Framework outlined agency responsibility and plans for coordination.

---

[246]FDA (1992), p. 22,982. The definition of modification continues, "Modifications may be minor, such as a single mutation that affects one gene, or major alterations of genetic material that affect many genes. Most, if not all, cultivated food crops have been genetically modified."

[247]Peck (2017), p. 322.

[248]See FDA (2018a).

[249]Keefe (2015). The biotechnology consultation was completed soon after USDA had granted nonregulated status.

An independent study requested by the Biotechnology Working Group complemented these policy documents. The National Academies of Sciences, Engineering, and Medicine assembled a committee of experts to study future products of biotechnology, including genetic engineering, genome engineering, and targeted manipulation of genetic information, with a focus on risks (or lack of risks) and risk assessment.[250] That document, published in January 2017, asserted that "there was no regulation, law, or statute to mandate a central review of biotechnology products or to develop an oversight system that is coordinated among agencies, minimizes gaps and redundancies in product review, provides more certainty for product developers as to the regulatory path, and embraces the principles of anticipation, participation, responsiveness, and transparency."[251]

The National Academies report viewed the updated Coordinated Framework and the National Strategy as starting points, but recommended "a consistent, efficient, and effective decision-making framework that continues to balance innovation and safety."[252] Indeed, although the Coordinated Framework is flexible and could govern many types of biotechnology products, the statutory jurisdiction of the FDA, USDA, and EPA may result in "gaps or redundancies in regulatory oversight."[253] Not all products of innovative biotechnologies will fit within existing agency jurisdictions, so regulators may need to use "the flexibility available under their statutes to minimize gaps in jurisdiction" and choose the most suitable regulatory system for new products.[254] Agencies may be unprepared, in capacity and expertise, for the risk analysis required for an increased number of new products, including complex products with fewer comparators.[255] Moreover the US regulatory system is complex and fragmented, posing uncertainty and other difficulties for product developers and the general public. Regulatory oversight that is too complex will deter innovation; instead, an oversight framework must match "the scope, scale, complexity, and tempo of future technological developments and increase[] public confidence in the safety of products entering the marketplace."[256]

The National Academies report identified two significant challenges to regulators in light of developments (including new products beyond agriculture) in

---

[250]NASEM (2017).

[251]NASEM (2017), p. 141.

[252]NASEM (2017), p. 11.

[253]NASEM (2017), p. 6. Other agencies may also have jurisdiction over some new products, but the 2017 Coordinated Framework does not outline the roles of those agencies.

[254]NASEM (2017), p. 11.

[255]NASEM (2017), pp. 8, 172. Risk assessment for GE crops often uses a nonbiotechnology comparator. With new technology (for example, a product with only synthetic DNA), however, finding a nonbiotech comparator may be difficult, so "the idea of 'comparator' may need to expand to include similar existing biotechnology products with which regulatory agencies already have experience" (p. 155). One commentator suggested that in the future regulators could require "omics" data to prove substantial equivalence of crops developed with innovative biotechnology to traditional crops. Enríquez (2017b), pp. 533–534.

[256]NASEM (2017), p. 11.

biotechnology: "To find jurisdiction under existing statutes to regulate all the products that may pose risks to consumers and ... [t]o utilize the best available risk-analysis tools consistent with agency authorities to provide nuanced oversight that protects consumers while fostering beneficial innovation."[257] The report set out conclusions and recommendations about the ability of the current regulatory system to cope with new developments in the next 5–10 years, in light of the flexibility, complexity, and fragmentation of the current regulatory scheme. USDA, FDA, and EPA will be challenged by these developments and should (among other recommendations) increase scientific capacity, keep informed of new developments, support internal and external research, and work together to ensure effective risk assessment and regulation.[258]

The National Academies committee concluded that a beneficial approach would be "a single point of entry into the regulatory system with a decision-making structure aimed to assess and manage product risk, to direct products to their appropriate regulatory agencies, and to increase transparency for developers and society."[259] This approach would help to ensure that products are evaluated consistently and efficiently. Moreover, it would also help to "triage" new products, so that familiar products take less regulatory effort, leaving regulators free to evaluate products with novel traits that may require complex risk assessments.[260]

The National Academies report made recommendations consistent with the updated Coordinated Framework and the current US regulatory system. It relied in part on earlier National Academies reports. For example, a 2016 National Academies report recommended that regulatory focus be on products of biotechnology, rather than the process. Regulation should consider the risk of novel characteristics to health or environment, the extent of uncertainty about severity of harm, and potential for exposure. If a new product has no intended traits or alterations that raise health or environment concerns, it should face no further testing. Products with potential for health or environmental effects or with differences from comparators that cannot be interpreted should face further safety testing.[261]

A regulatory system focused on appropriate risk evaluation will be important for health and environment, but may also help to encourage consumer acceptance of new technologies. Consumer awareness about GE crops in agriculture is low in the United States, and many consumers do not know that GE food is available in grocery stores. Nonetheless although many consumers believe that GE foods pose a health hazard, a majority would buy GE produce, especially if that produce had been

[257]NASEM (2017), p. 11.

[258]NASEM (2017), pp. 171–185.

[259]NASEM (2017), pp. 174–175.

[260]NASEM (2017), pp. 176–177.

[261]NASEM (2016a), pp. 26–27, 513. The National Academies also recommended that policymakers address socio-economic, as well as scientific, issues and facilitate communication with the public.

treated with fewer pesticides.[262] A food and health survey published in 2017 indicated that fewer than 20% of respondents identified biotechnology as an important food safety issue.[263] Yet a 2014 food technology survey indicated that 37% of respondents knew that GE foods were available in supermarkets (with considerable confusion about which foods), but only 28% had a favorable view of food biotechnology.[264] It seems likely that consumer awareness of innovative biotechnology, including gene editing, is even lower than awareness of older GE technology; literature seems to focus on consumer concern about human applications of gene editing. Consumer education will be critical for acceptance of products from innovative biotechnology. USDA and FDA are cooperating on an initiative to prepare and communicate science-based educational information about agricultural biotechnology for consumers.[265]

Because innovative plant breeding techniques do not always fit within existing "product definitions, regulatory frameworks and risk assessment approaches" used for GE products,[266] a number of commentators have outlined characteristics of more effective regulation. In 2012, for example, after USDA had decided not to regulate products of several innovative technologies, but before USDA had considered crops from CRISPR/Cas9 technology, scholars indicated, first, that regulatory frameworks must remain "fit for purpose," noting that narrow definitions of GE products may not apply to plants developed with new technologies. Second, regulation and risk assessment must be "proportional to the level of risk" to health and environment, especially when products of new technology (for example, cisgenesis) resemble conventionally-bred plants. Third, regulatory measures should prevent harm, but also stimulate innovation and encourage consumer acceptance of new products. Fourth, regulatory frameworks must facilitate international harmonization to avoid trade disputes and damage caused by asynchronous authorizations. Finally, regulatory systems must ensure that products with "equal potential to cause harm" face similar risk assessments that focus on characteristics of the plant, not the process of development.[267]

Similarly, other scholars recommended a risk-based system focused on the products of biotechnology. Starting from the premise that "there are high-risk organisms, but no high-risk techniques," they argue for regulatory measures, grounded in data and experience and focused on the end product, to assess new technologies and their risks.[268] Risk analysis should start with the identification of the object of protection—for example, food safety, avoidance of harm to agriculture

[262]NASEM (2016a), pp. 48, 303.

[263]IFIC (2017). Only about 5% of consumers ranked biotechnology as their first food safety concern; about 13%, their second or third concern.

[264]IFIC (2014).

[265]FDA (2018b).

[266]Podevin et al. (2012), p. 1057.

[267]Podevin et al. (2012), pp. 1058–1060.

[268]Conko et al. (2016), p. 498.

or ecosystem, prevention of weediness, protection of biodiversity—coupled with awareness of benefits of GE crops and products. The likelihood of hazard (from very low to very high) and the magnitude of harm (from negligible to major) help to determine risk.[269] Neither the type of breeding techniques (conventional or molecular) nor the source of DNA (related or distant organism) determines risk.[270] These scholars believe that most GE crops will pose only negligible risk, and they assert that regulatory requirements should be related to the risk of the GE product, with case-by-case review, or even prohibition, reserved for products posing significant risks.

Moreover, another scholar suggested that a new legislative framework, also focused on risk, could "build on existing agency experience and expertise; better align regulatory oversight with risk; simplify the approval process for developers; address legitimate health, safety, and environmental concerns; adapt seamlessly to changes in technology; and provide federal funding for monitoring and research on the greatest areas of scientific uncertainty or known deleterious impacts."[271] An important component of this recommended approach is the identification of a single lead agency to coordinate oversight for regulatory approvals—USDA for plants, FDA for animals, and EPA for microbes or insects—with a mandate to consult with other agencies when evaluation calls for additional expertise. The recommended framework would require agencies to define risk classifications by regulation, classify new varieties by risk, and regulate those varieties according to their risk classification.[272]

7.10 Conclusion

As the discussion above indicates, US regulation of plants developed with GE technology is complex, fragmented, time-consuming, and expensive, with developers subject to regulation in two or even three administrative agencies.[273] For

[269]Conko et al. (2016), pp. 498–499. Risk is "an arithmetic function of the likelihood that the genetic modification will lead to harm and the magnitude of the resulting harm, conventionally stated as: Risk = Hazard × Exposure" (p. 499).

[270]Conko et al. (2016), p. 501. Most regulatory systems, including the US, "are neither scientifically defensible nor justifiable: all too often, they lead to the plants of lowest risk being subject to the highest degree of scrutiny. The result is a massive waste of limited resources, huge disincentives to innovation in a time of great need and no increase in public or environmental safety."

[271]Peck (2017), pp. 328–329.

[272]Peck (2017), p. 333. Peck recommended creation of a Biosafety Clearinghouse to help developers identify the appropriate regulatory agency for their product and to facilitate agency consultations; EPA environmental monitoring of risky varieties; and federally-funded research to encourage innovation and ensure health and safety (pp. 335–339).

[273]Indeed, a recent criticism characterized regulation of GE crops as "a scientifically unjustified barrier to agricultural innovation." CAST (2018), p. 16.

innovative biotechnologies, the scope of regulation is unclear. Regulatory definitions, established in light of rDNA technology, are narrow and exclude some crops developed with innovative biotechnologies, including gene editing. USDA has declined to regulate several crops developed with CRISPR/Cas9, as well as crops developed with other new technologies. Lacking plant pest material, they do not fit with the agency's regulatory authority. Moreover, USDA has formalized its decision not to regulate many crops developed with innovative GE technologies. Regulatory uncertainty for innovative technologies exists in other jurisdictions, including the European Union.

Innovative biotechnologies, like CRISPR/Cas9, may prove critical in providing food and feed for growing global populations. CRISPR, for example, is less costly and more precise than some other technologies, making it attractive to a wide range of enterprises. Regulation of these crops to avoid risks to health and environment is critical, but excessive regulation, with its high compliance costs, may exclude small enterprises and public research bodies. Public institutions, including universities, were critical to early development of agricultural biotechnology, but the cost of regulatory compliance often precluded universities and small private companies from the agricultural biotechnology market.[274] Indeed, for small companies and public institutions, the "unnecessarily complicated, onerous, and unscientific regulatory system presents a near insurmountable barrier" to commercialization, so their investments do not reach farmers or benefit consumers and the environment.[275] Only companies that develop crops with potentially high global sales will be able to afford "burdensome and costly regulatory requirements," and these companies may be reluctant to develop crops needed by farmers in developing countries.[276] Thus, as the United States develops new regulatory policy, requirements should be flexible, adaptable to the rapid development of technology, and related closely to the level of risk presented by products of innovative GE technologies.[277]

One thoughtful commentator, who believes that innovative biotechnology could help to ensure global food security, asserted that "genomics is at roughly the stage GM technology was at its point of first commercialization, where expectations pertaining to its potential were very high. Thus the important question is whether genomics will suffer a similar fate to GMOs, or will the international regulatory and

[274]See CAST (2018), pp. 2–3, 11–12, 15.

[275]CAST (2018), p. 16. Small-market, specialty, and perennial crops are particularly affected by high costs and data demands of the regulatory process (p. 12).

[276]Conko et al. (2016), pp. 501–502. These large crops are often annual field crops. CAST (2018), p. 12.

[277]See Conko et al. (2016), p. 502. See also Bergkamp (2017), pp. 62–63: "[R]egulators should think twice before regulating the risks associated with new technologies. Innovation is important to any society, and innovation requires risk-taking." Instead, regulators should ask "whether we are better off without the proposed restrictions, or with a more modest regime. Not all risks can and should be regulated; some risks are well worth taking."

trade systems allow this new technology to reach its full potential to assist in achieving food security goals over the near future?"[278]

References[279]

Agricultural Marketing Service (AMS), USDA (n.d.) GMO disclosure & labeling. https://www.ams.usda.gov/rules-regulations/gmo. Accessed 3 Feb 2018

AMS, USDA (2018a) Proposed rule: National Bioengineered Food Disclosure Standard. Fed Regist 83:19,860–19,889 (4 May)

AMS, USDA (2018b) Final rule: national bioengineered food disclosure standard. Fed Regist 83:65,814–65,876, codified at 7 Code of Federal Regulations part 66

Agriculture and Agri-Food Canada (AAFC) (2017) Low-Level Presence. http://www.agr.gc.ca/eng/industry-markets-and-trade/agri-food-trade-issues/technical-trade-issues-in-agriculture/low-level-presence/?id=1384370877312. Accessed 8 Feb 2018

Alliance for Bio-Integrity v. Shalala (2000) 116 F. Supp. 2d 166 (D DC)

American Association for the Advancement of Science (AAAS) (2012) Statement by the AAAS Board of Directors on Labeling of Genetically Modified Foods. https://www.aaas.org/sites/default/files/AAAS_GM_statement.pdf. Accessed 5 Jan 2108

American Medical Association (AMA) (2012) Policy H-480.958: bioengineered (genetically engineered) crops and foods. Link from https://policysearch.ama-assn.org/policyfinder. Accessed 12 Feb 2018

American Soybean Association, Ron Moore, President (2017) Comment submitted to the FDA in response to Docket No. FDA-2-16-N-4389

Animal and Plant Health Inspection Service (APHIS), USDA (2007) APHIS policy on responding to the low-level presence of regulated genetically engineered plant materials. Fed Regist 72:14,649–14,651 (29 Mar)

APHIS (2016) Environmental impact statement; introduction of the products of biotechnology. Fed Regist 81:6225–6228 (5 Feb)

APHIS (2017a) Revisions to USDA-APHIS 7 CFR part 340 Regulations governing the importation, interstate movement, and environmental release of genetically engineered organisms. Draft Programmatic Environmental Impact Statement - January, 2017. USDA, Washington, DC

APHIS (2017b) Proposed rule; importation, interstate movement, and environmental release of certain genetically engineered organisms. Fed Regist 82:7008–7039 (19 Jan)

APHIS (2017c) Proposed rule; withdrawal, importation, interstate movement, and environmental release of certain genetically engineered organisms. Fed Regist 82:51,582–51,583 (7 Nov)

APHIS (2018a) BRS interstate/release and release permits and notifications. https://www.aphis.usda.gov/aphis/ourfocus/biotechnology/permits-notifications-petitions/sa_permits/status-update/release-permits. Accessed 29 Dec 2018

APHIS (2018b) Regulated articles letters of inquiry. https://www.aphis.usda.gov/aphis/ourfocus/biotechnology/am-i-regulated/regulated_article_letters_of_inquiry/regulated_article_letters_of_inquiry. Accessed 20 Dec 2018

APHIS (2018c) Details on USDA plant breeding innovations. https://www.aphis.usda.gov/aphis/ourfocus/biotechnology/brs-news-and-information/pbi-details

APHIS (2018d) Petitions for determination of nonregulated status (last updated 29 Mar). https://www.aphis.usda.gov/aphis/ourfocus/biotechnology/permits-notifications-petitions/petitions/petition-status. Accessed 27 Apr 2018

[278]Kerr (2017), pp. 67, 72 (quotation, p. 67).

[279]Statutes and regulations are cited in footnotes and not included here.

Baumann M (2016) CRISPR/Cas9 genome editing -- new and old ethical issues arising from a revolutionary technology. Nanoethics 10:139–159
Bergeson LL (2017) Enlisting modern technologies to ensure a safe food supply. Nat Resour Environ 31(3):31–34
Bergkamp L (2017) The reality of risk regulation. Eur J Risk Regul 8(1):56–63
Biology Fortified (2018) GENetic Engineering Risk Atlas (GENERA). http://genera.biofortified.org/wp. Accessed 21 Feb 2018
Biotechnology Regulatory Services, APHIS, USDA (2011) User's guide: general document preparation guidelines for submission to BRS. https://www.aphis.usda.gov/brs/pdf/Doc_Prep_Guidance.pdf. Accessed 8 Feb 2018
BRS (2016) Request to extend nonregulated status from a previous determination: extension guidance for developers. https://www.aphis.usda.gov/brs/aphisdocs/guidance_ext_nonreg.pdf. Accessed 8 Feb 2018
BRS (2017a) Regulatory status under 2017 withdrawn proposed 340 rule (last modified, 14 June; archived page). https://www.aphis.usda.gov/aphis/ourfocus/biotechnology/biotech-rule-revision/2016-340-rule/reg-status-table. Accessed 29 Dec 2018
BRS (2017b) Am I Regulated under 7 CFR part 340? https://www.aphis.usda.gov/aphis/ourfocus/biotechnology/am-i-regulated. Accessed 8 Feb 2018
BRS (2017c) Permit User's Guide with special guidance for ePermits. https://www.aphis.usda.gov/biotechnology/downloads/permit_guidance.pdf. Accessed 8 Feb 2018
Brookes G, Barfoot P (2017a) GM crops: global socio-economic and environmental impacts 1996–2015. PG Economics Ltd, Dorchester
Brookes G, Barfoot P (2017b) Environmental impacts of genetically modified (GM) crop use 1996–2015: impacts on pesticide use and carbon emissions. GM Crops Food 8(2):117–147. https://doi.org/10.1080/21645698.2017.1309490
Brookes G, Barfoot P (2017c) Farm income and production impacts of using GM crop technology 1996–2015. GM Crops Food 8(3):156–193. https://doi.org/10.1080/21645698.2017.1317919
Brookes G, Barfoot P (2018) Farm income and production impacts of using GM crop technology 1996–2016. GM Crops Food 9(2):58–89. https://doi.org/10.1080/21645698.2018.1464866
Codex Alimentarius Commission (CAC) (2003a) Principles for the risk analysis of foods derived from modern biotechnology. Doc CAC/GL/44-2003. World Health Organization and Food and Agriculture Organization, Rome
Codex Alimentarius Commission (CAC) (2003b) Guideline for the conduct of food safety assessment of foods derived from recombinant-DNA plants. Doc CAC/GL/45-2003 (Annex 3, 2008). World Health Organization and Food and Agriculture Organization, Rome
Conko G, Kershen DL, Miller H, Parrott WA (2016) A risk-based approach to the regulation of genetically engineered organisms. Nat Biotechnol 34(5):493–503
Council for Agricultural Science and Technology (CAST) (2016) The impact of asynchronous approvals for biotech crops on agricultural sustainability, trade, and innovation (QTA2016-2). CAST, Ames
CAST (2018) Regulatory barriers to the development of innovative agricultural biotechnology by small businesses and universities (IP59). CAST, Ames
Deloitte (2017) Study of electronic or digital link disclosure: a third-party evaluation of challenges impacting access to bioengineered food disclosure. USDA, Washington, DC
Enríquez P (2017a) Genome editing and the jurisprudence of scientific empiricism. Vanderbilt J Entertain Technol Law 19(3):603–695
Enríquez P (2017b) CRISPR GMOs. N C J Law Technol 18(4):432–539
Environmental Protection Agency (EPA) (2001) Regulations under the Federal Insecticide, Fungicide, and Rodenticide Act for Plant-Incorporated Protectants (Formerly Plant-Pesticides). Fed Regist 66:37,772, codified at 40 CFR parts 152 and 174
EPA (2007) PRN 2007-2: guidance on small-scale field testing and low-level presence in food of plant-incorporated protectants (PIPs). https://www.epa.gov/pesticide-registration/prn-2007-2-guidance-small-scale-field-testing-and-low-level-presence-food. Accessed 8 Feb 2018

EPA (2018) Withdrawal of proposed rules. Fed Regist 83:20,004–20,008 (7 May)
European Commission (2010) A Decade of EU-Funded GMO Research (2001–2010) (EUR 24473-EN). https://ec.europa.eu/research/biosociety/pdf/a_decade_of_eu-funded_gmo_research.pdf. Accessed 12 Feb 2018
Feeley J, Fisk MC (2017) Syngenta agrees to pay more than $1.4 billion in corn accord. www.bloomberg.com. Accessed 8 Feb 2018
Fernandez-Cornejo J, Wechsler S, Livingston M, Mitchell L (2014) Genetically engineered crops in the United States (ERR 162). ERS, USDA, Washington, DC
Fernandez-Cornejo J, Wechsler S, Milkove D (2016) The adoption of genetically engineered alfalfa, canola, and sugarbeets in the United States (EIB 163). ERS, USDA, Washington, DC
Food and Drug Administration (FDA) (1992) Statement of policy: foods derived from new plant varieties. Fed Regist 57:22,984–23,005 (29 May)
FDA (1997) Guidance on consultation procedures under FDA's 1992 statement of policy - foods derived from new plant varieties. https://www.fda.gov/Food/GuidanceRegulation/GuidanceDocumentsRegulatoryInformation/ucm096126.htm. Accessed 8 Feb 2018
FDA (2006) Guidance for industry: recommendations for the early food safety evaluation of new non-pesticidal proteins produced by new plant varieties intended for food use. https://www.fda.gov/Food/GuidanceRegulation/GuidanceDocumentsRegulatoryInformation/ucm096156.htm. Accessed 8 Feb 2018
FDA (2015) Guidance for industry: voluntary labeling indicating whether foods have or have not been derived from genetically engineered plants. https://www.fda.gov/Food/GuidanceRegulation/GuidanceDocumentsRegulatoryInformation/ucm059098.htm. Accessed 8 Feb 2018
FDA (2016) Final rule. Substances generally recognized as safe. Fed Regist 81:54,960–55,055, amending 21 CFR parts 170, 570, and others
FDA (2017) Notification. Genome editing in new plant varieties used for foods: request for comments. Fed Regist 82:6564–6566 (19 Jan)
FDA (2018a) Biotechnology consultations on food from GE plant varieties. https://www.accessdata.fda.gov/scripts/fdcc/?set=Biocon. Accessed 20 Dec 2018
FDA (2018b) Agricultural biotechnology education and outreach initiative. https://www.fda.gov/Food/ResourcesForYou/Consumers/ucm579348.htm. Accessed 28 Feb 2018
FDA (2018c) New protein consultations (early food safety evaluation). https://www.accessdata.fda.gov/scripts/fdcc/index.cfm?set=npc. Accessed 20 Dec 2018
Foreign Agricultural Service (FAS), USDA (2012) International Statement on Low Level Presence. https://www.fas.usda.gov/international-statement-low-level-presence. Accessed 8 Feb 2018
Giddings V (2018) Gene editing, government regulation, and greening our future. https://itif.org/publications/2018/04/03/gene-editing-government-regulation-and-greening-our-future. Accessed 23 Apr 2018
Gleim S, Smith SJ, Phillips PWB (2016) Regulatory system impacts on global GM crop adoption patterns. Estey J Int Law Trade Policy 17(2):96–116
Global Alliance for Ag Biotech Trade (GAABT) (2015) Practical approach to address Low Level Presence (LLP) of agricultural biotechnology-derived plant products in food, feed, and grain for processing (FFP). GAABT, Washington, DC. https://croplife.org/wp-content/uploads/2015/04/GAABT-LLP-Solutions-QA-FINAL-21-April-2015.pdf. Accessed 8 Feb 2018
Greene C, Wechsler SJ, Adalja A, Hanson J (2016) Economic issues in the coexistence of organic, genetically engineered (GE), and non-GE crops (EIB 149). ERS, USDA, Washington, DC
Gregoire M (2010) Biotechnology Regulatory Services, APHIS, USDA, Letter to Gary W. Rudgers (Dow AgroScience LLC), Re: APHIS review as to whether *Zea mays* plants with the IPK1 gene deleted using zinc nuclease technology is regulated by APHIS (26 May)
Gregoire MC (2012) Letter to Gary W. Rudgers, Re: APHIS review as to whether plant varieties created using zinc-finger nucleases techniques for targeted gene deletion are regulated by APHIS (8 Mar)

Grossman MR (2007) The coexistence of GM and other crops in the European Union. Kansas J Law Public Policy 16(3):325–392
Grossman MR (2010) Monsanto Co. v. Geertson Seed Farms: US Supreme Court decides GM alfalfa case. Eur Food Feed Law Rev 5(4):216–221
Grossman MR (2012) Genetically modified crops and their products in the United States: a review of the regulatory system. Jahrbuch des Agrarrechts XI:69–96
Grossman MR (2016a) The United States Bioengineered Food Disclosure Standard: labels for genetically engineered food. Eur Food Feed Law Rev 11(6):502–507
Grossman MR (2016b) Genetic technology and food security: a view from the United States. In: Norer R (ed) Genetic technology and food security. Springer, Cham, pp 289–332
Grossman MR (2017) US FDA enacts final rule for GRAS substances. Eur Food Feed Law Rev 12 (2):169–172
Grossman MR (2018) Agricultural biotechnology: regulation in the United States and the European Union. In: Bremmers H, Purnhagen K (eds) Regulating and managing food safety in the EU – a legal-economic perspective. Springer, Cham, pp 331–380
Hicklin T (2014) The start of a New Genomic Era. In: CRISPR-CAS9: Engineering A Revolution in Gene Editing (Science special report), p 2. http://www.sciencemag.org/collections/crispr-cas9-engineering-revolution-gene-editing. Accessed 15 Jan 2018
Holdren JP et al (OSTP) (2011) Principles for Regulation and Oversight of Emerging Technologies: Memorandum for Heads of Executive Departments and Agencies. https://obamawhitehouse.archives.gov/sites/default/files/omb/inforeg/for-agencies/Principles-for-Regulation-and-Oversight-of-Emerging-Technologies-new.pdf. Accessed 8 Feb 2018
Holdren JP et al (OSTP) (2015) Modernizing the Regulatory System for Biotechnology Products, Memorandum for Heads of Food and Drug Administration, Environmental Protection Agency, and Department of Agriculture. https://obamawhitehouse.archives.gov/sites/default/files/microsites/ostp/modernizing_the_reg_system_for_biotech_products_memo_final.pdf. Accessed 8 Feb 2018
International Food Information Council (IFIC) (2014) Consumer perceptions of food technology survey. Link from surveys at http://www.foodinsight.org. Accessed 28 Feb 2018
IFIC (2017) 2017 Food and health survey. Link from surveys at http://www.foodinsight.org. Accessed 28 Feb 2018
International Service for the Acquisition of Agri-Biotech Applications (ISAAA) (2017) Global status of commercialized Biotech/GM crops: 2016. Report and executive summary (ISAAA brief no. 52). ISAAA, Ithaca
ISAAA (2018a) GM approval database. http://www.isaaa.org/gmapprovaldatabase/default.asp. Accessed 8 Feb 2018
ISAAA (2018b) Substantial equivalence of GM and Non-GM crops (Pocket K No. 56). http://isaaa.org/resources/publications/pocketk/56/default.asp. Accessed 26 Apr 2018
ISAAA (2018c) Global status of commercialized Biotech/GM crops in 2017: Biotech crop adoption surges as economic benefits accumulate in 22 years. Report and executive summary (ISAAA brief no. 53). ISAAA, Ithaca
Jaffe G (2017) Biotech Blog: USDA should establish a science-based regulatory system to address genetically engineered and gene-edited crops. https://cspinet.org/news/biotech-blog-usda-should-establish-science-based-regulatory-system-address-genetically. Accessed 12 Feb 2018
Kahrmann J, Bömeke O, Leggewie G (2017) Aged GMO legislation meets new genome editing techniques. Zeitschrift für Europäisches Umwelt- und Planungsrecht 2:176–182
Keefe DM (FDA) (2015) Letter to Neal Carter, Okanagan Specialty Fruits, Inc. Biotechnology Consultation Agency Response Letter BNF 000132 (20 Mar)
Kerr WA (2017) Genomics, international trade and food security. Estey J Int Law Trade Policy 18 (2):63–77
Kniss A (2017) Long-term trends in the intensity and relative toxicity of herbicide use. Nat Commun 8(24865):1–7. https://doi.org/10.1038/ncomms14865

Kuzma J (2016) A missed opportunity for U.S. biotechnology regulation. Science 353:1211–1213. https://doi.org/10.1126/science.aai7854
Mahaffey H, Taheripour F, Tyner WE (2016) Evaluating the economic and environmental impacts of a global GMO ban. J Environ Protect 7:1522–1546
Monsanto Co. v. Geertson Seed Farms (2010), 561 US 139 (2010) (US Supreme Court decision)
National Academies of Sciences, Engineering, and Medicine (NASEM) (2016a) Genetically engineered crops: experiences and prospects. National Academies Press, Washington, DC
NASEM (2016b) Gene drives on the horizon: advancing science, navigating uncertainty, and aligning research with public values. National Academies Press, Washington, DC
NASEM (2017) Preparing for future products of biotechnology. National Academies Press, Washington, DC
National Academy of Sciences (NAS) (2018) 2018 NAS Prize in Food and Agriculture Sciences. http://www.nasonline.org/programs/awards/2018/Barrangou.html. Accessed 12 Feb 2018
National Agricultural Statistics Service (NASS), USDA (2017) Acreage. http://usda.mannlib.cornell.edu/usda/current/Acre/Acre-06-30-2017.pdf. Accessed 23 Feb 2018
NASS (2018) Acreage. http://usda.mannlib.cornell.edu/usda/current/Acre/Acre-06-29-2018.pdf. Accessed 29 June 2018
National Institutes of Health (2016) NIH guidelines for research involving recombinant or synthetic nucleic acid molecules (NIH guidelines). https://osp.od.nih.gov/wp-content/uploads/2013/06/NIH_Guidelines.pdf. Accessed 13 Feb 2018
Obama B (2011) Executive Order 13563 -- improving regulation and regulatory review. Code Fed Regulat 3:13,563 (18 Jan)
Office of Science and Technology Policy (OSTP) (1986) Coordinated framework for regulation of biotechnology products. Fed Regist 51:23,302–23,367 (26 June)
OSTP (1992) Exercise of federal oversight within scope of statutory authority: planned introductions of biotechnology products into the environment. Fed Regist 57:6753–6762 (27 Feb)
OSTP (2002) Proposed federal actions to update field test requirements for biotechnology derived plants and to establish early food safety assessments for new proteins produced by such plants. Fed Regist 67:50,578–50,580 (2 Aug)
Peck A (2017) Re-framing biotechnology regulation. Food Drug Law J 72:314–340
Podevin N, Devos Y, Davies HV, Nielsen KM (2012) Transgenic or not? No simple answer! New biotechnology-based plant breeding techniques and the regulatory landscape. EMBO Rep 13 (12):1057–1061
Redick TP (2017) Liability prevention for agricultural biotechnology. Drake J Agric Law 22 (1):31–64
Schaart J et al (2015) Opportunities of new plant breeding techniques. Edepot.wur.nl/357723. Accessed 3 May 2018
Smart RD, Blum M, Wesseler J (2017) Trends in approval times for genetically engineered crops in the United States and the European Union. J Agric Econ 68(1):182–198
Smyth SJ (2017) Genetically modified crops, regulatory delays, and international trade. Food Energy Secur 6:78–86. https://doi.org/10.1002/fes3.100
StarLink Corn Products Liability Litigation (In re) (2000) 212 F. Supp. 2d 828 (ND Ill)
Sunstein CR (2017) On mandatory labeling, with special reference to genetically modified foods. Univ Pa Law Rev 165(5):1043–1095
Task Force on Agriculture and Rural Prosperity, USDA (2018) Report to the President of the United States from the Task Force on Agriculture and Rural Prosperity. USDA, Washington, DC
Tidgren KA (2018) Corn Farmers May Begin Filing Claims in Syngenta Settlement Next Month; Plaintiffs Seek Preliminary Approval of $1.51 Billion Syngenta Settlement. Link from https://www.calt.iastate.edu/blogpost. Accessed 26 Apr 2018
Trump DJ (2017) Executive order 13790. Promoting agriculture and rural prosperity in America. Fed Regist 82:20,237–20,239 (28 Apr)
US Department of Agriculture (USDA) (1986) Advanced notice of proposed USDA guidelines for biotechnology research. Fed Regist 51:23,367–23,393 (25 June)

USDA (1991) Proposed USDA guidelines for research involving the planned introduction into the environment of organisms with deliberately modified hereditary traits. Fed Regist 56:4134–4151 (1 Feb)

USDA (2018) Secretary Perdue Issues USDA Statement on Plant Breeding Innovation (Press Release No. 0070-18, 28 Mar 2018). https://www.usda.gov/media/press-releases/2018/03/28/secretary-perdue-issues-usda-statement-plant-breeding-innovation. Accessed 24 Apr 2018

USDA, Advisory Committee on Biotechnology and 21st Century Agriculture (AC21) (2016) A framework for local coexistence discussions. https://www.usda.gov/sites/default/files/documents/ac21-report-final-local-coexistence.pdf. Accessed 4 Feb 2018

US Government Accountability Office (US GAO) (2016) Genetically Engineered Crops: USDA Needs to Enhance Oversight and Better Understand Impacts of Unintended Mixing with Other Crops (GAO-16-241)

US White House (2016) National Strategy for Modernizing the Regulatory System for Biotechnology Products (Biotechnology Working Group, Emerging Technologies Interagency Policy Coordination Committee)

US White House (2017) Modernizing the Regulatory System for Biotechnology Products: Final Version of the 2017 Update to the Coordinated Framework for the Regulation of Biotechnology

Waltz E (2016) CRISPR-edited crops free to enter market, skip regulation. Nat Biotechnol 34(6):582

Waltz E (2018) With a free pass, CRISPR-edited plants reach market in record time. Nat Biotechnol 36(1):6–7

Wolt JD, Wang K, Yang B (2016) The regulatory status of genome-edited crops. Plant Biotechnol J 14:510–518

Zhang F (2014) CRISPR: the democratization of gene editing. In: CRISPR-CAS9: engineering a revolution in gene editing (Science special report), p 3. http://www.sciencemag.org/collections/crispr-cas9-engineering-revolution-gene-editing. Accessed 15 Jan 2018

Chapter 8
Comparative Analysis: The Regulation of Plants Derived from Genome Editing in Argentina, Australia, Canada, the European Union, Japan and the United States

David Hamburger

Abstract A comparison of the cultivation of genetically modified organism (GMOs) and consumption of their products (Sect. 8.2) reveals the distinctness of each examined country's approach towards GMOs. Not surprisingly, this finds its continuation in diverging and differing legal frameworks for their regulation. The diversity of approaches is not only reflected in different regulatory triggers and point of entries into the regulatory regime (Sect. 8.3), but also by varying labelling (Sect. 8.5) and coexistence provisions (Sect. 8.6). When taking a closer look at the regulatory status of genome edited plant varieties and the products derived from them, it becomes apparent that the differences of the regulatory frameworks manifest in the legal classification of those plants and their produce. Consequently, genome edited organisms (GEOs) are treated vastly differently by the examined legal regimes (Sect. 8.4). However, it should be borne in mind that some of the examined countries are currently working on a revision of their regulations (Sect. 8.7).

8.1 Introduction

The reports on the regulatory status of genome edited crops in Argentina, Australia, Canada, the European Union (EU), Japan and the United States (USA) have identified anticipated problems, innovative solutions and existing regulatory gaps. In order to benefit from them beyond the mere assessment of the respective national status quo, it is imperative to interrelate their findings with each other.

Against this background, a comparative analysis is a multipurpose instrument to gain additional insights into the regulation of genetically modified plants and the products derived from them.

D. Hamburger (✉)
Faculty of Law, University of Passau, Passau, Germany
e-mail: david.hamburger@uni-passau.de

H.-G. Dederer, D. Hamburger (eds.), *Regulation of Genome Editing in Plant Biotechnology*, https://doi.org/10.1007/978-3-030-17119-3_8

By identifying differences and similarities of socio-economic factors and comparing them to the regulatory outcome, a determination on their influence on the regulation of agricultural biotechnology can be made.

In addition, a comparative analysis can be used to uncover the weaknesses and strengths of the respective regulations. Since the effectiveness of a system can be assessed more easily when it is put in relation to alternatives, instead of determining it in an absolute manner, a comparative approach is very promising.

Furthermore, concepts of other legal systems can be used as guidance or benchmark for own reform efforts. However, this requires not only an assessment of how effective the respective regulatory concept is, but it must also be determined whether transferability exists at all. Such transferability can, however, only be guaranteed if the underlying conditions of the regulatory systems are at least comparable.

Apart from an analysis of the substantive regulations, it is a comparison of the different procedures used to update the existing regulatory regime for genome edited plants that can give an additional valuable insight into the mechanisms which render a reform effort a success or a failure.

To this end, a comparative analysis is carried out on the basis of the country reports[1] of this contributed volume.

8.2 Country Overview

8.2.1 Cultivation of GM Crops

8.2.1.1 Approved GM Plants or GM Events

A comparison of the number of approved GM plants is for several reasons rather complicated and burdensome. First, it must be distinguished between the different uses an approval has been granted for. Common categories in that regard are contained use, field trials, cultivation and marketing. Second, a distinction must be made between the number of approved GM plant varieties and the number of approved GM events. Since a single plant can harbour several (so called stacked) GM events, the number referring to GM plant varieties and that referring to GM events can differ significantly from each other. The same is true the other way round because the same single GM event can be used in different plant varieties. Third, attention must be paid to whether the authorisation is still in force, has been revoked or is expired.

Figure 8.1 illustrates the number of GM plant events that are currently approved for commercial cultivation in Argentina, Australia, Canada, the EU, Japan and the USA. Based on the given numbers the countries can be divided into three groups.

---

[1]In the following any abstract reference to countries also includes the EU, even though strictly speaking it does not constitute a nation state.

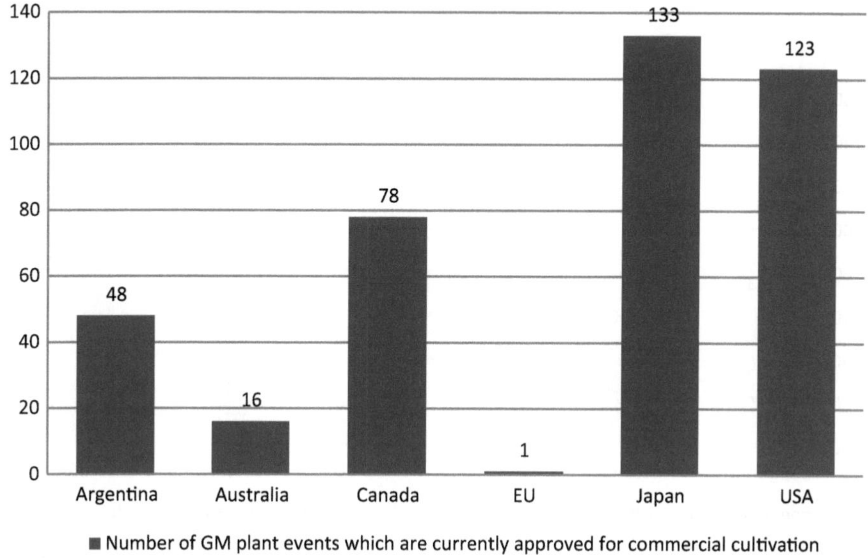

Fig. 8.1 Number of GM plant events which are currently approved for commercial cultivation. With regard to Argentina the figure indicates the number of approved GM crops for cultivation. The Argentinian Ministry of Agriculture provides a constantly updated list of approved GM events; cf. Ministerio de Agroindustria (2018). In the case of Australia the number refers to GM plants which have been authorized for commercial release; cf. Australian Government Department of Health and Office of the Gene Technology Regulator (2018). The data for Canada displays the approval for unconfined release into the environment of plants with novel traits which are considered to be LMOs as defined by the Cartagena Protocol on Biosafety; cf. Canadian Food Inspection Agency (2018). Since the term plant with novel traits encompasses GM as well as non-GM varieties this distinction has to be made to ensure comparability. Overall, 98 plants with novel traits are currently approved for environmental release in Canada; cf. Canadian Food Inspection Agency (2018). In the EU, the only GM crop, that is currently approved for cultivation, is maize genetically modified with the event Mon810; see Chap. 5 (Country Report on the EU), Sect. 5.1 at fn.3. With regard to Japan it is referred to the number of GM plant varieties approved for cultivation, distribution, and import; cf. Chap. 6 (Country Report on Japan), Sect. 6.1, at fn.7. In case of the USA the figure shows the number of GM crop varieties which have received the status of an unregulated article by USDA; cf. United States Department of Agriculture (2018b). The different sources of the used data do not adhere to a uniform terminology concerning "plant variety" and "GM event". Therefore, some of the sources use the term "plant variety" to describe the number of events. The importance of this distinction is evident in the case of the EU: While only one event is approved (MON810), the Common EU Catalogue of Varieties lists over 50 maize varieties modified with that event

The USA and Japan form the upper tier with over 100 events which are currently approved for cultivation. The lower tier constitutes of Australia and the EU with under 20 approved GM events. Argentina and Canada are positioned approximately in the middle with 40 and 78 approved plant events.

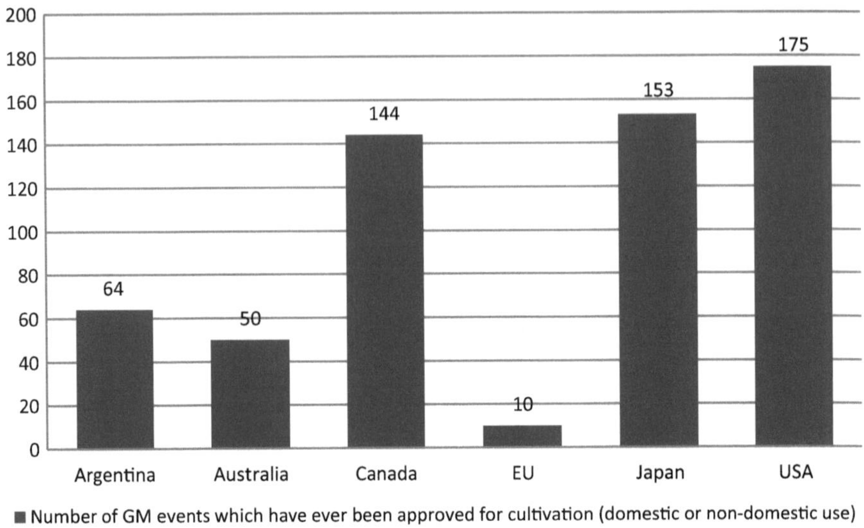

Fig. 8.2 Number of GM events which have ever been approved for cultivation (domestic or non-domestic use) (International Service for the Acquisition of Agri-biotech Applications 2018c)

The widely cited[2] numbers of approved GM events given by the ISAAA database are used for Fig. 8.2. However, the data should be treated with caution. First of all, the database seems only to consider new approvals but not whether a once granted approval is still in force.[3] Consequently, expired or revoked approvals are still listed what inflates the numbers artificially. Secondly, the accuracy of the information can at least be called into question. Concerning the EU, ten events are listed as approved for cultivation.[4] However, some of the events have actually never been approved for cultivation.[5]

[2]Cf. Blair and Regenstein (2015), p. 85; Newton (2014), p. 202; Kuntz (2018), p. 183; Spielman and Zambrano (2013), p. 184; Juma (2016), p. 245; Capalbo and Suzuki (2017), p. 270.

[3]Cf. International Service for the Acquisition of Agri-biotech Applications (2018d). This website gives an overview of the changes made to the database since July 2014. However, only new approvals are listed here. On the individual website for each event it is, however, indicated if the approval is still in force.

[4]Those are seven events regarding ornamental flowers (carnations), MON810, Liberty Link Maize (T25) and the Amflora Potato; cf. International Service for the Acquisition of Agri-biotech Applications (2018e).

[5]The database indicates that ornamental carnations with the event 'Moonlight' (event code 123.2.38) were approved for cultivation in 2007; International Service for the Acquisition of Agri-biotech Applications (2014). However, the corresponding decision of the European Commission states clearly that '[t]he product may be put to ornamental use only, with the exception of cultivation'; European Commission (2007), Art.3. The 'Moonberry' (event code IFD-25958-3) and 'Moonvelvet' event (event code IFD-264Ø7-2) are listed as approved for cultivation and only on the website of the specific event, it is indicated that they have only been approved for import.

Consequently, Fig. 8.2 shows all GM events which have ever been approved for cultivation. Due to revoked or expired approvals, the number is higher than those stated in Fig. 8.1.

When comparing Figs. 8.1 and 8.2, the relative increase in the case of Australia and the EU compared to the other countries is noteworthy. The numbers stated in Fig. 8.2 rise compared to Fig. 8.1 by 33.3%, 84.6%, 15.0%, 42.3% for Argentina, Canada, Japan and the USA respectively. However, with regard to Australia and the EU, the numbers are spiking up 312.5% and 1000% respectively. Keeping in mind that Fig. 8.1 uses the current number of approved GM varieties while Fig. 8.2 states the number of all events that had been approved at any point in time, the big discrepancy is a testimony of a shift in policy. It is a hint that the regulatory framework or at least its application used to be more permissive in the past then it is nowadays. While this is certainly not a new insight into the EU's attitude towards GMOs, it is interesting to note that such a trend can also be observed regarding Australia. However, the sharp percentage increase in the case of Australia and the EU is of course also due to the fact that there have been comparatively few approved GM events in absolute numbers to begin with.

8.2.1.2 Current Cultivation of GM Crops

A substantial cultivation of GMOs is taking place in Canada, Argentina and especially the United States (cf. Fig. 8.3). The USA is worldwide by far the biggest producer of GM crops with almost 73 million hectares followed by Brazil, Argentina

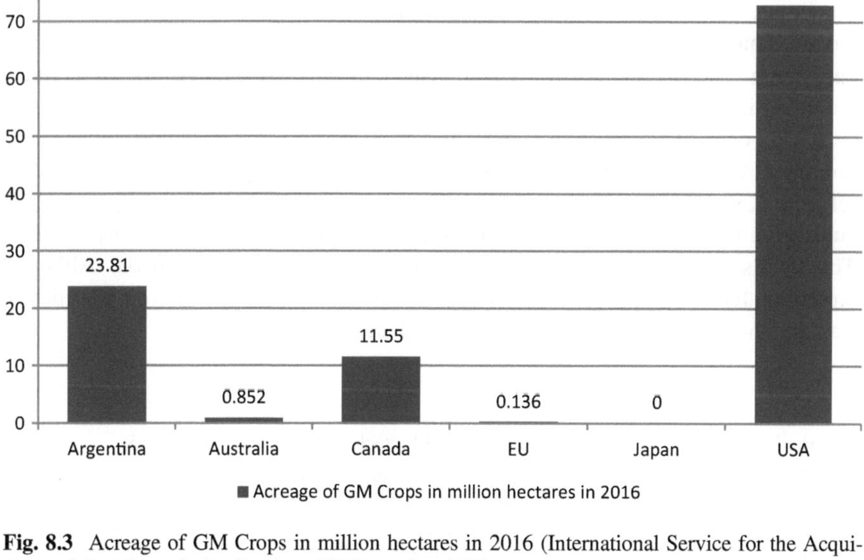

Fig. 8.3 Acreage of GM Crops in million hectares in 2016 (International Service for the Acquisition of Agri-biotech Applications 2016, pp. 5 and 73)

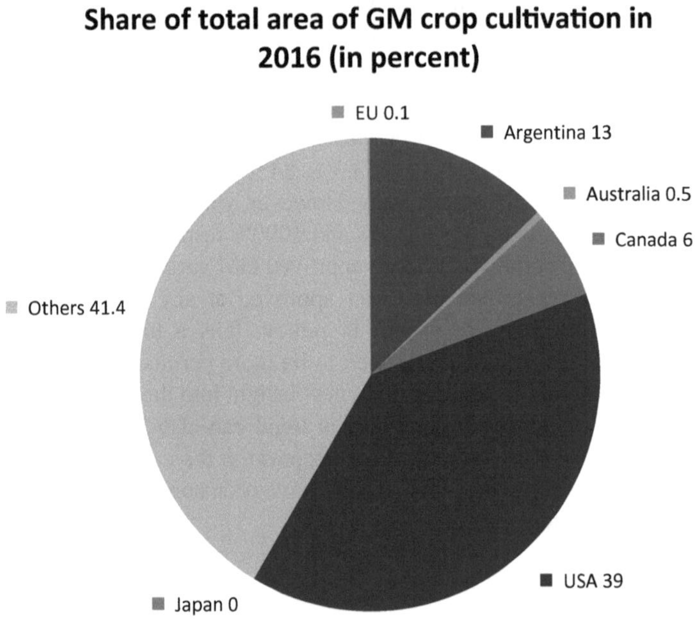

Fig. 8.4 Share of total area of GM crop cultivation in 2016 (Calculated on the basis of the data given in International Service for the Acquisition of Agri-biotech Applications (2016), p. 5)

and Canada with 49.1, 23.8, and 11.6 million hectares respectively.[6] Australia has a rather insignificant number of GM acreage. Whereas the EU has virtually no cultivation and Japan has actually no cultivation at all.

The six countries examined in detail in this contributed volume account together for almost 60% of the worldwide GM acreage (cf. Fig. 8.4). Argentina and the United States alone are responsible for more than half of the total world GM crop cultivation. The EU and Australia with a share of only 0.1% and 0.5% respectively are rather insignificant while Canada is at least in the single digit region (6%).

As the case of Japan shows, there is not necessarily a correlation between the number of GM varieties approved for cultivation and the degree of actual cultivation (cf. Fig. 8.5). While Japan is the country with the most GM varieties approved for cultivation (compared to the other six countries), it does not cultivate any of them. This is also confirmed by a comparison of Argentina with Canada. Although Argentina has only approved just slightly over half as many GM crop varieties as Canada, the area under cultivation is more than twice as large.[7]

[6]International Service for the Acquisition of Agri-biotech Applications (2016), p. 5.

[7]The less extensive cultivation of GM plant varieties in Canada could be explained with a smaller agricultural area. However, the total area of arable land in Canada was in 2015 even slightly higher than that of Argentina (43.6 million ha compared to 39.2 million ha); cf. Food and Agricultural Organization of the United Nations (2018), Section Land Use—arable land.

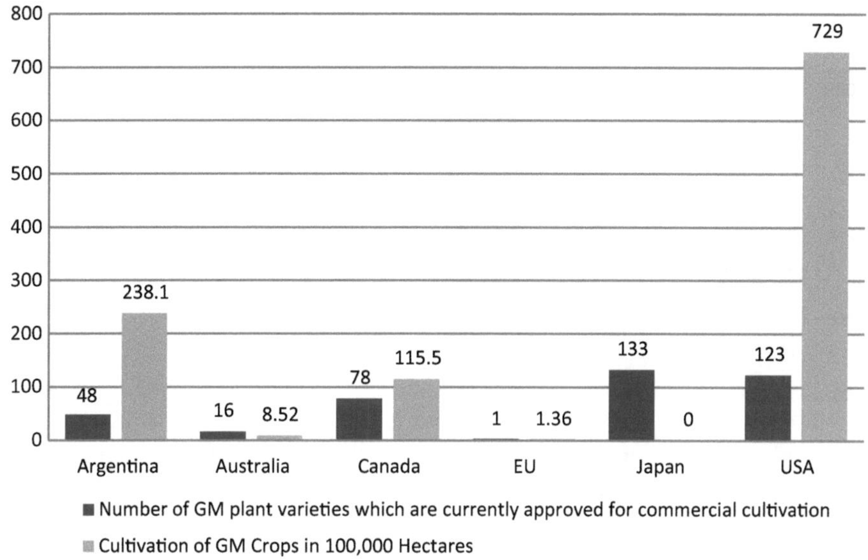

Fig. 8.5 Correlation between the numbers of GM varieties approved for cultivation and the actual cultivation in 2016

However, absolute numbers, as in Fig. 8.5, have the disadvantage that, for example, the different sizes of arable land in each country are not taken into account.

In order to remedy this situation and to ensure a better overview, it is, therefore, advisable to have recourse to relative numbers. One viable option is the adoption rate of GM crops as illustrated in Fig. 8.6. The comparable low adoption rate in the EU and Australia can be explained by the geographically limited cultivation in both countries. In 2016, only the EU member states Spain, Portugal, Czech Republic and Slovakia cultivated GM crops.[8] In Australia, GM varieties are banned from cultivation in Australian Capital Territory, South Australia and Tasmania.[9] However, those geographical limitations do not cause the low adoption rate alone. The GMO ban in the three regions of Australia, for instance, prevents the transport of GM crops or seeds through these states.[10] As a result, the transport costs for GM products produced in other regions of Australia rise what leads to an increase of the overall production cost and a decrease of competitiveness. Furthermore, it can be assumed that the negative regulatory and public attitude towards GMOs partly discourages farmers from using them.

[8]International Service for the Acquisition of Agri-biotech Applications (2016), p. 73.

[9]See Chap. 3 (Country Report on Australia), Sect. 3.3.1.1.

[10]See Chap. 3 (Country Report on Australia), Sect. 3.3.1.1.

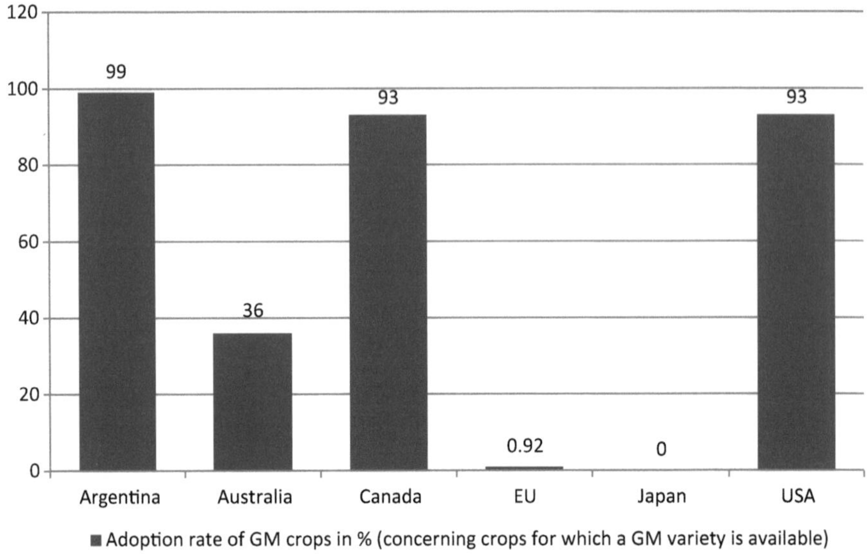

Fig. 8.6 Adoption rate of GM crops in 2016 (in percent) concerning crops for which a GM variety is available (International Service for the Acquisition of Agri-biotech Applications (2016). The adoption rate for the EU is not mentioned in the ISAAA report but can be calculated. In 2016 the total area of GM crop cultivation in the EU amounted to 136,363 hectares; International Service for the Acquisition of Agri-biotech Applications (2016), p. 74. Since only the maize line Mon810 is currently cultivated in the EU, this figure must be put into relation with the total area planted with maize in the EU. In 2016, 8,563,110 ha of grain maize and 6,251,190 ha of green maize were cultivated in the EU; Eurostat (2018a, b). The 136,363 GM maize hectares constitute 0.92% of the total EU maize acreage of 14,814,300 ha)

Argentina, Canada, and the USA on the other hand have an adoption rate of over 90%, which is a rather common adoption rate in countries that allow the cultivation of GM crops.[11]

A good impression of how widespread the use of GM crop varieties is, can also be obtained when one looks closely at the share GM cultivation has of the total national acreage (cf. Fig. 8.7). This data is especially interesting in the case of Australia. Figure 8.7 illustrates that when it comes to actual cultivation of GM crops Australia is with a share of GM cultivation of only 1.52% more comparable to the EU than to the other examined GM cultivating countries.

However, solely based on Fig. 8.7 no conclusion should be drawn regarding the openness towards GM technology in agriculture. The use of GM varieties is limited to certain crops since only a comparable small number of plant species have a GM variety.[12] If these varieties are not among the main crops of the respective country, it

[11]Cf. International Service for the Acquisition of Agri-biotech Applications (2016), p. 2.

[12]In detail, these are: alfalfa, apple, bean, canola, carnation, chicory, cotton, creeping bentgrass, eggplant, eucalyptus, flax, maize, melon, papaya, petunia, plum, poplar, potato, rice, rose, soybean,

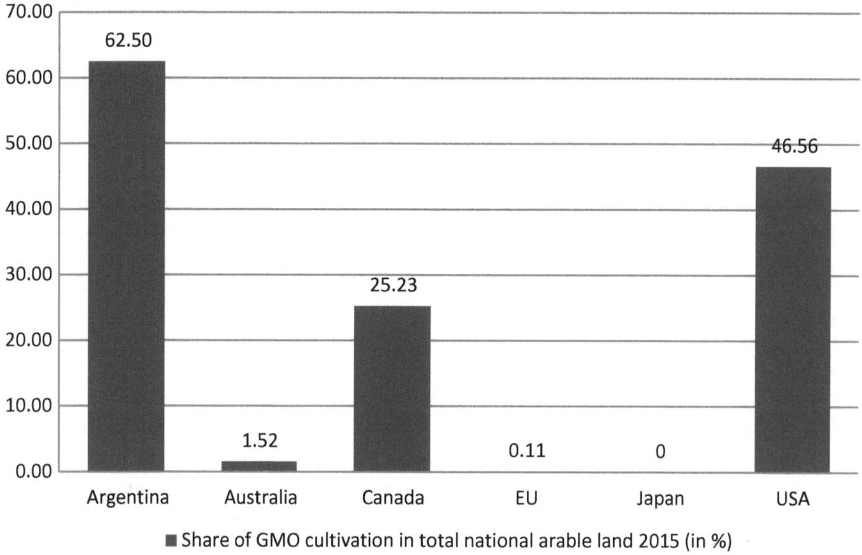

Fig. 8.7 Share of GMO cultivation in total national arable land in 2015 (The figures given are calculated based on data provided by the FAO and ISAAA; James (2015), pp. 15 and 198; Food and Agricultural Organization of the United Nations (2018). The FAO is the source of the data on the agricultural land and the ISAA of the area currently under cultivation with GMOs. Since the most recent data the FAO provides refers to the year 2015, the ISAAA data pertaining to the year 2015 has been used as well to ensure comparability)

is not surprising that the total share of GM cultivation is also rather small. For instance, in Australia only GM varieties of canola, cotton and carnation are currently approved for cultivation.[13] Canola and cotton, however, were only harvested on an area of 2,637,422 ha in 2016 compared to a total harvested area of agricultural crops of 22,728,026 ha.[14] The two major crops in Australia have been barley and wheat with 4,107,648 ha and 11,282,202 ha respectively harvested in 2016. For barley no GM alternative exists and a wheat GM variety is not approved for cultivation in Australia or elsewhere in the world.[15] If GM varieties are not available for the two main crops and the adoption rate of the existing varieties is rather low (cf. Fig. 8.6), it

squash, sugar beet, sugarcane, sweet pepper, tobacco, tomato, wheat. Cf. International Service for the Acquisition of Agri-biotech Applications (2018b).

[13]Australian Government Department of Health and Office of the Gene Technology Regulator (2018).

[14]280,422 ha cotton seed and 2,357,000 ha canola (rapeseed); cf. Food and Agricultural Organization of the United Nations (2018), Section Crops—Australia—area harvested.

[15]International Service for the Acquisition of Agri-biotech Applications (2018a). The non-approval of this glyphosate-resistant wheat variety is explained on the one hand by the lack of acceptance by the public; cf. Rao (2015), p. 343. On the other hand, however, it is also due to the fact that the pollen of wheat drifts much further than, for example, that of maize; cf. Gustafson (2014), p. 96. Detailed on this, using Canada as an example, Magnan (2016), pp. 151–157; Eaton (2013).

is not surprising that the overall share of GM varieties in Australia is comparatively low as well.

The situation in Argentina differs largely from the Australian case. Here the major crops harvested in 2016 are maize, wheat and soybeans with 5,346,593 ha, 5,629,213 ha and 19,504,648 ha respectively. Regarding maize and soybeans, GM varieties are approved for cultivation. Since those two alone are accounting for 74.9% of the total harvested area in 2016[16] and their adoption rate is at almost 100% (cf. Fig. 8.6), the high share of arable land (cf. Fig. 8.7) can be easily explained.

8.2.1.3 Conclusion

With regard to the cultivation of GM crops, the analysed countries paint a very heterogeneous picture. In the end, it is mixture of countries that do not cultivate GMOs (Japan), those that rely heavily on genetic engineering (Argentina, Canada, USA) and those that are rather reluctant to make use of GMOs (Australia, EU) when it comes to cultivation or environmental release.

8.2.2 Consumption of GMOs and Products Derived from Them

8.2.2.1 Approved GM Plant Events for Food and Feed

A closer look at the number of approved GM plant events for the use in food or feed (cf. Fig. 8.8) reveals a slightly different situation compared to the cultivation of GM crops. Japan, which does not cultivate GM crops, approved at the same time the most GM events for use in food and feed compared to the other five countries. A similar situation can be described with regard to the EU. Even though almost no cultivation takes place, nearly 100 GM events are approved with respect to food and feed. This discrepancy can be explained by the fact that while both countries are reluctant to allow the cultivation of GM plants within their territory, they depend on the importation of food and feed which contains GM material. This opportunistic approach ultimately leads to a situation where no correlation between the aversion to cultivation and the actual consumption of GM products exists. With regard to the EU, it is particularly noticeable that currently only one GM event is approved for cultivation but almost 100 for use in food or feed. This calls into question whether there is a scientific-based correlation between the risk and the approval of GM

[16]In 2016 the total harvested area of agricultural crops in Argentina amounted to 36,826,764 ha; cf. Food and Agricultural Organization of the United Nations (2018), Section Crops—Argentina—area harvested.

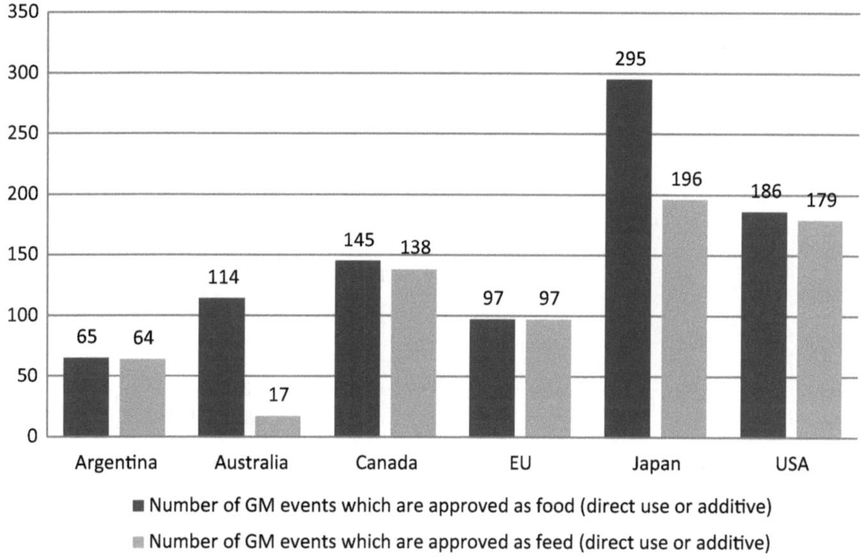

Fig. 8.8 Number of GM events which are approved as food or feed (International Service for the Acquisition of Agri-biotech Applications (2018c). The caveat regarding the accuracy of these numbers expressed at fn.3 applies here as well)

products or whether this discrepancy is politically motivated. It seems at least difficult to comprehend why products are deemed safe enough for consumption but not for cultivation. However, it must be acknowledged that the potential risks associated with cultivation are different from those associated with the consumption of GM products. Differences in treatment are therefore not a priori unjustifiable.

In the case of Australia, it is striking that the number of approved events for use in food is significantly higher than that concerning animal feed. However, considering that Australia meets most of its demand for stockfeed by domestic production[17] and is, therefore, not dependent on importation, this number is far less surprising. A higher number of approved GM event for feed would only make sense if there were a considerable higher demand for the import of (GM) feed.

8.2.2.2 Import of GM Crops

There exists no reliable data concerning the amount of agricultural products derived from GMOs that are imported by each of the examined countries.

In order to allow at least an approximation, the quantity of imported soybeans can be used. Seventy eight percent of the worldwide planted soybeans in 2016 were genetically modified.[18] Consequently, GM soy can hardly be avoided during import.

[17]Office of the Gene Technology Regulator (2008).

[18]International Service for the Acquisition of Agri-biotech Applications (2016), p. 90.

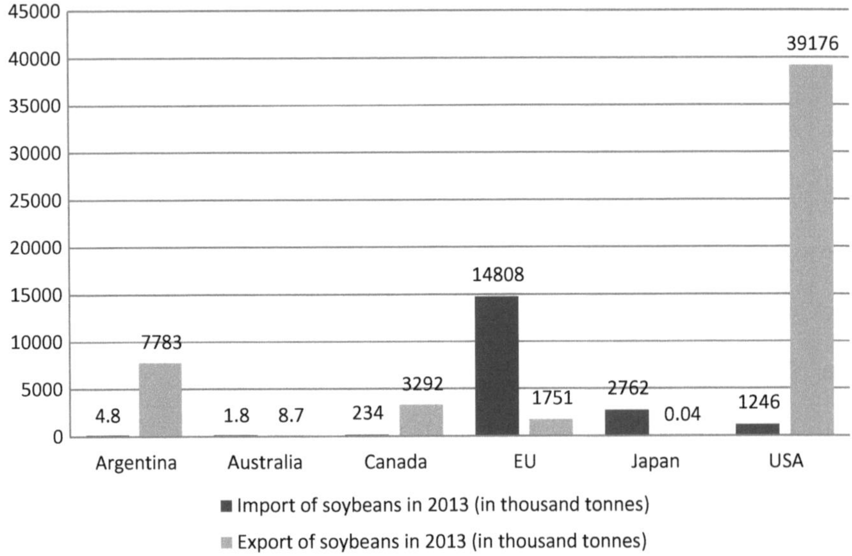

Fig. 8.9 Import and export of soybeans in 2013 (Food and Agricultural Organization of the United Nations (2018), Category Crops and livestock products)

Additionally, GM soybeans are approved for importation in all six analysed countries.[19] This suggests that the ratio of imports of soybeans at least approximately reflects the ratio of imported GM soybeans.

Figure 8.9 shows that while Australia is more or less self-sufficient regarding its soy supply, the other countries can be divided into soy net importing (EU, Japan) and net exporting countries (Argentina, Canada, USA). The exact same separation can also be made with regard to the cultivation of GM soybeans. In the exporting countries, GM soybeans are approved for cultivation, in the importing countries they are approved for importation but not cultivated.[20] This once more highlights the already mentioned discrepancy between the consumption and cultivation of GMOs.

8.2.2.3 Conclusion

Regarding the actual consumption and use of GM material in food and feed, both the handling of the approval of GMO products and their importation shows that a negative attitude towards cultivation has a far lesser impact on the consumption than one would anticipate.

[19]Cf. International Service for the Acquisition of Agri-biotech Applications (2018c).

[20]In the EU GM soybean is not approved for cultivation. In the case of Japan GM soybeans are approved for cultivation but currently not cultivated. Cf. International Service for the Acquisition of Agri-biotech Applications (2018c).

This is most likely due to the fact that the majority of GM commodities are used as feedstuff and not as food.[21] Since the use of GMOs in animal feed is viewed less critically by consumers than the direct use in products for human consumption, the pressure on the legislator to regulate GM feed more restrictively is at the same time less prevalent. However, especially in the EU there is a consumer demand for animal products that have been produced without having recourse to genetically engineered feed. That this has a comparatively low impact on the use and import of GM feed might be related to the labelling requirements. There is no obligation to specifically label animal products in the EU if the animals have been fed genetically modified feed.[22] Accordingly, farmers do not have to fear any losses that could result from consumers spurning their products because of GM labelling.

However, the use of genetically modified feed can also simply be explained as an economic necessity, since the adoption rate of individual GM varieties is so high that their use is in part without alternatives.

The different treatment of cultivation and consumption can, therefore, be explained by a melange of public opinion, economic factors and political interests.

8.2.3 Status Quo of Genome Edited Plants and Products Derived from Them

The way and extent to which genome edited plants are already used, varies from country to country significantly.

In Argentina, active research on genome edited crops takes place[23] and a field trial of a genome edited soybean by Calyxt has already been conducted.[24] Additionally, a few genome edited plant varieties have gone through the relevant consultation process in order to determine their regulatory status. Most of them were not regarded as GMOs but as new varieties of conventional plants. However, no genome edited plants are currently cultivated or marketed in Argentina.[25]

---

[21]It is estimated that about 70–90% of the worldwide harvested GM crops are consumed as feedstock by food-producing animals; Flachowsky et al. (2012), p. 180; Lucht (2015), p. 4255.

[22]However, on the national level voluntary negative labelling exist. In the case of Germany, the Federal Ministry of Food and Agriculture introduced the label “ohne Gentechnik” (engl. “without gene technology”), which indicates that - in the case of a food or food ingredient of animal origin—no genetically modified feed has been administered to the animal from which the food was obtained; cf. § 3a (4) EG-Gentechnik-Durchführungsgesetz (EGGenTDurchfG). However, it is still possible to feed such animals with genetically modified feed. For example, in the case of poultry, pigs or cattle, it is sufficient for the labelling “without gene technology” if they have not received any genetically modified feed 10 weeks, 4 months or 12 months respectively before slaughter. Cf. Appendix to § 3a (4) Sentence 2 EGGenTDurchfG.

[23]Cf. La Capital (2018).

[24]Calyxt, Inc. (2016).

[25]For further details see Chap. 2 (Country Report on Argentina), Sect. 2.5.

Genome edited plants have not yet been authorized either for contained use, field trials or cultivation in Australia and no such approvals are currently pending. The situation is the same for products derived from genome edited plants which are currently neither approved for marketing nor is an approval pending.[26]

Canadian researchers are engaged in plant breeding using genome editing techniques.[27] Field trials of a genome edited plant variety took already place in Canada prior to its approval by the Canadian authorities.[28] Two genome-edited plant varieties are currently approved for commercial cultivation.[29] In 2018 Cibus introduced its genome edited SU Canola to the market for cultivation.[30] However, there is not yet a marketing of food containing products derived from genome edited crops.

In the EU, a contained use takes place in the context of laboratory research. Various field tests have been carried out in the past or are currently underway. However, there is right now no cultivation or marketing of GEOs or products derived from them.[31]

Japanese researchers have already made use of the genome editing technique at laboratories in compliance with containment requirements. The first field trial of a genome edited plant variety took place in 2017. Neither cultivation nor marketing of agricultural GEOs or products derived from them takes currently place in Japan.[32]

Comparable extensive research is conducted on genome edited plant varieties in the United States.[33] At the same time, it is unclear to what extent field trials with genome edited plant varieties take place in the USA. Since USDA determined many genome edited crops as not regulated articles under its biotechnology regulations,[34] field trials do not need an approval by USDA, what makes it difficult to keep track of them. Nonetheless, Calyxt, a Minnesota based biotechnology company, indicates on its website that with regard to several genome edited crop varieties field tests have been or are currently being performed.[35] Cibus' genome edited SU Canola is cultivated in the USA since 2017.[36] Calyxt launched its genome edited high oleic soybean variety on the US market in 2018.[37] The situation with regard to marketing as food or feed is rather unclear since manufacturers can declare their products to be

[26] For further details see Chap. 3 (Country Report on Australia), Sect. 3.5.

[27] See for example PlantForm (2015).

[28] Cf. For example Canadian Food Inspection Agency (2013). The second approved genome edited plant variety has never been tested in Canada, but the results of field trials conducted in the USA have been used during the approval procedure, cf. Canadian Food Inspection Agency (2014).

[29] For further details see Chap. 4 (Country Report on Canada), Sect. 4.5.

[30] Ayers (2018), Pratt (2018) and Cibus (2018).

[31] For further details see Chap. 5 (Country Report on the EU), Sect. 5.5.

[32] For further details see Chap. 6 (Country Report on Japan), Sect. 6.4.

[33] Cf. Ricroch et al. (2017), p. 178; Lusser et al. (2012), p. 233.

[34] For further details see Chap. 7 (Country Report on the USA), Sect. 7.8.2.2.

[35] Calyxt (2018a).

[36] Cibus (2018) and Pratt (2018).

[37] Calyxt (2018b).

Table 8.1 Status quo of genome edited plants and products derived from them

	Argentina	Australia	Canada	EU	Japan	USA
Contained use	✓	✗	✓	✓	✓	✓
Field trial	✓	✗	✓	✓	✓	✓
Cultivation	✗	✗	✓	✗	✗	✓
Marketing as food or feed	✗	✗	✗	✗	✗	✓

GRAS on their own volition and prevent so a premarket review by the FDA. Since a GRAS notification to the FDA is voluntary, the FDA might not even know when genome edited food enters the market. It has been estimated that food containing material of Calyxt's genome edited soybean could make its way on the market until end of 2018.[38] After 17.000 acres of Calyxt's genome edited soy bean variety have been planted in the United States in 2018,[39] the first food and feed products produced from that harvest to be sold on the US market have been reported in 2019.[40]

Summing up, the market entry of GEOs and products derived from them is most advanced in Canada and the USA. In Argentina, the EU and Japan GEOs have not yet progressed beyond field trials while Australia did not yet make use of genome edited plants at all (Table 8.1).

8.3 The Point of Entry into the Regulatory Framework for Genetically Modified Organisms (GMOs): Regulatory Triggers

To be able to determine whether the current regulatory regime for GMOs is applicable to GEOs, it is crucial to identify the trigger for the applicability of the regulatory framework.

Not only use different countries different regulatory triggers, but in some instances, different triggers are also used by the same country.

Consequently, the different regulatory triggers must be examined country-by-country and according to their individual scope of application.

In order to facilitate a clear comparison, only the triggers for regulating the cultivation of genetically modified plants and sale of foodstuff derived from them are examined below. Information on the regulatory trigger for other possible applications of GEOs—such as animal feed—can be found in the respective country reports.

[38]Dewey (2018).

[39]Calyxt (2019a).

[40]Calyxt (2019b).

8.3.1 *Argentina*[41]

Art.2 Resolution 763[42] stipulates an authorization requirement for the release of not commercially authorized GMOs. Pursuant to Art.1 Resolution 763 the regulation applies to all activities involving GMOs that could at least have potential use in an agricultural context. Therefore, the trigger of the Argentine GMO regulatory framework does not distinguish between different types of use like field testing, cultivation or marketing.

Consequently, the decisive criterion for the applicability of the regulatory framework is the existence of a GM plant, defined in Art.2 Nr.27 Resolution 701[43] as a "plant organism that has a combination of genetic material obtained through the application of modern biotechnology". Modern biotechnology is understood as "a) the application of in vitro nucleic acid techniques, including recombinant nucleic acid into cells or organelles, or b) fusion of cells beyond the taxonomic family, that overcome natural physiological reproductive or recombination barriers and that are not techniques used in traditional breeding and selection" (Art.2 Nr.8 Resolution 701).

This trigger is a combination of a product-based approach ("combination of genetic material") and a process-based approach ("obtained through the application of modern biotechnology").

8.3.2 *Australia*[44]

The Australian regulatory regime prohibits any dealings with GMOs without a licence.[45] "Dealing" is very broadly defined,[46] so that in the end any activities related to GMOs are covered by the scope of application.

The decisive criterion is, therefore, the presence of a GMO as defined by the Gene Technology Act in Para.10: "genetically modified organism means: (a) an organism that has been modified by gene technology; or (b) an organism that has inherited particular traits from an organism (the initial organism), being traits that occurred in the initial organism because of gene technology; or (c) anything declared by the

---

[41]For this section see Chap. 2 (Country Report on Argentina), Sect. 2.3.1.

[42]Ministerio de Agroindustria (2011).

[43]Servicio Nacional de Sanidad y Calidad Agroalimentaria (2011).

[44]For this section see Chap. 3 (Country Report on Australia), Sect. 3.3.1.

[45]Gene Technology Act 2000, Para. 32.

[46]Pursuant to Para. 10 Gene Technology Act 2000 "deal with, in relation to a GMO, means the following: (a) conduct experiments with the GMO; (b) make, develop, produce or manufacture the GMO; (c) breed the GMO; (d) propagate the GMO; (e) use the GMO in the course of manufacture of a thing that is not the GMO; (f) grow, raise or culture the GMO; (g) import the GMO; (h) transport the GMO; (i) dispose of the GMO; and includes the possession, supply or use of the GMO for the purposes of, or in the course of, a dealing mentioned in any of paragraphs (a) to (i)."

regulations to be a genetically modified organism, or that belongs to a class of things declared by the regulations to be genetically modified organisms; (...)".[47]

Accordingly, the application of the regulatory framework depends mainly on whether "gene technology" has been used. Gene technology is defined as "any technique for the modification of genes or other genetic material, but does not include: (a) sexual reproduction; or (b) homologous recombination; or (c) any other technique specified in the regulations for the purposes of this paragraph" (Gene Technology Act 2000, Para. 10).[48]

The trigger is a combination of a process-based and a product-based approach. While the requirement that the organism "has been modified by gene technology" is emphasized in part (a) of the GMO definition, part (b) refers to "inherited particular traits".

However, Australia has different rules for the sale of food produced using gene technology.[49] Those food related regulations do not use the trigger of the Gene Technology Act 2000 but provide a distinctive point of entry: "food produced using gene technology".[50] This is defined as "food which has been derived or developed from an organism which has been modified by gene technology".[51] And "gene technology" is understood as "recombinant DNA techniques that alter the heritable genetic material of living cells or organisms".[52] Of particular importance is here the divergent definition of "gene technology". This could lead to a situation where an organism is considered to be a GMO under the Gene Technology Act 2000 but food containing that GMO is not considered to be "food produced using gene technology"—or vice versa.

In contrast to the cultivation, the trigger for the GM regulations on food is—apart from the inevitable genetic modification—purely process-based, since it does not take into account any product's characteristics.

---

[47]Gene Technology Act 2000, Para. 10.

[48]For a detailed illustration of the Australian GMO definition, which considers exceptions as well, see Chap. 3 (Country Report on Australia), Sect. 3.3.1.1.

[49]See Standard 1.5.2 of the Australia New Zealand Food Standards Code and the further explanations in Chap. 3 (Country Report on Australia), Sect. 3.3.1.2.

[50]Cf. Section 3 of Standard 1.5.2 of the Australia New Zealand Food Standards Code.

[51]Cf. Standard 1.1.2 of the Australia New Zealand Food Standards Code.

[52]Cf. Standard 1.1.2 of the Australia New Zealand Food Standards Code.

8.3.3 *Canada*[53]

With regard to the unconfined cultivation of seeds,[54] a notification and a corresponding authorisation by the competent authority is required.[55] This requirement applies to all seed in the same way, regardless of the breeding technique or physical characteristics, as long as the seed was not grown in Canada prior to December 1996.[56]

Relevant with regard to the cultivation of genetically modified plants are the additional requirements which are imposed on so called plants with a novel trait.[57] A plant with a novel trait is described as "a plant containing a trait not present in plants of the same species already existing as stable, cultivated populations in Canada, or is present at a level significantly outside the range of that trait in stable, cultivated populations of that plant species in Canada".[58]

It should be emphasised here that it is not the plant with a novel trait which triggers the authorization requirement, but that a novel trait triggers only additional formal and substantive requirements within the approval process.

Canada makes use of a different trigger when it comes to the marketing of food. With regard to the marketing of so called "novel food", a notification and a corresponding authorisation by the competent authority is required.[59] Novel food is inter alia defined as "a food that is derived from a plant, animal or microorganism that has been genetically modified such that (i) the plant, animal or microorganism exhibits characteristics that were not previously observed in that plant, animal or microorganism, (ii) the plant, animal or microorganism no longer exhibits characteristics that were previously observed in that plant, animal or microorganism, or (iii) one or more characteristics of the plant, animal or microorganism no longer fall within the anticipated range for that plant, animal or microorganism".[60] While part (i)–(iii) takes a product-based approach by referring to the characteristics of the food

[53]For this section see Chap. 4 (Country Report on Canada), Sect. 4.3.1.

[54]The Seeds Act, R.S.C., 1985, c. S-8 and the Seeds Regulations, C.R.C.,c. 1400 refer only to seeds and not to plants. However, seed is very broadly defined as "any plant part of any species belonging to the plant kingdom, represented, sold or used to grow a plant"; Seeds Act, R.S.C., 1985, c. S-8, Para.2. Therefore, the corresponding plant is considered to be covered as well.

[55]Seeds Act, R.S.C., 1985, c. S-8, Para. 3 (1) (b) in conjunction with Seeds Regulations, C.R.C.,c. 1400, Part V, Para 109 (1).

[56]This is a rather simplified description of the exemption rules. For the detailed specification of exempted seeds see Seeds Regulations, C.R.C.,c. 1400, Part V, Para.108.

[57]Cf. Seeds Regulations, C.R.C.,c. 1400, Para.110 (d).

[58]Directive 94-08, Assessment Criteria for Determining Environmental Safety of Plants With Novel Traits, Sec.1. This is, however, not the legal definition used in Seeds Regulations, C.R.C.,c. 1400, Para.107 (1).

[59]Food and Drug Regulations, C.R.C., c. 870, Para. B.28.002.

[60]Food and Drug Regulations, C.R.C., c. 870, Para. B.28.001.

in question, the first part of the definition ("has been genetically modified"[61]) is solely linked to the production process. The notion that the Canadian GMO framework is based solely on a product-based approach is, therefore, a common misconception. However "novel food" can also be "a substance, including a microorganism, that does not have a history of safe use as a food [...] [or] a food that has been manufactured, prepared, preserved or packaged by a process that (i) has not been previously applied to that food, and (ii) causes the food to undergo a major change".[62] Therefore, the novel food regulatory regime is not limited to GMOs, but applies also in general to any food with certain characteristics whether or not it is genetically modified.

8.3.4 EU[63]

While the European legal framework regulates contained use, field trials, cultivation and marketing as food or feed differently, the common denominator is that the GMO framework applies if these activities are carried out in relation to genetically modified organisms.[64]

The GMO definition for the European regulatory framework for environmental release is laid down in Art. 2 (2) in connection with Annex IA and Annex IB of Directive 2001/18/EC. The definition applies to the regulation of field trials, cultivation and the placing on the market of GMOs. Here, a GMO is defined as "an organism, with the exception of human beings, in which the genetic material has been altered in a way that does not occur naturally by mating and/or natural recombination". Required is in essence (1) a genetic alteration (2) in a way that does not occur naturally. The definition is supplemented by three lists, one of which gives examples of techniques resulting in GMOs, the second lists techniques not resulting in GMOs and the last one determines techniques that are resulting in GMOs but are exempted from the scope of the GMO framework.[65]

---

[61] "genetically modify" is defined as "to change the heritable traits of a plant, animal or microorganism by means of intentional manipulation"; Food and Drug Regulations, C.R.C., c. 870, Para. B.28.001.

[62] Food and Drug Regulations, C.R.C., c. 870, Para. B.28.001.

[63] For this section see Chap. 5 (Country Report on the EU), Sect. 5.3.1.

[64] Cf. Art.4 (1) Directive 2001/18/EC; Art.3 (1) Regulation 1829/2003. Strictly speaking Directive 2009/41/EC, which is regulating the contained use, does, pursuant to Art.2 (b), not apply to GMOs but to genetically modified micro-organisms (so called GMMs). While every GMM is also a GMO within the meaning of the EU regulatory framework, not every GMO is at the same time a GMM.

[65] For a detailed illustration of the European GMO definition see Chap. 5 (Country Report on the EU), Sect. 5.3.1.

Whether part (2) of this GMO definition has a product-based component, has been contentious among scholars.[66] However, the European Court of Justice implied in its judgment on 25 July 2018 that part (2) is to be interpreted as process-based.[67] With part (1), i.e. the required alteration of genetic material, a product-based component is still present but the ECJ seems to let every change suffice without making special demands on the kind of genetic alteration.[68]

8.3.5 Japan[69]

The Japanese regulatory framework for GMOs in form of the so called Cartagena Law applies to the use of a living modified organism (LMO).[70] Therefore, for the regulations to be applicable there must (1) take place a use and (2) the use must relate to a living modified organism.

"Use" is defined very broadly as "use for provision as food, animal feed or other purposes, cultivation and other growing, processing, storage, transportation and disposal, and other acts attendant with these".[71] Consequently, this part of the trigger serves as a single point of entry into the regulatory framework, since it encompasses virtually every kind of interaction with genetically modified plants.

"Living modified organism" is understood as "an organism that possesses nucleic acid, or a replicated product thereof, obtained through use of the [sic!] any of the [stipulated technologies]".[72]

This is, again, a twofold trigger with a product-based aspect ("possesses nucleic acid") and a process-based criterion ("obtained through").

With regard to GM food the Ministry of Health, Labour and Welfare (MHLW) imposed a mandatory safety assessment based on its competence under the Food Sanitation Law.[73] A food or food additives are subject to the risk assessment if they consist at least in part of organisms "produced by recombinant DNA techniques".[74]

---

[66]This aspect is further explained in see Chap. 5 (Country Report on the EU), Sect. 5.3.2.

[67]ECJ, Case C-528/16 *Confédération paysanne and Others* (2018), ECLI:EU:C:2018:583, para. 29.

[68]ECJ, Case C-528/16 *Confédération paysanne and Others* (2018), ECLI:EU:C:2018:583, para. 28.

[69]For this section see Chap. 6 (Country Report on Japan), Sect. 6.3.1.

[70]Cf. Cartagena Law 2003, Art.4 and Art.12. The English translation of the official title of the Cartagena Law is "Act on the Conservation and Sustainable Use of Biological Diversity through Regulations on the Use of Living Modified Organisms (Act No. 97 of 2003)". This is commonly referred to as "Cartagena Law 2003". The Cartagena Law uses the term "living modified organism" instead of "genetically modified organism", as it is an act of implementation of the Cartagena Protocol which uses the term LMO instead of GMO.

[71]Cartagena Law 2003, Art.2 (3).

[72]Cartagena Law 2003, Art.2 (2).

[73]Ministry of Health, Labour and Welfare (2018, 2019).

[74]Ministry of Health and Welfare (2000a).

"Recombinant DNA techniques" is defined as "a technology that DNA cleaved or recombined by, for example, by enzymes, is transferred to living cells for proliferation [sic!]".[75] Even though the regulatory trigger requires only the use of a recombinant DNA technique, it is not a purely process-based one. Since the legal definition of "recombinant DNA techniques" involves the transfer of DNA, a product-based component is present as well.

The legal relationship between the Cartagena Law and the mandatory safety assessment based on the Food Sanitation Law remains to some extent ambiguous. One could make a *lex specialis* assumptions where the mandatory safety assessment supersedes the Cartagena Law. However, the Cartagena Law came into force after the mandatory safety assessment based on the Food Sanitation Law had already been in place and it explicitly applies to the "use for provision as food". Hence, it stands to reason that both legal provisions apply next to each other with regard to food and the mandatory safety assessment stipulates only additional requirements.

8.3.6 USA[76]

With regard to cultivation, the Plant Protection Act allows the Secretary of Agriculture to "prohibit or restrict the importation, entry, exportation, or movement in interstate commerce of any plant [...] if [...] [this] is necessary to prevent the introduction [...] or the dissemination of a plant pest".[77] The department of Agriculture has made use of this competence by issuing specific regulations on the introduction of organisms and products altered or produced through genetic engineering.[78] These regulations require a permit for the introduction of any regulated article.[79] Introduction is rather broadly determined as "[t]o move into or through the United States, to release into the environment, to move interstate, or any attempt thereat".[80] Consequently "introduction" encompasses at least the importation, cultivation and interstate transportation.

A "regulated article" is, inter alia, defined as "[a]ny organism which has been altered or produced through genetic engineering, if the donor organism, recipient organism, or vector or vector agent belongs to any genera or taxa designated in § 340.2 and meets the definition of plant pest [...] or any other organism or product altered or produced through genetic engineering which the Administrator, determines is a plant pest or has reason to believe is a plant pest".[81] This trigger is a

[75]Ministry of Health and Welfare (2000b).

[76]For this section see Chap. 7 (Country Report on the USA), Sect. 7.3.2.

[77]Plant Protection Act, 7 U.S.C. §§ 7701–7786, § 7712 (a).

[78]7 CFR Part 340.

[79]7 CFR Part 340, § 340.0 (a).

[80]7 CFR Part 340, § 340.1.

[81]7 CFR Part 340, § 340.1. This is a significantly shortened version of the definition. For the full definition, see the legal text or Chap. 7 (Country Report on the USA), Sect. 7.3.2.1, fn.64.

combination of a process-based approach ("through genetic engineering"[82]) and a product-based approach ("plant pest"), what makes apparent that the United States regulates gene technology specifically and not just plant pests in general.

Additional requirements apply to genetically modified plants that express pesticidal substances, since it is prohibited to "distribute or sell to any person any pesticide that is not registered".[83] With regard to genetically modified plants, that prohibition applies to such GMOs which have been altered to express insecticides like bacillus thuringiensis (Bt). Plants with so called "plant-incorporated protectants"[84] require, therefore, an additional authorization by the Environmental Protection Agency (EPA).

Pursuant to the Federal Food, Drug, and Cosmetic Act (FD&C Act) "[t]he introduction or delivery for introduction into interstate commerce of any food [. . .] that is adulterated or misbranded"[85] is prohibited. A food is, inter alia, considered to be adulterated "if it bears or contains any added poisonous or added deleterious substance"[86] or "if it is or if it bears or contains any food additive that is unsafe".[87]

With regard to genetically modified plants, it is the transferred genetic material that is considered to be a food additive[88] and can, therefore, be subject to the food additive regulation.[89] Food additives are deemed to be unsafe and therefore prohibited if they have not been granted an approval or are exempt from approval.[90] Excluded from the definition of "food additive" and as a consequence exempt from approval are substances that are "generally recognized as safe" (GRAS).[91] The competent Food and Drug Administration (FDA) generally considers material from genetically modified plants as GRAS.[92] Consequently, food containing

[82]"Genetic engineering" is in that context defined as "genetic modification of organisms by recombinant DNA techniques"; 7 CFR Part 340, § 340.1.

[83]Federal Insecticide, Fungicide, and Rodenticide Act, 7 USC §§ 136-136y, §136a (a).

[84]"Plant-incorporated protectant" is defined as "a pesticidal substance that is intended to be produced and used in a living plant, or in the produce thereof, and the genetic material necessary for production of such a pesticidal substance. It also includes any inert ingredient contained in the plant, or produce thereof"; 40 C.F.R. Part 174, § 174.3.

[85]Federal Food, Drug, and Cosmetic Act, 21 U.S.C. §§ 301 - 399h, § 331 (a).

[86]Federal Food, Drug, and Cosmetic Act, 21 U.S.C. §§ 301 - 399h, § 342 (a) (2) (A).

[87]Federal Food, Drug, and Cosmetic Act, 21 U.S.C. §§ 301 - 399h, § 342 (a) (2) (C) (i). The definition has been shortened considerably. For the complete definition, see the legal text.

[88]"Food additive" is defined as "*any substance the intended use of which results or may reasonably be expected to result, directly or indirectly, in its becoming a component or otherwise affecting the characteristics of any food [. . .], if such substance is not generally recognized [. . .] to be safe under the conditions of its intended use*"; Federal Food, Drug, and Cosmetic Act, 21 U.S.C. §§ 301 - 399h, § 321 (s).

[89]Cf. Food and Drug Administration (1992), p. 22990.

[90]Federal Food, Drug, and Cosmetic Act, 21 U.S.C. §§ 301 - 399h, § 348 (a).

[91]Federal Food, Drug, and Cosmetic Act, 21 U.S.C. §§ 301 - 399h, § 321 (s).

[92]Food and Drug Administration (1992), p. 22990.

materials of genetically modified plants are not subject to a mandatory premarket review by the FDA.

However, food is also considered to be adulterated "if it bears or contains a pesticide chemical residue that is unsafe".[93] This becomes relevant for food that is produced from genetically modified plants with plant-incorporated protectants (see above). Such food can only be sold if a tolerance level is met or an exemption is granted by the EPA.[94]

8.3.7 Comparative Analysis

When comparing the different regulatory frameworks, it becomes apparent that most of them rely as a trigger on varying definitions of genetically modified organism. The only noteworthy exceptions are the Canadian regulations for the cultivation of plants and the US trigger for food (cf. Table 8.2).

When examining the regulatory frameworks from the perspective of whether a product-based or a process-based regulatory trigger is used, only two kinds of systematic approaches can be identified: (1) a combination of a product- and a process-based trigger or (2) a solely product-based trigger (cf. Table 8.3)

However, if one considers the regulatory regime as a whole, it becomes obvious that none of the examined legal frameworks is purely a product-based one. Eventually, each of the analysed regulatory concepts will also fall back on the process of genetic modification to determine whether additional regulatory requirements apply.

Consequently, none of the analysed regulatory frameworks abstains from using regulations specifically tailored for GMOs. Only the United States and Canada provide in part a regulatory regime which is generally applicable to GMOs and non-GMOs or products derived from them (cf. Table 8.4).

It is hard to draw any definitive conclusion from the kind of regulatory trigger used. In principle, it can be assumed that a general or product-based regulatory approach is advantageous for the adoption of GMOs, since in that case no stricter general rules can apply to GMOs compared to conventionally bred plants. This notion is backed by the rather embracing attitude of Canada and the USA towards gene technology in agriculture.[95]

However, this is only an indicator, since the high adoption rate of GMOs in Argentina[96] shows that a process-based or GM specific approach can be designed in a permissive way as well.

This shows that based alone on the type of regulatory trigger no meaningful conclusions can be made with regard to the regulatory regime. The common used

---

[93]Federal Food, Drug, and Cosmetic Act, 21 U.S.C. §§ 301 - 399h, § 342 (a) (2) (B).

[94]Federal Food, Drug, and Cosmetic Act, 21 U.S.C. §§ 301 - 399h, § 346 (a).

[95]Cf. Sect. 8.2.

[96]Cf. Fig. 8.4 (Sect. 8.2.1.2), Fig. 8.6 (Sect. 8.2.1.2) and Fig. 8.7 (Sect. 8.2.1.2).

Table 8.2 Used regulatory trigger

	Argentina	Australia	Canada	EU	Japan	USA
Cultivation	GMO definition	GMO definition	Novel trait	GMO definition	LMO[a] definition	1) Genetic engineering + Plant pest definition 2) Plant-incorporated protectants
Food	GMO definition	Gene technology definition	Novel food	GMO definition	1. LMO definition 2. Recombinant DNA techniques	Adulterated food

[a]Cf. the explanation in Fn.66

distinction between a process-based or product-based approach lacks, therefore, considerably in value.

More or less the same can be said about the distinction between a GM specific regulatory regime and a general framework. This is mainly due to the fact that a process-based approach is at the same time always GM specific by nature and only a purely product-based regulatory regime can be classified as generally applicable.

A different kind of insight into the regulatory regime provides a closer look at the fact whether it has the same trigger for different forms of use of GMOs (single point of entry) or whether different triggers apply to different uses (multiple points of entry). At first glance it can be assumed that a single point of entry regulatory regime is advantageous for the adoption of GMOs, since it provides a more streamlined regulatory process (cf. Argentina). However, a single point of entry regulatory regime can also serve as a bottleneck to prevent the widespread adoption of GMOs. This is indicated by the fact that with Japan and the European Union two reluctant adopters of GMOs make use of a single point of entry system (cf. Table 8.5). Even though in the case of Japan, an additional trigger applies to the regulation of food, the uniform trigger of the Cartagena Law is still applicable, what justifies qualifying it as single point of entry.

It can, therefore, finally be concluded that the type of trigger can have a comparatively low impact on the actual effect the regulatory framework has on the adoption of GMOs. If it shall be explained why the adoption of GMOs in agriculture differs so much from country to country, other factors like the actual application of the regulatory framework or the political and social environment have to be taken into account as well.

8.4 Regulatory Status of Genome Edited Plants

After identifying the respective regulatory triggers, it is now possible to determine how plants derived from genome editing are classified by each regulatory framework.

Table 8.3 Product- or process-based regulatory trigger

	Argentina	Australia		Canada		EU	Japan	USA	
		Cultivation	Food	Cultivation	Food			Cultivation	Food
Process-based	✓	✓	✓	×	✓	✓	✓	✓	×
Product-based	✓	✓	×	✓	✓	✓	✓	✓	✓

Table 8.4 Specific GM or general trigger

	Argentina	Australia	Canada		EU	Japan	USA	
			Cultivation	Food[a]			Cultivation[b]	Food
Specific	✓	✓	×	✓	✓	✓	✓	×
General	×	×	✓	✓	×	×	✓	✓

[a]Of the three independent and distinct definitions of "novel food" only one refers specifically to a genetic modification. Therefore, while the definition of "novel food" as trigger is overall general in nature, it comprises at the same time a specific provision for GMOs
[b]With regard to a restriction of plant cultivation, the US regulatory framework uses as general trigger the existence of a plant pest or the threat thereof. However, within this general approach a specific trigger for genetically modified plants exists. Cf. Sect. 8.3.6

Table 8.5 Single point of entry or multiple point of entry

	Argentina	Australia	Canada	EU	Japan	USA
Single point of entry	✓	×	×	✓	✓	×
Multiple points of entry	×	✓	✓	×	(✓)	✓

The comparative analysis will again concentrate on genome edited plants intended for cultivation and food containing genome edited material.

8.4.1 Argentina[97]

The method how Argentine regulators determine whether a plant derived from genome editing is considered a GMO under the current regulatory framework is illustrated in Fig. 2.1 (Sect. 2.3.2).

As a result, plants derived from ODM, SDN-1 and SDN-2 are not considered to be GMOs. In general the application of SDN-3 will result in a GMO but with the one exception of a "perfect allelic replacement".

Since the Argentine GMO regulatory framework does not distinguish between different types of use, the same applies to food (Table 8.6).

8.4.2 Australia[98]

With regard to plants, the Australian regulator (Gene Technology Regulator, GTR) has adopted an interim approach, until the 2019 amendments to the regulatory framework will come into force in October 2019. Plant varieties derived from

[97]For this section see Chap. 2 (Country Report on Argentina), Sect. 2.3.2.
[98]For this section see Chap. 3 (Country Report on Australia), Sect. 3.3.2.

Table 8.6 Classification of plants and food derived from ODM, SDN-1, SDN-2 or SDN-3 in Argentina

	ODM	SDN-1	SDN-2	SDN-3
GMOs	×	×	×	✓

Table 8.7 Interim classification of plants derived from ODM, SDN-1, SDN-2 or SDN-3 in Australia

	ODM	SDN-1	SDN-2	SDN-3
GMO plants	✓	×	✓	✓

Table 8.8 Interim classification of food derived from ODM, SDN-1, SDN-2 or SDN-3 in Australia

	ODM	SDN-1	SDN-2	SDN-3
Food produced using gene technology	×	×	×	✓

SDN-1 are considered to be non-GMOs while an application of the ODM, SDN-2 and SDN-3 techniques results in GMOs (cf. Table 8.7).

With regard to food derived from genome edited plants, the responsible Australian regulator (Food Standards Australia New Zealand, FSANZ) has not yet made an official determination. However, FSANZ has indicated that it will take into account the proposed interim approach of two scientific workshops held in 2012 and 2013. As part of that interim approach, only SDN-3 is considered to lead to food that fulfils the definition of "produced using gene technology" (cf. Table 8.8).

8.4.3 *Canada*[99]

Since the Canadian regulatory trigger regarding the cultivation of plants is the novel trait of a new plant variety, no definitive determination can be made concerning individual genome editing techniques. Whether the stricter rules for plants with novel traits apply, does not depend on the technique used but whether a novel trait is expressed in the new plant variety.

With regard to the marketing of food containing material of genome edited plants the Canadian regulator has not yet made any determinations, since market approval for such products has not yet been sought. Keeping in mind that the trigger of Canadian regulation of novel food has a process-based component ("has been genetically modified"), it does not seem completely implausible that certain genome

[99]For this section see Chap. 4 (Country Report on Canada), Sect. 4.3.2.

editing techniques are in general not covered by this trigger. This will finally depend on the interpretation of "genetically modify"[100] and whether the genome editing technique in question falls within the scope of that definition.

8.4.4 *EU*[101]

Since the ruling of the European Court of Justice (ECJ) in July 2017[102] at least the regulatory classification of genome edited crops created via ODM and SDN-1 is clear-cut.[103] Due to the process-based interpretation of the ECJ and by means of an inductive reasoning, it can be assumed that all genome editing techniques result in GMOs—as long as they cause any genetic alteration—and trigger, therefore, the European regulatory framework for genetically modified organisms.[104]

The same applies to food derived from genome edited plants because of the single-point of entry the European GMO framework uses (Table 8.9).

8.4.5 *Japan*

Since Japan's regulator had not made a legal classification regarding genome edited crops, their regulatory status was characterised by legal uncertainty.[105] Based on the

[100]"Genetically modify" is defined as "to change the heritable traits of a plant, animal or microorganism by means of intentional manipulation"; Food and Drug Regulations, C.R.C., c. 870, Sec. B.28.001.

[101]For this section see Chap. 5 (Country Report on the EU), Sect. 5.3.2.

[102]ECJ, Case C-528/16 *Confédération paysanne and Others* (2018), ECLI:EU:C:2018:583.

[103]Strictly speaking, only herbicide-resistant plants created via ODM or SDN-1 have been subject to the ruling of the ECJ. The ruling of the ECJ refers to "techniques/methods of mutagenesis such as those at issue in the main proceedings"; ECJ, Case C-528/16 *Confédération paysanne and Others* (2018), ECLI:EU:C:2018:583, Para. 28. In the main proceedings it is only referred to ODM and SDN-1; Conseil d'État, n°388649 *Confédération paysanne et autres* (2016), Para. 23. It should be noted, however, that ODM and SDN-1 are here only mentioned as examples for modern directed mutagenesis using genetic engineering techniques (cf. the use "*notamment*"; engl. "including" or "in particular"). The opinion of the Advocate General referred also only exemplary ("such as") to ODM and SDN-1; Opinion of the Advocate General Bobek, Case C-528/16 *Confédération paysanne and Others* (2018), Para. 46. Therefore, it can be argued that SDN-2 is directly covered as well by the judgment since SDN-2 can also be understood as a mutagenesis technique due to its close resemblance to ODM (both cause small changes to the DNA based on a template without incorporating foreign DNA into the genome).

[104]Cf. the general assessment of the EU's GMO definition above (Sect. 8.3.4). Since the ECJ ruled in paragraph 29 of the judgment in the case 528/16 that at least SDN-1 and ODM "alter the genetic material of an organism in a way that does not occur naturally" just based on the process used, the same is true *a fortiori* for SDN-2 and SDN-3 when applying the Court's reasoning.

[105]For this section see Chap. 6 (Country Report on Japan), Sect. 6.3.2.

Table 8.9 Classification of plants and food derived from ODM, SDN-1, SDN-2 or SDN-3 in the EU

	ODM	SDN-1	SDN-2	SDN-3
GMOs	✓	✓	✓	✓

suggestions of two expert meetings on the handling of genome editing technology held in August 2018 by Japan's Ministry of Environment[106] and the conclusions drawn by an advisory panel of Japan's Ministry of Environment,[107] the ministry released its policy regarding the regulation of genome edited crops in February 2019.[108] This policy is, however, not an amendment or update of the regulatory framework but only an authoritative interpretation of the current Cartagena Law (and the related provisions) with respect to genome edited plant varieties. Pursuant to that policy, genome edited crops are not considered to be LMOs and therefore not subject to stricter regulation if no nucleic acid that was processed extracellularly has been inserted or no such nucleic acid is anymore present in the final organism.[109] Furthermore, excluded are still processes using nucleic acid of an organism belonging to the same species as that of the target organism or nucleic acid of an organism belonging to a species that exchanges nucleic acid with the species of the target organism.[110] Considering all this, conclusions can be drawn for the individual genome editing techniques. Keeping in mind that the genomic changes induced by means of SDN-1, SDN-2 and ODM are indistinguishable from naturally occurring alterations, plants obtained in that way will not be considered to be LMOs as long as no foreign nucleic acid, e.g. a vector's DNA or a nuclease's sgRNA, is integrated into the host genome. With regard to SDN-3 the classification will depend on the kind of DNA that has been inserted into the host genome. If nucleic acid of an organism belonging to the same species or of an organism belonging to a species that exchanges nucleic acid with the species of the target organism has been integrated into the genome, the resulting plant will not be classified as LMO. Again, this assumption is only valid in case that no foreign nucleic acid was incorporated during the genome editing process. If SDN-3 is, however, used to integrate a foreign gene into the host's genome, the resulting organism is considered is to be a LMO (cf. Table 8.10).

This interpretation applies, however, not to the additional regulatory requirements applicable to food produced via "recombinant DNA techniques". To clarify whether those provisions apply to genome editing as well, the competent Ministry of Health, Labour and Welfare started a similiar process as described above. Based on suggestions of the Research Sub-Committee for Genetically Modified Food[111] and after a

[106] Sato (2018a, 2018b).

[107] Kurai and Sato (2018).

[108] Sato (2019a).

[109] Sato (2019a), p. 2.

[110] Cf. Chap. 6 (Country Report on Japan), Sect. 6.2.

[111] Sato (2018c, d, e).

Table 8.10 Assumed classification of plants and food derived from ODM, SDN-1, SDN-2 or SDN-3 in Japan

	ODM	SDN-1	SDN-2	SDN-3
GMOs according to the Cartagena Law	×	×	×	✓/×
GMOs according to the GM food regulations	×	×	×	✓

public comment period,[112] the ministry released its policy regarding food derived from genome editing in March 2019. In general, genome edited food containing transgenic genes is considered to be "produced by recombinant DNA techniques" (cf. Sec. 8.3.5) and therefore subject to the GM food regulation of the MHLW.[113] However, if there are no transgenic genes present in the final food product and only (1) a base-pair deletion, (2) a substitution, (3) a naturally occurring gene deletion and/or (4) insertion of one to several base pairs has been induced via genome editing, the MHLW's GM food regulations do not apply.[114] In the end, these criteria are less different from those applied by the Ministry of Environment (cf. paragraph above) than they appear at first glance, since "transgenic" is only a paraphrase for the fact that the gene belongs to a species that in nature does not exchanges nucleic acid with the species of the target organism. Accordingly, the only difference is the additional limitation to the four specific genome editing results mentioned above. A closer look at these limitation reveals that SDN-1 (base deletions, substitutions, insertions) and SDN-2 or ODM (insertions of up to several base pairs) seem to fall within their scope while SDN-3 (gene insertions) does not. This has the consequence that even if a gene of a sexually compatible species is inserted via SDN-3 the GM food regulations apply. However, even if the genome edited food in question is not subject to the GM food regulations, it can be subject to a safety review like a GM product if the developer cannot confirm that there is no production of an allergenic or toxic substance due to off-target mutations.[115]

8.4.6 USA[116]

Whether genome edited plants are considered to be regulated articles under USDA regulations, depends on whether they fulfill the corresponding definition of a "regulated article". In a nut-shell, the classification of genome edited plants depends on whether (1) the organism has been altered or produced through genetic engineer-

[112]Sato (2019b).

[113]Sato (2019c), p. 3.

[114]Sato (2019c), pp. 3–4.

[115]Sato (2019c), p. 5.

[116]For this section see Chap. 7 (Country Report on the USA), Sect. 7.8.

Table 8.11 Classification of plants derived from ODM, SDN-1, SDN-2 or SDN-3 in the USA[a]

	ODM	SDN-1	SDN-2	SDN-3	
Regulated article	×	×	×	✓	×

[a]This classification is only valid under the assumption made in Sect. 8.4.6

ing and (2) whether the donor organism, recipient organism, or vector or vector agent meets the definition of plant pest.

With regard to (1), the narrow definition of genetic engineering, which encompasses only "genetic modification of organisms by recombinant DNA techniques",[117] does not exclude genome edited plants in general, since some genome editing approaches make use of rDNA while others do not.[118]

Aspect (2) of the regulatory trigger does not allow an unambiguous determination, since genome editing can resort to various vectors (viral, non-viral and physical vectors).[119] As one of the non-viral vectors Agrobacterium can be used,[120] which is regarded as constituting a plant pest by USDA.

To tackle this regulatory uncertainty USDA published a statement[121] and further information on its website[122] concerning its handling of innovative plant breeding. As a general rule USDA stated that it does not regulate "plants that could otherwise have been developed through traditional breeding techniques as long as they are developed without the use of a plant pest as the donor or vector and they are not themselves plant pests."[123] This includes deletions, single base pair substitutions, insertions from compatible plant relatives and complete null segregants.[124]

Therefore, under the assumption that (1) neither the donor nor the recipient organism are plant pests, (2) that either no plant pest was used as a vector or that such a plant pest vector is no longer present in the final product and (3) that the plant has no pesticidal properties[125] the following classifications can be made: plants derived from ODM, SDN-1 and SDN-2 are not considered to be regulated articles, since the genetic alterations could also be induced by conventional breeding techniques or occur in nature. With regard to SDN-3 a classification on a case-by-case basis is required to determine whether the individual application of SDN-3 falls within the definition of a "regulated article" (Table 8.11).

Regarding the regulation of food, the FDA has not yet made a determination on the classification of foodstuff derived from genome edited plants and their products.

[117]7 CFR Part 340, § 340.1.

[118]Wolt et al. (2016), p. 511.

[119]Cf. Li et al. (2018), pp. 209–212; Lino et al. (2018), p. 1241; Li et al. (2015), pp. 453–458.

[120]Ma et al. (2017).

[121]United States Department of Agriculture (2018a).

[122]United States Department of Agriculture (2018c).

[123]United States Department of Agriculture (2018c).

[124]United States Department of Agriculture (2018c).

[125]Otherwise, the EPA could have regulatory competences. Cf. Chap. 7 (Country Report on the USA), Sect. 7.8.1.

Table 8.12 Assumed classification of food derived from ODM, SDN-1, SDN-2 or SDN-3 without plant-incorporated protectants in the USA

	ODM	SDN-1	SDN-2	SDN-3
Adulterated food	×	×	×	×

With regard to ODM, SDN-1 and SDN-2, it seems already questionable if FDA has the competence to regulate, since no stable integrated transferred genetic material, which could be considered to be a food additive, is present in the final end product.[126] Even if the FDA should not share this point of view, it is likely that food derived from genome editing is considered to be GRAS and therefore not subject to a mandatory premarket review by the FDA.

However, since food is also considered to be adulterated "if it bears or contains a pesticide chemical residue that is unsafe",[127] food that is made out of genome edited plants with plant-incorporated protectants can be considered as adulterated (Table 8.12).

8.4.7 *Comparative Analysis*

When comparing the different approaches (cf. Table 8.13), it becomes apparent that the Canadian regime has a special status among them. Due to the solely product-based approach with regard to the cultivation of plants, an ex ante classification of a genome edited plant based on the used technique is not possible in the case of Canada.

Having a closer look at the spectrum of classifications, it is striking that while at the one end the EU classifies all genome editing techniques as leading to GMOs, the USA at the other end considers only certain applications of SDN-3 as covered by the GMO framework.

Besides that, the other countries adopt an approach which lies somewhere in-between of those two.

Since the regulatory framework for GMOs is usually more restrictive than that for non-GMOs, the classification of genome edited plants as non-GMOs is an indicator for the permissiveness towards new breeding technologies. So it comes as no surprise that it is the regulatory landscape of Argentina and the USA which does not impose the stricter rules for traditional GMOs on most genome edited plants and products derived from them. This could be expected since the USA and Argentina are among the countries with the highest acreage of cultivated GMOs[128] and it stands to reason that they are open to adopt GEOs on a wide scale as well.

[126] However, foreign genetic material is present in the plant in the intermediate steps.

[127] Federal Food, Drug, and Cosmetic Act, 21 U.S.C. §§ 301 - 399h, § 342 (a) (2) (B).

[128] Cf. Fig. 8.4.

Table 8.13 Classification of plants derived from ODM, SDN-1, SDN-2 or SDN-3 as GMOs

	Argentina[a]	Australia		Canada		EU	Japan[c]	USA[b]	
		Cultivation	Food	Cultivation	Food			Cultivation	Food
ODM	×	✓	×	–	?	✓	×	×	×
SDN-1	×	×	×	–	?	✓	×	×	×
SDN-2	×	✓	×	–	?	✓	×	×	×
SDN-3	✓	✓	✓	–	?	✓	✓/×	× ✓	×

[a]In the case of Argentina, a "perfect allelic replacement" via SDN-3 is likely to be not considered to lead to a GMO. Cf. Chap. 2 (Country Report on Argentina), Sect. 2.3.2

[b]With regard to the classification in the USA, assumptions have been made which limit the scope of application of the regulatory regime. For those assumptions see Sect. 8.4.6

[c]Concerning Japan only the classification pursuant to the Cartagena Law is displayed here

Table 8.14 Countries in which an official decision on the regulatory status of genome edited plants has already been made

	Argentina	Australia	Canada	EU	Japan	USA
Decision made	✓	✓	✓	✓	✓	✓

Table 8.15 Countries in which an official decision on the regulatory status of food derived from genome edited plants has already been made

	Argentina	Australia[a]	Canada	EU	Japan	USA
Decision made	✓	×	×	✓	✓	×

[a]The Australian interim approach, which has been illustrated above, is not based on an official decision but on the outcomes of two scientific workshops. FSANZ indicated, however, that it would base its decision on the outcome of the workshops

Of the examined countries, all have made an official determination concerning the status and legal classification of genome edited plants (cf. Table 8.14).

However, only Argentina, the EU and Japan have so far made a decision on the legal classification of food which contains material that is derived from genome edited plants (cf. Table 8.15). This is not surprising, as these countries use a single point of entry into their regulatory regime (cf. Table 8.5). Therefore, a decision on the regulatory status of genome edited plants also applies to food.

Since Australia, Canada and the USA use a multiple point of entry into their regulatory frameworks (cf. Table 8.5), the classification made with regard to genome edited plants cannot be extrapolated to food derived from genome edited plants. A determination of the status of such foodstuff is, therefore, likely to happen as soon as marketing is imminent.

8.5 Labelling

Whether products containing genetically modified or genome edited substances have to be labelled, is very differently regulated in the examined countries. In order to ensure a clearer comparison, only the labelling regulations for foodstuff are analyzed below. How other product categories are handled with regard to labelling is described in the individual country reports.

8.5.1 Argentina[129]

There is no legal obligation to label GMOs in Argentina. Since GMO products are treated as being equivalent to their conventional counterparts, labelling of those products is considered to be potentially misleading.

[129]For this section see Chap. 2 (Country Report on Argentina), Sect. 2.8.

8.5.2 *Australia*[130]

In Australia the labelling of all "genetically modified food" is mandatory unless an exemption applies. Interestingly a different trigger applies to labelling ("genetically modified food") than to the marketing authorization of food derived from genetically modified crops ("food produced using gene technology"). Decisive for the labelling requirement is whether the food contains "DNA or protein which, as a result of the use of gene technology, is different in chemical sequence or structure from DNA or protein present in counterpart food that has not been produced using gene technology".[131]

Since that trigger is narrower than the trigger for marketing authorization of food, it can be the case that food which is subject to the GMO framework is at the same time not subject to mandatory labelling. Conversely, foodstuff to which the GMO regulatory framework does not apply are by no means subject to mandatory labelling. Since only food containing material from plants derived from certain applications of SDN-3 is likely to be covered by the GMO regulatory framework, it stands to reason that only food containing material from plants derived from SDN-3 that fulfills the definition of "genetically modified food" and is not exempted has to be labelled.

Food can contain up to 1% of unintended presence of otherwise approved GM material without requiring labelling.

8.5.3 *Canada*[132]

Canada has no mandatory labelling system for GMOs in place and does, therefore, not require that GEOs are labelled.

8.5.4 *EU*[133]

In the EU, the labelling of food produced from or containing GMOs is mandatory. "GMO" is defined in exactly the same way as with regard to the authorization of cultivation or the marketing of food.[134] Therefore, the trigger for labelling is the same as for the authorization of food. Consequently, every authorized GM food has to be labelled.

---

[130]For this section see Chap. 3 (Country Report on Australia), Sect. 3.8.

[131]Australia New Zealand Food Standards Code, Standard 1.5.2, Sec. 4 (5).

[132]For this section see Chap. 4 (Country Report on Canada), Sect. 4.8.

[133]For this section see Chap. 5 (Country Report on the EU), Sect. 5.8.

[134]Cf. Regulation (EC) No 1829/2003, Art.12 (1) in conjunction with Art.2.

Food can contain up to 0.9% of unintended presence of otherwise approved GM material without requiring labelling.

8.5.5 *Japan*[135]

Only GM food that is explicitly mentioned in the labelling provisions is subject to mandatory labelling while other GM foodstuff is exempted from labelling requirements.[136] However, even the listed food products can contain up to 5% of unintended presence of otherwise approved GM material without requiring labelling.

Whether food containing material from genome edited plants is subject to a mandatory labelling, depends, therefore, on whether the genome edited plant is considered to be a GMO and whether the crop or processed product thereof is listed as requiring labelling.

8.5.6 *USA*[137]

In 2016, the USA passed a mandatory labelling law for GM food called National Bioengineered Food Disclosure Standard. Previously there was no labelling requirement for food containing material from genetically modified plants at the federal level. In December 2018 USDA implemented the disclosure standard by issuing final regulations.[138] As part of the implementation USDA was inter alia tasked to specify the definitions used and to set a threshold of unintended GM presence up to which labelling is not required.

Whether and, if so, which food derived from genome edited plants is going to be covered by the labelling standard, depends on the interpretation of the labelling trigger: "bioengineered food". "Bioengineered food" is defined in the disclosure standard as "food (A) that contains genetic material that has been modified through in vitro recombinant deoxyribonucleic acid (DNA) techniques; and (B) for which the modification could not otherwise be obtained through conventional breeding or found in nature".[139] The relation of part A to part B ("and") seems to be cumulative in nature,[140] so that a food must fulfill part A and B at the same time to be considered as "bioengineered food". With regard to part A USDA clarified that food with

[135]For this section see Chap. 6 (Country Report on Japan), Sect. 6.6.

[136]For a list of the crops and the processed products made out of them that are subject to mandatory GMO labelling see Sato (2016), p. 23.

[137]For this section see Chap. 7 (Country Report on the USA), Sect. 7.6.

[138]United States Department of Agriculture and Agricultural Marketing Service (2018).

[139]National Bioengineered Food Disclosure Standard, 7 USC §§ 1639-1639c, § 1639 (1).

[140]Since part B refers explicitly to "the modification" described in part A, it is clear that both parts of the definition must be read together.

undetectable modified genetic material does not constitute bioengineered foods.[141] This does not say anything specifically about the applicability to food derived from genome edited plants, since the genetic modification is always detectable, even though the source of such a change is not always identifiable.[142]

Against the background of genome edited plants, it is the interpretation of part B which will be decisive for the applicability of the labelling requirement. Unfortunately, the USDA explicitly refrained from any clarifications on the meaning of "found in nature" because it found "it unnecessary to define the term".[143] With regard to genome editing genetic alterations induced by means of ODM, SDN-1 and SDN-2 can theoretically also be obtained through conventional breeding or found in nature.[144] Therefore, it is likely that food containing material of such genome edited plants does not require labelling. With regard to SDN-3, a case-by-case assessment is probably required to determine whether the modification can be obtained through conventional breeding as well (cf. part B of the definition) and whether rDNA was used in the process (cf. part A of the definition).

Exempted from the labelling requirement is food with an inadvertent or technically unavoidable GM presence of up to five percent for each ingredient.[145]

8.5.7 Comparative Analysis

The majority of the examined countries has a mandatory labelling system for GM food in place (cf. Table 8.16). Only Argentina and Canada abstain from the use of a labelling requirement for genetically modified foodstuff.

Generally speaking, those countries that embrace the cultivation of GMOs refrain from a mandatory labelling requirement of GM food. This can be interpreted as an indicator that mandatory labelling is prevalent in countries where a negative attitude towards GMOs exists. However, the USA is now an important exception, after a decision was taken in 2016 to introduce a labelling obligation.

Since Argentina and Canada do not require a mandatory labelling, there is also no threshold for unintended GM presence that could apply (cf. Table 8.17). Noteworthy is the case of Japan and the USA that allow up to 5% of adventitious presence of GM material in food without requiring labelling as GM food. Compared to the threshold

---

[141] United States Department of Agriculture and Agricultural Marketing Service (2018), p. 65816.

[142] For the distinction between "detection," "identification," and "traceability" see Hamburger (2018), section "Coexistence Measures and Identity Preservation Systems". Nonetheless, a genetic alteration might be no longer detectable if the produce is processed or refined in a certain way. However this is not an issue that is specifically linked to genome edited plants but applies to all genetically modified crops and their products.

[143] United States Department of Agriculture and Agricultural Marketing Service (2018), p. 65818.

[144] Voigt and Klima (2017), p. 321.

[145] United States Department of Agriculture and Agricultural Marketing Service (2018), p. 65872.

Table 8.16 Mandatory labelling of food containing genetically modified substances

	Argentina	Australia	Canada	EU	Japan	USA
GMO mandatory labelling	×	✓	×	✓	✓	✓

Table 8.17 Threshold of unintended presence of GM material

	Argentina	Australia	Canada	EU	Japan	USA[a]
Labelling threshold	–	1%	–	0.9%	5%	5%

[a]The threshold has not yet been set, but the competent authority is currently checking various options. Cf. Chap. 7 (Country Report on the USA), Sect. 7.6

used in other jurisdictions, this high tolerance could make it possible for an above-average number of products to avoid being labelled.[146]

The various threshold levels in place hinder the free circulation of goods between national markets and is, therefore, an obstacle to free trade, since, for example, a product that contains GM material below the threshold of the Japanese regulatory system might still be subject to European mandatory labelling due to the far lower tolerance level. This was probably also one of the reasons why the US struggled to define its tolerance threshold for the newly introduced labelling requirement.

The existence of a mandatory GM labelling requirement raises the question whether food derived from genome edited crops is subject to those labelling provisions as well. In the case of Argentina and Canada the situation is clear-cut, since no mandatory labelling exists at all. An equally unambiguous but opposite approach pursues the EU where every food containing genome edited material is subject to mandatory labelling. The other countries opted to apply a differentiating path by requiring that only certain genome edited food is labelled (Table 8.18).

8.6 Identity Preservation System (Coexistence)

The terms "identity preservation" and "coexistence" are used here only to refer to measures that are deemed to ensure farmers', manufacturers', sellers' and consumers' freedom of choice. They do not encompass the segregation of GM and non-GM crops due to security considerations (e.g. special rules for field trials), but only address crops which are approved and therefore considered to be safe.

To ensure the freedom of farmers to choose which kind of crop to grow on their field, the freedom of manufacturers which kind of produce to process, the freedom of sellers which products to sell and the freedom of consumers which products to buy, it must be prevented that GMO and non-GMO product lines mix with each other. This

[146]There exist, however, indicators that a higher threshold of 5% has only a rather small impact on the number of products which are requiring labelling. Cf. Viljoen and Marx (2013), p. 389; Oh and Ezezika (2014), p. 11.

Table 8.18 Mandatory labelling of food containing genome edited substances

	Argentina	Australia	Canada	EU	Japan		USA	
ODM	×	×	×	✓	×		×	
SDN-1	×	×	×	✓	✓	×	×	
SDN-2	×	×	×	✓	✓	×	×	
SDN-3	×	✓	×	✓	✓	×	✓	×

could happen during cultivation by cross-pollination, through wind or bees, during harvest by contaminated equipment and during processing or transportation by (un) intentional mixture.

The term ‘coexistence’ is used hereinafter only for measures applied in the period from sowing to harvest and intended to ensure the coexistence of different plant organisms. Coexistence measures are, for instance, isolation distances or buffer zones between different crops, a required approval from neighboring farmers if the minimum isolation distance is not respected, information duties (registration of areas in database, prior information to authorities or neighbors), staggered sowing (different plant cycles and rotation intervals of sexually compatible GM and non-GM crops), and the cleaning/separation of equipment or obligatory insurances.[147]

An identity preservation system, as understood here, ensures that the segregation established by coexistence measures is maintained after the harvest until the product reaches the end-consumer. This is achieved, inter alia, with the help of an end-to-end paper trail, segregated production facilities, separate storage, and testing procedures.[148]

8.6.1 Argentina[149]

In Argentina no mandatory coexistence or identity preservation provisions exist. As soon as a GM crop variety is approved for cultivation and marketing, they are considered as safe and any mandatory segregation measures consequently regarded as unnecessary.

Voluntary coexistence or identity preservation measures are possible. However, the implementation is the exclusive responsibility of those who benefit from them and the profiteers must also bear the costs.

In concrete terms, this means that conventional farmers, who want to protect their products from mixture with GM material, must take steps on their own to prevent admixture, since they have an interest in doing so in order to ensure a price premium on the market. The same holds true for GMO farmers that require a certain purity of

[147]Schenkelaars and Wesseler (2016), pp. 6–8; Lee (2014), p. 244; Beckmann et al. (2014), p. 376.

[148]Kumar and Sopory (2008), p. 306; Smyth et al. (2004), p. 140; Wiseman (2009), p. 257. This and the previous paragraph was taken from Hamburger (2018).

[149]For this section see Chap. 2 (Country Report on Argentina), Sect. 2.9.

their products—for example, because the special trait of the GM crop is used for pharmaceutical manufacturing.

8.6.2 Australia[150]

Australia has no general rules on coexistence or identity preservation measures. Instead, individually required measures to prevent the unintended presences of GM material are laid down in the specific licence to grow a GM crop.[151] At the state and territory level, further mandatory requirements can be imposed on GM crop adopters.

8.6.3 Canada[152]

Canada has no mandatory coexistence or identity preservation system in place. If a producer wishes to preserve the purity of a produce, he has to take the necessary measures himself and bears the financial burden.[153]

8.6.4 EU[154]

In the EU, the member states retained the competence to set the rules for coexistence and identity preservation measures. Therefore, the individual requirements can vary significantly from member state to member state. The European Commission, however, issued a general guidance for the design of appropriate measures and several plant-specific guidance documents.

Responsible for the implementation of these measures are the users of GMOs and not the individual profiteer.

---

[150]For this section see Chap. 3 (Country Report on Australia), Sect. 3.9.

[151]Crothers (2017), p. 20.

[152]For this section see Chap. 4 (Country Report on Canada), Sect. 4.9.

[153]Danielson and Watters (2017), pp. 15–16.

[154]For this section see Chap. 5 (Country Report on the EU), Sect. 5.9.

8.6.5 Japan

The central government of Japan issued only segregation rules for field trials.[155] Since those are rather based on security concerns during the trial stage, the primary objective is not to ensure freedom of choice. Therefore, those guidelines do not stipulate coexistence or identity preservation measures, as they are understood here.

However, at the local level a wide variety of different coexistence and identity preservation measures are in place. In some regions, farmers who wish to grow GM crops must inform their neighbours prior to any cultivation and in a few instances neighbouring farmers are even required to consent. Other regional governments impose burdensome administrative requirements on a GM farmer and require additional distance spaces.[156]

8.6.6 USA[157]

In the United States, no mandatory obligations exist to ensure coexistence or identity preservation of agricultural products. Consequently, any measures to prevent the commingling of GMOs with non-GMOs are voluntarily and must be carried out by those who want to benefit from them.

8.6.7 Comparative Analysis

While Australia, the EU, and Japan have mandatory coexistence or identity preservation provisions in place, Argentina, Canada, and the USA rely on voluntary measures (cf. Table 8.19).

The same allocation of the examined countries can be made when looking at the cultivation of GM crops (cf. Fig. 8.7 in Sect. 8.2.1.2). This at least indicates that there is a connection between mandatory coexistence measures and the extent to which GM crops are cultivated. That correlation can be explained on the one hand by the fact that farmers refrain from using GM crops because required buffer zones cannot be maintained[158] and the liability risk for contamination of neighboring fields is perceived as too high. On the other hand, coexistence and identity preservation requirements are proven to increase the production costs.[159] Since a mandatory

155 Ministry of Agriculture, Forestry and Fisheries (2008) (in Japanese only).

156 For an overview over the different regional requirements see Sato (2016), pp. 17–21.

157 For this section see Chap. 7 (Country Report on the USA), Sect. 7.5.

158 Lee (2014), p. 244.

159 Schenkelaars and Wesseler (2016), p. 9; Falck-Zepeda (2006), p. 1204; Gabriel and Menrad (2015), pp. 482, 484.

Table 8.19 Mandatory coexistence measures or identity preservation system in place

	Argentina	Australia	Canada	EU	Japan	USA
Coexistence	×	✓	×	✓	✓	×

coexistence system usually imposes the costs on the GM farmer (cf. Table 8.20), this might lead to a situation where the additional revenue from growing GMOs does not outweigh the extra cost of coexistence measures.[160]

Interestingly all three jurisdictions, that have a mandatory coexistence system in place, impose not just the obligations to take measures but also the duty to bear the costs of these measures on the GM farmer. This may be explained by a common underlying reasoning of those countries that conventional agriculture must be protected from a contamination by GM material. Even if not explicitly stated, it seems reasonable to assume that an implied "polluter pays principle" seems to be at work here.

Since the mandatory coexistence or identity preservation measures imposed by Australia, the EU (member states) and Japan only apply to GMO users, the classification of GEOs as GMOs or non-GMOS will decide whether they are also subject to them. Consequently, the classification of GEOs is not only decisive for their market approval but also for the rules which apply to their handling in general.

8.7 Reform Efforts

Decisive for the future regulation of GEOs is not only the applicability and interpretation of the current status quo but also the intent and content of possible reform efforts. While several interest groups or industry efforts may exist, official reform efforts are more promising to give an idea of how likely it is that a regulatory framework will be revised.

8.7.1 Argentina[161]

In the case of Argentina, efforts to update the regulatory regime for new breeding technologies have already led to an amendment of the regulatory framework. Due to the flexibility of the current system, the competent authorities are confident that they are well equipped for future technological innovations and approval requests. Consequently, no efforts to change the legal regime, which is currently in place, are made.

[160] Venus et al. (2017), p. 421.

[161] For this section see Chap. 2 (Country Report on Argentina), Sect. 2.6.

Table 8.20 Responsibility for coexistence or identity preservation measures

	Argentina	Australia	Canada	EU	Japan	USA
GMO users	×	✓	×	✓	✓	×
Profiteers	✓	×	✓	×	×	✓

8.7.2 Australia

Regarding the regulations for environmental release of GMOs an official review of the current regime was carried out between 2016 and 2019. In April 2019 the Gene Technology Amendment Regulations 2019 have been adopted.[162] However, they will not enter into force before October 2019. The most relevant changes of this update of the current Australian regulatory regime with regard to genome editing are that organisms modified by certain genome editing techniques are now explicitly exempted from the GMO definition while others are now explicitly covered.[163] A priori considered to be a GMO are "[a]n organism that has had its genome modified by oligonucleotide-directed mutagenesis" and "[a]n organism modified by repair of single-strand or double-strand breaks of genomic DNA induced by a site-directed nuclease, if a nucleic acid template was added to guide homology-directed repair".[164] While the former description refers to the ODM technique, the latter one encompasses SDN-2 and SDN-3. Instead, excluded from the GMO definition is "[a]n organism modified by repair of single-strand or double-strand breaks of genomic DNA induced by a site-directed nuclease, if a nucleic acid template was not added to guide homology-directed repair".[165] This is in the end a paraphrase of the SDN-1 technique. Consequently, only genome edited plants bred via SDN-1 are exempted of the scope of application of the Gene Technology Act, while plants modified via SDN-2, SDN-3 and ODM are covered. It must be noted, however, that the classification of organisms modified by those different techniques cannot be made as precise as the legislator intended. According to the wording SDN-2 and SDN-3 is only explicitly covered if it was used to guide homology-directed repair (HR). However, SDN-3 can also be used to guide non-homologous end joining

[162]Gene Technology Amendment (2019 Measures No. 1) Regulations 2019. https://www.legislation.gov.au/Details/F2019L00573/. Accessed 7 June 2019.

[163]Strictly speaking this is to some degree an oversimplification. Under certain circumstances an exempted organism could still be regarded as GMO and an organism here classified as GMO could be exempted by a different provision. For more details on this see Explanatory Statement: Select Legislative Instrument 2019 No. XX. Gene Technology Amendment (2019 Measures No. 1) Regulations 2019. https://www.legislation.gov.au/Details/F2019L00573/. Accessed 7 June 2019, pp. 8–9.

[164]Gene Technology Amendment (2019 Measures No. 1) Regulations 2019. https://www.legislation.gov.au/Details/F2019L00573/. Accessed 7 June 2019, Sect. 25.

[165]Gene Technology Amendment (2019 Measures No. 1) Regulations 2019. https://www.legislation.gov.au/Details/F2019L00573/. Accessed 7 June 2019, Sect. 26

(NHEJ).[166] This SDN-3-NHEJ technique would even be covered by the wording of the SDN-1 exemption since a nucleic acid template is not added to guide HR. Furthermore, more recently genome editing is no longer limited to either HR or NHEJ but can also employ microhomology-mediated end joining (MMEJ) as well.[167] Therefore, there remains regulatory uncertainty with regard to certain applications of genome editing.

Furthermore, the food regulator considers an update of the regulatory framework applicable to GM food since 2017. While several update proposals for suitable regulations of new breeding techniques are currently under discussion, no definitive decisions have been made so far.

8.7.3 Canada[168]

Canadian authorities have indicated no plan for revising the regulatory framework in the near future. This is hardly surprising, since applications for environmental release of genome edited plants have already been processed under the umbrella the product-based regime without any prior amendments needed.

8.7.4 EU[169]

In anticipation of the ECJ's judgment on the regulatory status of certain genome edited crops, the European Commission's initiatives regarding the regulation of new breeding techniques[170] did not result in any legislative proposals. However, after the court issued its ruling there is no longer a reasonable justification for this watchful waiting. In view of the elections to the EU Parliament end of May 2019 and the resulting reappointment of the Commission, it can be assumed that legislative initiatives or official reform efforts will be pursued in the future.[171]

8.7.5 Japan

After a fairly long time of hesitation to clarify the regulation of genome edited crops,[172] the Ministry of Environment and the Ministry of Health, Labour and

---

[166] European Food Safety Authority (2012), p. 13.

[167] Ata et al. (2018).

[168] For this section see Chap. 4 (Country Report on Canada), Sect. 4.6.

[169] For this section see Chap. 5 (Country Report on the EU), Sect. 5.6.

[170] Cf. European Commission (2017).

[171] Similar Lappin (2018), p. 3.

[172] Cf. Chap. 6 (Country Report on Japan), Sect. 6.5.

Welfare adopted the aforementioned policies regarding genome edited plants and food derived from them.[173] Reform efforts exist currently only with respect to the labelling of food products derived from genome edited crops. In that regard the Consumer Affairs Agency held a first meeting in May 2019 to examine how food products derived from genome edited plants could be labelled.[174]

8.7.6 USA

Neither the National Strategy for Modernizing the Regulatory System for Biotechnology Products published in 2016 nor the 2017 Update of the Coordinated Framework addressed specific regulatory concepts or proposals for an update of the current legal regime in the light of new breeding technologies.

However, in June 2019 the USDA published a proposal to revise their current regulations.[175] According to the USDA's own statement, this "would mark the first comprehensive revision of the regulations since they were established in 1987".[176] In respect of genome edited plant varieties significant changes have been proposed in the form of the complete exemption of certain techniques of genome editing from the scope of the regulations. Namely, the regulations would not apply to plants modified such that "(1) The genetic modification is solely a deletion of any size; or (2) The genetic modification is a single base pair substitution; or (3) The genetic modification is solely introducing nucleic acid sequences from within the plant's natural gene pool or from editing of nucleic acid sequences in a plant to correspond to a sequence known to occur in that plant's natural gene pool; or (4) The plant is an offspring of a GE plant that does not retain the genetic modification in the parent."[177] Since "the natural gene pool of a plant is determined by those plants with which the plant is sexually compatible",[178] the exemption applies as long as no transgene has been inserted. Considering all this, plants bred via SDN-1, SDN-2 and ODM would be exempted. Plant varieties produced via SDN-3 would be exempted as long as the inserted gene belongs to the plant's natural gene pool, i.e. is no transgene. However, this assumption seems to be only valid as long as no transgene nucleic acid, e.g. a vector's DNA or a nuclease's sgRNA, is present in the final end product. The fact that this classification is congruent with that in Table 8.1 is intended, because "[t]he exemptions reflect the Secretary of Agriculture's March 28, 2018, statement that

173Cf. Sect. 8.4.5.

174Sato (2019d).

175United States Department of Agriculture (2019).

176United States Department of Agriculture (2019), p. 26514.

177United States Department of Agriculture (2019), p. 26537.

178United States Department of Agriculture (2019), p. 26520.

Table 8.21 Current official efforts to update the regulatory regime for GEOs

	Argentina	Australia	Canada	EU	Japan	USA
Reform efforts	×	✓	×	×	✓	✓

USDA does not plan to regulate plants that could otherwise have been developed through traditional breeding techniques."[179]

8.7.7 *Comparative Analysis*

Currently only Australia, Japan (to a minor extent) and the USA undergo an official process to update their regulatory framework for genetically modified organism to ensure its suitability for GEOs (cf. Table 8.21).

The absence of reform efforts are, however, not an indicator for the actual or perceived fitness of the current legal regime with regard to new breeding techniques. While it is true that Argentina and Canada made no reform efforts because an update has already been made or is generally not considered to be necessary, the situation in the EU is quite different.[180] Within the EU, there seems to be predominant consensus on the shortcomings and unsuitability of the regulatory approach, but at the same time there is momentarily a lack of political will to change the status quo.

8.8 Conclusion

Since the degree, form and extent of the actual use of genetically modified crops and their products is so decidedly different in the examined countries (Sect. 8.2), it is not surprising that the corresponding regulatory frameworks are equally diverse.

This diversity is not only reflected in different regulatory triggers and point of entries (Sect. 8.3), but consequently also in a varying classification of GEOs (Sect. 8.4). These disparities are underlined and reinforced additionally by the different concepts and strategies when it comes to labelling (Sect. 8.5) and coexistence measures (Sect. 8.6).

This indicates that, from a regulatory point of view, GEOs are likely to meet the same fate as traditional GMOs: legal fragmentation causes non-tariff barriers to trade, research and development investments focus on countries with a permissive regulatory approach and legal hurdles slow down considerably the adoption of new plant varieties.

[179]United States Department of Agriculture (2019), p. 26519.

[180]Cf. Chap. 7 (Country Report on the USA), Sect. 7.9.

However, it should also be borne in mind that reform efforts are still underway in several countries (Sect. 8.7). Therefore, there is still the opportunity not to repeat the mistakes of the past.

References

Ata H, Ekstrom TL, Martínez-Gálvez G, Mann CM, Dvornikov AV, Schaefbauer KJ, Ma AC, Dobbs D, Clark KJ, Ekker SC, Burgess SM (2018) Robust activation of microhomology-mediated end joining for precision gene editing applications. PLoS Genet 14(9):e1007652

Australian Government Department of Health, Office of the Gene Technology Regulator (2018) Table of authorisations for commercial releases of GM plants (subset of list of licences involving Intentional Release). http://www.ogtr.gov.au/internet/ogtr/publishing.nsf/Content/cr-1. Accessed 20 Apr 2018

Ayers K (2018) Plant breeder expands offerings for Western Canada. https://www.farms.com/ag-industry-news/plant-breeder-expands-offerings-for-western-canada-244.aspx. Accessed 23 July 2018

Beckmann V, Soregaroli C, Wesseler J (2014) Coexistence. In: Smyth SJ, Phillips PWB, Castle D (eds) Handbook on agriculture, biotechnology and development. Edward Elgar, Cheltenham, pp 372–391

Blair R, Regenstein JM (2015) Genetic modification and food quality: a down to earth analysis. Wiley Blackwell, Chichester

Calyxt (2018a) Products in our development pipeline. http://www.calyxt.com/products/products-in-our-development-pipeline/. Accessed 19 July 2018

Calyxt (2018b) Calyxt exceeds farmer adoption milestone for high-oleic soybean product launch. http://www.calyxt.com/calyxt-exceeds-farmer-adoption-milestone-for-high-oleic-soybean-product-launch/. Accessed 19 July 2018

Calyxt (2019a) Calyxt doubles 2018 high oleic soybean acres. https://www.calyxt.com/calyxt-doubles-2018-high-oleic-soybean-acres/. Accessed 22 May 2019

Calyxt (2019b) First commercial sale of Calyxt high oleic soybean oil on the U.S. market. https://www.calyxt.com/first-commercial-sale-of-calyxt-high-oleic-soybean-oil-on-the-u-s-market/. Accessed 23 May 2019

Calyxt, Inc. (2016) Calyxt completes production of 30 tons of its high oleic soybean product in Argentina. http://www.calyxt.com/wp-content/uploads/2016/05/PR-5.24.16_Calyxt_Argentina_Soybean.pdf. Accessed 19 Sept 2018

Canadian Food Inspection Agency (2013) DD 2013-100: determination of the safety of Cibus Canada Inc.'s Canola (Brassica napus L.) Event 5715. http://www.inspection.gc.ca/plants/plants-with-novel-traits/approved-under-review/decision-documents/dd-2013-100/eng/1427383332253/1427383674669. Accessed 19 July 2018

Canadian Food Inspection Agency (2014) Decision document DD2014-101 determination of the safety of BASF Canada Inc.'s Canola (Brassica napus) Event CLB-1. http://www.inspection.gc.ca/plants/plants-with-novel-traits/approved-under-review/decision-documents/dd2014-101/eng/1454107718197/1454107776755#a31. Accessed 19 July 2018

Canadian Food Inspection Agency (2018) Plants with Novel Traits (PNT) and novel feeds from plant sources approved in Canada. http://inspection.gc.ca/active/netapp/plantnoveltraitpnt-vegecarnouvcn/pntvcne.aspx. Accessed 9 May 2018

Capalbo DMF, Suzuki MT (2017) The importance of Bacillus thuringiensis in the context of genetically modified plants in Brazil. In: Fiuza LM, Polanczyk RA, Crickmore N (eds) Bacillus thuringiensis and Lysinibacillus sphaericus: characterization and use in the field of biocontrol. Springer International Publishing, Cham, pp 259–271

Cibus (2018) Products. https://www.cibus.com/products.php. Accessed 23 July 2018

Crothers L (2017) Australia agricultural biotechnology annual: 2017 agricultural biotechnology report. https://gain.fas.usda.gov/Recent%20GAIN%20Publications/Agricultural%20Biotechnology%20Annual_Canberra_Australia_11-27-2017.pdf. Accessed 20 Sept 2018

Danielson E, Watters A (2017) Canada: agricultural biotechnology annual 2017. https://gain.fas.usda.gov/Recent%20GAIN%20Publications/Agricultural%20Biotechnology%20Annual_Ottawa_Canada_12-21-2017.pdf. Accessed 20 Sept 2018

Dewey C (2018) The future of food: scientists have found a fast and cheap way to edit your foods' DNA. https://www.washingtonpost.com/news/business/wp/2018/08/11/feature/the-future-of-food-scientists-have-found-a-fast-and-cheap-way-to-edit-your-edibles-dna/?noredirect=on&utm_term=.3b4e6be63099. Accessed 2 Oct 2018

Directive 94-08, Assessment Criteria for Determining Environmental Safety of Plants With Novel Traits

Eaton E (2013) Growing resistance: Canadian farmers and the politics of genetically modified wheat. University of Manitoba Press, Winnipeg

European Commission (2007) Commission Decision of 23 May 2007 concerning the placing on the market, in accordance with Directive 2001/18/EC of the European Parliament and of the Council, of a carnation (Dianthus caryophyllus L., line 123.2.38) genetically modified for flower colour (2007/364/EC). Official Journal of the European Union L 138/50

European Commission (2017) Modern biotechnologies in the agri-food sector. https://ec.europa.eu/food/plant/gmo/modern_biotech_en. Accessed 28 Sept 2018

Eurostat (2018a) Grain maize and corn-cob-mix by area, production and humidity. http://ec.europa.eu/eurostat/web/products-datasets/-/tag00093. Accessed 22 May 2018

Eurostat (2018b) Green maize by area, production and humidity. http://ec.europa.eu/eurostat/web/products-datasets/-/tag00101. Accessed 22 May 2018

Falck-Zepeda J (2006) Coexistence, genetically modified biotechnologies and biosafety: implications for developing countries. Am J Agric Econ 88:1200–1208

Flachowsky G, Schafft H, Meyer U (2012) Animal feeding studies for nutritional and safety assessments of feeds from genetically modified plants: a review. J Verbr Lebensm 7:179–194. https://doi.org/10.1007/s00003-012-0777-9

Food and Agricultural Organization of the United Nations (2018) FAOSTAT. http://www.fao.org/faostat/en/#data/RL. Accessed 28 May 2018

Food and Drug Administration (1992) Statement of policy: foods derived from new plant varieties. Fed Regist 57:22983–23005

Gabriel A, Menrad K (2015) Cost of coexistence of GM and non-GM products in the food supply chains of rapeseed oil and maize starch in Germany. Agribusiness 31:472–490. https://doi.org/10.1002/agr.21415

Gustafson E (2014) We the eaters: if we change dinner, we can change the World. Rodale

Hamburger D (2018) Normative criteria and their inclusion in a regulatory framework for new plant varieties derived from genome editing. Front Bioeng Biotechnol 6

International Service for the Acquisition of Agri-biotech Applications (2014) Event Name: 123.2.38 (40644). http://www.isaaa.org/gmapprovaldatabase/event/default.asp?EventID=31. Accessed 18 May 2018

International Service for the Acquisition of Agri-biotech Applications (2016) Global Status of Commercialized Biotech/GM Crops: 2016: ISAAA Brief No. 52

International Service for the Acquisition of Agri-biotech Applications (2018a) GM Approval Database, MON71800. http://www.isaaa.org/gmapprovaldatabase/event/default.asp?EventID=237&Event=MON71800. Accessed 7 June 2018

International Service for the Acquisition of Agri-biotech Applications (2018b) GM Crops List. http://www.isaaa.org/gmapprovaldatabase/cropslist/default.asp. Accessed 7 June 2018

International Service for the Acquisition of Agri-biotech Applications (2018c) GM Approval Database. http://www.isaaa.org/gmapprovaldatabase/default.asp. Accessed 9 May 2018

International Service for the Acquisition of Agri-biotech Applications (2018d) GM Approval Updates. http://www.isaaa.org/gmapprovaldatabase/updates/default.asp. Accessed 9 May 2018

International Service for the Acquisition of Agri-biotech Applications (2018e) GM Approval Database - EU. http://www.isaaa.org/gmapprovaldatabase/advsearch/default.asp?CropID=Any&TraitTypeID=Any&DeveloperID=Any&CountryID=EU&ApprovalTypeID=3. Accessed 18 May 2018

James C (2015) Global status of commercialized biotech/GM crops: 2015 ISAAA brief No. 51: 20th anniversary (1996 to 2015) of the global commercialization of biotech crops and biotech crop highlights in 2015. ISAAA, Manila

Juma C (2016) Innovation and its enemies: why people resist new technologies. Oxford University Press, New York

Kumar A, Sopory SK (eds) (2008) Recent advances in plant biotechnology and its applications: Prof. Dr. Karl-Hermann Neumann commemorative volume. I.K. International Pub. House, New Delhi

Kuntz M (2018) Transgenic plants and beyond. Advances in botanical research, vol 86. Academic Press, San Diego

Kurai T, Sato S (2018) Environment ministry proposes policy for regulating genome editing. https://gain.fas.usda.gov/Recent%20GAIN%20Publications/Environment%20Ministry%20Proposes%20Policy%20for%20Regulating%20Genome%20Editing%20_Tokyo_Japan_9-25-2018.pdf. Accessed 29 May 2019

La Capital (2018) INTA y dos alimentos del futuro: "superpapas" y leche no alergénica. http://www.lacapitalmdp.com/inta-y-dos-alimentos-del-futuro-superpapas-y-leche-no-alergenica/. Accessed 20 July 2018

Lappin J (2018) EU Court extends GMO directive to new plant breeding Techniques_Brussels. https://gain.fas.usda.gov/Recent%20GAIN%20Publications/EU%20Court%20Extends%20GMO%20Directive%20to%20New%20Plant%20Breeding%20Techniques_Brussels%20USEU_Belgium%20EU-28_7-27-2018.pdf. Accessed 19 Sept 2018

Lee M (2014) EU environmental law, governance and decision-making, 2nd ed. Modern studies in European law, vol 43. Hart, Oxford

Li L, He Z-Y, Wei X-W, Gao G-P, Wei Y-Q (2015) Challenges in CRISPR/CAS9 delivery: potential roles of nonviral vectors. Hum Gene Ther 26:452–462. https://doi.org/10.1089/hum.2015.069

Li L, Hu S, Chen X (2018) Non-viral delivery systems for CRISPR/Cas9-based genome editing: challenges and opportunities. Biomaterials 171:207–218. https://doi.org/10.1016/j.biomaterials.2018.04.031

Lino CA, Harper JC, Carney JP, Timlin JA (2018) Delivering CRISPR: a review of the challenges and approaches. Drug Deliv 25:1234–1257. https://doi.org/10.1080/10717544.2018.1474964

Lucht JM (2015) Public acceptance of plant biotechnology and GM crops. Viruses 7:4254–4281. https://doi.org/10.3390/v7082819

Lusser M, Parisi C, Plan D, Rodriguez-Cerezo E (2012) Deployment of new biotechnologies in plant breeding. Nat Biotechnol 30:231–239. https://doi.org/10.1038/nbt.2142

Ma J, Xiang H, Donnelly DJ, Meng F-R, Xu H, Durnford D, Li X-Q (2017) Genome editing in potato plants by agrobacterium-mediated transient expression of transcription activator-like effector nucleases. Plant Biotechnol Rep 11:249–258. https://doi.org/10.1007/s11816-017-0448-5

Magnan A (2016) When wheat was king: the rise and fall of the Canada-UK wheat trade. UBC Press, Vancouver

Ministerio de Agroindustria (2011) Resolución-763/2011. http://www.senasa.gob.ar/normativas/resolucion-763-2011-ministerio-de-agroindustria. Accessed 24 July 2018

Ministerio de Agroindustria (2018) OGM Comerciales. https://www.agroindustria.gob.ar/sitio/areas/biotecnologia/ogm/. Accessed 18 May 2018

Ministry of Agriculture, Forestry and Fisheries (2008) Field trail guidelines for genetically modified crops. http://www.naro.affrc.go.jp/archive/nias/gmo/indicator20080731.pdf. Accessed 22 Sept 2018

Ministry of Health and Welfare (2000a) Announcement no. 232. https://www.mhlw.go.jp/english/topics/food/3-2.html. Accessed 31 May 2019

Ministry of Health and Welfare (2000b) Announcement no. 233. https://www.mhlw.go.jp/english/topics/food/3-3.html. Accessed 31 May 2019

Ministry of Health, Labour and Welfare (2018) Procedure for safety assessment. https://www.mhlw.go.jp/english/topics/foodsafety/dna/01.html. Accessed 31 May 2019

Ministry of Health, Labour and Welfare (2019) Food produced by recombinant DNA techniques. https://www.mhlw.go.jp/english/topics/foodsafety/dna/index.html. Accessed 31 May 2019

Newton ED (2014) GMO food: a reference handbook. Contemporary world issues. ABC-CLIO, Santa Barbara

Office of the Gene Technology Regulator (2008) GM crops and stockfeed - fact sheet. http://www.ogtr.gov.au/internet/ogtr/publishing.nsf/Content/gmstockfeed-htm. Accessed 23 May 2018

Oh J, Ezezika OC (2014) To label or not to label: balancing the risks, benefits and costs of mandatory labelling of GM food in Africa. Agric Food Sec 3:8–15. https://doi.org/10.1186/2048-7010-3-8

PlantForm (2015) PlantForm, University of Calgary research collaboration awarded NSERC Engage Grant. http://www.plantformcorp.com/news-detail.aspx?id=c51ad109-a5d5-4449-a6c3-0d94539fc0e7#.W1WBe7guCUl. Accessed 23 July 2018

Pratt S (2018) New herbicide-tolerant options coming soon for canola growers. New Seed Variety Guide 2018:10–13

Rao VS (2015) Transgenic herbicide resistance in plants. CRC Press, Boca Raton

Ricroch A, Clairand P, Harwood W (2017) Use of CRISPR systems in plant genome editing: toward new opportunities in agriculture. Emerg Topics Life Sci 1:169–182. https://doi.org/10.1042/ETLS20170085

Sato S (2016) Japan: agricultural biotechnology annual. https://gain.fas.usda.gov/Recent%20GAIN%20Publications/AGRICULTURAL%20BIOTECHNOLOGY%20ANNUAL_Tokyo_Japan_11-30-2016.pdf. Accessed 11 Feb 2019

Sato S (2018a) Japan discusses genome editing technology. https://gain.fas.usda.gov/Recent%20GAIN%20Publications/Japan%20Discusses%20Genome%20Editing%20Technology_Tokyo_Japan_8-10-2018.pdf. Accessed 28 Sept 2018

Sato S (2018b) Japan holds second meeting to discuss genome editing technology. https://gain.fas.usda.gov/Recent%20GAIN%20Publications/Japan%20Holds%20Second%20Meeting%20to%20Discuss%20Genome%20Editing%20Technology_Tokyo_Japan_8-22-2018.pdf. Accessed 28 Sept 2018

Sato S (2018c) Health ministry begins discussion of genome edited foods. https://gain.fas.usda.gov/Recent%20GAIN%20Publications/Health%20Ministry%20Begins%20Discussion%20of%20Genome%20Edited%20Foods%20_Tokyo_Japan_9-26-2018.pdf. Accessed 6 June 2019

Sato S (2018d) Japanese health ministry continues discussions on genome edited food. https://gain.fas.usda.gov/Recent%20GAIN%20Publications/Japanese%20Health%20Ministry%20Continues%20Discussions%20on%20Genome%20Edited%20Food_Tokyo_Japan_12-4-2018.pdf. Accessed 5 June 2019

Sato S (2018e) Health ministry proposes draft genome edited food policy. https://gain.fas.usda.gov/Recent%20GAIN%20Publications/Health%20Ministry%20Proposes%20Draft%20Genome%20Edited%20Food%20Policy_Tokyo_Japan_12-19-2018.pdf. Accessed 5 June 2019

Sato S (2019a) Environment ministry finalizes policy for regulating genome editing. https://gain.fas.usda.gov/Recent%20GAIN%20Publications/Environment%20Ministry%20Finalizes%20Policy%20for%20Regulating%20Genome%20Editing_Tokyo_Japan_3-6-2019.pdf. Accessed 29 May 2019

Sato S (2019b) Health ministry invites comments on genome edited food policy. https://gain.fas.usda.gov/Recent%20GAIN%20Publications/Health%20Ministry%20Invites%20Comments%20on%20Genome%20Edited%20Food%20Policy_Tokyo_Japan_2-13-2019.pdf. Accessed 5 June 2019

Sato S (2019c) Japanese health ministry finalizes genome edited food policy. https://gain.fas.usda.gov/Recent%20GAIN%20Publications/Japanese%20Health%20Ministry%20Finalizes%20Genome%20Edited%20Food%20Policy_Tokyo_Japan_4-12-2019.pdf. Accessed 5 June 2019

Sato S (2019d) Japan initiates discussion of food labeling for genome edited foods. https://gain.fas.usda.gov/Recent%20GAIN%20Publications/Japan%20Initiates%20Discussion%20of%20Food%20Labeling%20for%20Genome%20Edited%20Foods_Tokyo_Japan_5-29-2019.pdf. Accessed 7 June 2019

Schenkelaars P, Wesseler J (2016) Farm-level GM coexistence policies in the EU: context, concepts and developments. EuroChoices 15:5–11. https://doi.org/10.1111/1746-692X.12112

Servicio Nacional de Sanidad y Calidad Agroalimentaria (2011) Resolución-701/2011. http://www.senasa.gob.ar/normativas/resolucion-701-2011-senasa-servicio-nacional-de-sanidad-y-calidad-agroalimentaria. Accessed 24 July 2018

Smyth S, Phillips PWB, Khachatourians GG (2004) Regulating the liabilities of agricultural biotechnology. CABI Publishing, Wallingford

Spielman DJ, Zambrano P (2013) Policy, investment, and partnerships for agricultural biotechnology research in Africa: emerging evidence. In: Falck-Zepeda J, Gruère G, Sithole-Niang I (eds) Genetically modified crops in Africa: economic and policy lessons from countries South of the Sahara. International Food Policy Research Institute, Washington, DC, pp 183–205

United States Department of Agriculture (2018a) Secretary perdue issues USDA statement on plant breeding innovation. https://www.usda.gov/media/press-releases/2018/03/28/secretary-perdue-issues-usda-statement-plant-breeding-innovation. Accessed 13 Aug 2018

United States Department of Agriculture (2018b) Petitions for determination of nonregulated status. https://www.aphis.usda.gov/aphis/ourfocus/biotechnology/permits-notifications-petitions/petitions/petition-status. Accessed 9 May 2018

United States Department of Agriculture (2018c) Details on USDA plant breeding innovations. https://www.aphis.usda.gov/aphis/ourfocus/biotechnology/brs-news-and-information/2018_brs_news/pbi-details. Accessed 13 Aug 2018

United States Department of Agriculture, Agricultural Marketing Service (2018) National bioengineered food disclosure standard: final rule. Fed Regist 83:65814–65876

United States Department of Agriculture (2019) Movement of certain genetically engineered organisms. Fed Regist 84:26514–26541

Venus TJ, Dillen K, Punt MJ, Wesseler JHH (2017) The costs of coexistence measures for genetically modified maize in Germany. J Agric Econ 68:407–426. https://doi.org/10.1111/1477-9552.12178

Viljoen CD, Marx GM (2013) The implications for mandatory GM labelling under the Consumer Protection Act in South Africa. Food Control 31:387–391. https://doi.org/10.1016/j.foodcont.2012.10.026

Voigt B, Klima J (2017) CRISPR-Plants & Co. – the quest for adequate risk regulation: modern plant breeding techniques and the current legal framework for risk regulation in the European Union. Zeitschrift für Europäisches Umwelt- und Planungsrecht 15:319–338

Wiseman G (2009) Real-time PCR: application to food authenticity and legislation. In: Logan J, Edwards K, Saunders N (eds) Real-time PCR: current technology and applications. Caister Academic Press, Norfolk, pp 253–267

Wolt JD, Wang K, Yang B (2016) The regulatory status of genome-edited crops. Plant Biotechnol J 14:510–518. https://doi.org/10.1111/pbi.12444

Appendix: Questionnaire

Country-Report: [Please state your Country]

Introduction

The introduction should make the reader aware of the significance of the respective country for the cultivation of genetically modified plants and the marketing of products derived from them (Table A.1).

The Regulatory Framework for Genetically Modified Organisms (GMOs): An Overview

Please give an overview of the regulatory framework concerning GMOs. What are the applicable laws and regulations? Which public authorities are responsible for the administration of the regulatory regime? In case there is more than one competent authority: what are their responsibilities and what is the relationship between the different authorities?

To allow the reader a quick understanding of the relevant actors and pertinent legislation, a tabular summary of your findings might be helpful (cf. the example for Canada below) (Table A.2).

H.-G. Dederer, D. Hamburger (eds.), *Regulation of Genome Editing in Plant Biotechnology*, https://doi.org/10.1007/978-3-030-17119-3

Table A.1 Facts that could be mentioned

• Since when are GM-plants cultivated? • Socio-economic effects of GMOs (*e.g.* profitability, farm size) • Acreage of GM-plants	• Share of GM-plants at total acreage of a variety (*e.g.* 90% of the cotton cultivation consists of GM-cotton) • Import/Export share of GM-plants and products derived from them at total national import/export of like products	• Do currently contained uses, field trials or cultivation of genome edited plants take place? • Are products derived from genome edited plants already on the market?

Table A.2 Actors and pertinent legislation

Authority	Area of responsibility	Pertinent legislation
Canadian Food Inspection Agency (CFIA)	Plants with novel traits Novel fertilizers and supplements Novel livestock feeds Veterinary biologics	Seeds Act Fertilizers Act Feeds Act Health of Animals Act
Health Canada	Novel foods Pest control products	Food and Drug Act Pest Control Products Act
Environment and Climate Change Canada (ECCC; former Environment Canada)	New Substances	Canadian Environmental Protection Act (1999)

Regulatory Status of Genome Edited Plants

Applicability of the Regulatory Framework for GMOs

Please describe the regulatory approach towards GMOs. Can it be described as product-based or process-based or is it some form of combination of both?

What criteria have to be fulfilled in order to trigger the application of the regulatory regime?

In case of a process-based approach, there seems to be usually a definition of 'genetically modified organism' (GMO) that must be met. If so, could you explain the definition and how it is interpreted and applied by the competent authority?

In case of a product-based regime, a different kind of trigger is used (e.g. "novelty" in Canada; "adulterated" or "regulated article" in the US). Please identify the respective trigger and describe its meaning and application.

Regulatory Classification of Genome Editing/Genome Edited Plants

Please analyse if the regulatory regime, which is applicable to GMOs, is applicable to genome edited plants and products derived from them as well.

The answer will most likely depend on whether the characteristics of the genome edited plant (*e.g.* 'transgenic' or not) or the technique used to modify the plant constitute a trigger as described in the previous chapter.

Table A.3 Suggestion for the subdivision of this chapter

Regulatory Classification of genome editing 1. SDN Type 1 2. SDN Type 2 3. SDN Type 3 4. ODM	Please note: This proposed subdivision is most likely unsuitable for a product-based regulatory framework. In that case, please adjust the subdivision accordingly In any case, you are welcome to choose the substructure which suits your analysis best

With regard to the different techniques, please differentiate, as possible, between SDN-1, SDN-2, SDN-3 and ODM (for definitions of these abbreviations see the glossary at the end of the questionnaire) (Table A.3). If possible and feasible, CRISPR/Cas9 should deserve special emphasis.

If the regulatory framework is product-based, a differentiation between the different techniques might be unsuitable. In that case, please explain the characteristics (of the plant or product), which are decisive for triggering the regulatory regime, and whether these characteristics can be achieved through one or several of the aforementioned techniques.

Regulatory Prerequisites for Activities Relating to Genome Edited Plants

Depending on the findings in the section "Regulatory Classification of Genome Editing/Genome Edited Plants" the regulatory regime for GMOs may be applicable (scenario 1) or may not be applicable (scenario 2) to genome edited plants or products derived from them.

In case of scenario 1: Please describe the administrative procedure, *e.g.* approval or notification procedure, in detail differentiating, if applicable, between contained use, field trials, cultivation and marketing (Table A.4).

In case of scenario 2: Is the use of genome edited plants or the marketing of products derived from such plants entirely unregulated? Do other rules for approval/ notification exist which might be still applicable?

It might be the case that scenario 1 and 2 are not mutually exclusive. In this case, the regulatory restrictions with regard to both scenarios should be part of the analysis.

Example: It might be the case that a genome edited plant altered by SDN-Type 1 does not trigger the regulatory regime which is applicable to GMOs whereas a plant modified by means of SDN-Type 3 does trigger those regulations (Table A.5).

Table A.4 Aspects that could be mentioned

• Are there any specifically adopted rules with regard to genome edited plants? • Which agency is responsible? • What are the requirements for a notification or for an application for approval? • What are the requirements for granting an approval? • What interests/factors are taken into consideration during the approval/notification procedure?	• Is the purpose of the end-use (*e.g.* as food, feed or drug) taken into account? • Are not risk related factors taken into account (*e.g.* new plant must have an economic and/or biological advantage; socio-economic concerns)? • Does a weighing of different/conflicting interests take place (*e.g.* risk *vs.* benefit)? • How are different risks weighed?	• What data is required for the risk assessment? • Who provides the data for the risk assessment (*e.g.* the applicant or independent researchers)? • Who carries out the risk assessment: applicant or authority? • How is dealt with uncertainty?

Table A.5 Suggestion for the subdivision of this chapter

Regulatory prerequisites 1. Contained use 2. Field trials 3. Cultivation 4. Marketing 4.1 Food 4.2 Feed 4.3 Medical products 4.4 Other products	Some regulatory frameworks do not differentiate as detailed between the requirements for the different activities as this outline does. In that case, please adjust the subdivision accordingly In any case, please choose the substructure which suits your analysis best

Status Quo of Genome Edited Plants and Products Derived from Them

If genome edited plants have already been approved/notified for contained use, field trials, or cultivation, could you name an example/examples (plant variety, the new trait of that plant and the technique used)? Are applications for approval of genome edited plants pending? Could you please state an example/examples? If approval/notification is not necessary, could you please name an example/examples of such an unregulated genome edited plant.

Are products derived from genome edited plants already on the market?

Reform Efforts

How suitable is the current regulatory regime for the regulation of genome edited plants? What are the shortcomings of the regulatory regime with respect to genome edited plants? What are the strengths?

Are there official plans (*e.g.* of the legislature or the executive) to update the current legal regime with regard to the regulation of genome edited plants and products derived from them? What are those proposals? How would you assess those proposals?

Do you have recommendations for an update of the current regulatory framework for genome edited plants and the products derived from them?

Low Level Presence

Is there a tolerance level for products containing genetically modified ingredients/ material up to which those products are exempt from approval/notification procedures or regulatory oversight?

How are products derived from genome edited plants treated in that regard? How is the problem addressed that certain genetic alterations by genome editing are untraceable? (Table A.6)

Labelling

Does an obligation exist to label products which contain genetically modified ingredients/material? Please describe the applicable regulations.

How are products derived from genome edited plants treated in that regard? How is the problem addressed that certain genetic alterations by genome editing are untraceable? (Table A.7)

Identity Preservation System (Coexistence)

Is there an identity preservation system in place? If so, please describe its components (*e.g.* separate cultivation areas or processing facilities; traceability; compensation in case of 'contamination').

Table A.6 Please note

In this chapter ("Low Level Presence") and the following ones ("Labelling", "Identity Preservation System (Coexistence)", "Liability") the mentioning of the rules for traditional GMOs has only the purpose to illustrate the regulatory matter at hand. However, the focus should be on genome edited plants and products derived from them. Therefore, the decisive question for this chapter is how the regulatory framework deals with the low level presence of ingredients derived from genome edited crops

Table A.7 Aspects that could be mentioned

• Mandatory or voluntary labelling? • What triggers the labelling obligation (*e.g.* novelty; genetic modification; distinctive characteristics from non-GM product)?	• Is there a tolerance level for low level presence exempting products from labelling requirements? • Is voluntary GM-free-labelling allowed?	• Up to what level can products containing GM-ingredients still be labelled as GM-free? Is there a zero-level-tolerance?

Table A.8 Aspects that could be mentioned

• Nature of liability rules: criminal, public or civil? • Damages: compensatory damages and/or punitive damages?	• Culpability: strict liability (liability without fault), intent or negligence (standard of care)?	• What activities are covered by liability laws (damage to the environment or human health; contamination of non-GM products; field trials or marketing without approval/notification)?

How are products derived from genome edited plants treated in that regard? How is the problem addressed that certain genetic alterations by genome editing are untraceable?

Liability

Are there specific liability provisions concerning the use of genetically modified plants and products derived from them?

How are activities relating to genome edited plants treated in that regard? (Table A.8)

Perception of Genome Editing

Position of Public Authorities

What is the attitude of the government or the competent authorities towards genome edited plants? Are they in favour of a strict regulation or are they embracing the new technology? Are the existing laws and regulations applied strictly or generously?

Public Opinion

Is there already an emerging public opinion on genome editing? Taking into account the public opinion on GMOs, what could the public position on genome editing be like?

Treatment of Other New Breeding Technologies

Genome editing is only a subset of a number of so called new breeding technologies. What is the regulatory status of other new breeding techniques such as Cisgenesis/Intragenesis, Grafting, Agro-infiltration, RNA-dependent DNA methylation (RdDM) or Reverse Breeding (for definitions of these techniques see the glossary at the end of the questionnaire)? (Table A.9)

Table A.9 Please note

This chapter shall only provide an overview. Please feel free to deal with this subject matter as brief as it suits you best. If no regulatory decision has been made yet, it is sufficient to state this. The same applies where the regulatory status is unclear

Conclusion

Please draw a conclusion regarding your findings.

Glossary

To ensure a uniform understanding among the country rapporteurs the glossary provides working definitions for those technical terms which have not yet received a clear-cut definition:

- **SDN** (Site-Directed Nucleases): a generic term for different nuclease techniques like Meganucleases (MN), Zinc Finger Nucleases (ZFNs), Transcription Activator-Like Effector Nucleases (TALENs) and Clustered Regularly Interspaced Short Palindromic Repeats (CRISPR/Cas)
  These techniques can be used in three ways[1]:
  - **SDN Type 1 (SDN-1)**: "generates site-specific random mutations (changes of single base pairs, short deletions and insertions) by non-homologous end-joining (NHEJ). During SDN-1, no repair template is provided to the cells together with the SDN. The double-strand DNA break (DSB) is repaired by NHEJ which is a natural DNA-break repair mechanism in the cell. This often (though not always) results in a single or a few base substitutions or small localized deletions or insertions. In the case of insertions, the inserted material is derived from the organism's own genome i.e. it is not exogenous. The DNA end (from the strand break) may also become joined to a completely unrelated site, which results in chromosomal translocation."
  - **SDN Type 2 (SDN-2)**: "generates site-specific desired point mutations by DNA repair processes through homologous recombination (HR). During SDN-2, a continuous stretch of DNA is delivered to the cells simultaneously with the SDN. This template DNA is homologous to the targeted area, spanning a few kilo base pairs, and overlaps the region of the DSB. The template DNA contains the specific base pair alteration(s) to be introduced into the target DNA or chromosome. It is then used by the cell to repair the DSB."

[1]The subsequent definitions are adopted literally from *European Food Safety Authority*, Scientific opinion addressing the safety assessment of plants developed using Zinc Finger Nuclease 3 and other Site-Directed Nucleases with similar function, adopted on 18 October 2012, pp. 5–6 (available at http://onlinelibrary.wiley.com/doi/10.2903/j.efsa.2012.2943/epdf).

- **SDN Type 3 (SDN-3)**: "a large stretch of donor DNA (up to several kilobases) is introduced together with the SDN complex to target DNA insertion into a predefined genomic locus. The predefined locus may or may not have extensive similarity to the DNA to be inserted. The insertion can take place either by HR or by NHEJ. Donor DNA can come from any species and it is delivered to the cell, along with the SDN, and it is targeted to the desired site of the genome and inserted into the DSB site."

- **ODM, Cisgenesis/Intragenesis, Grafting, Agro-infiltration, RNA-dependent DNA methylation (RdDM), Reverse breeding**: The understanding of the Joint Research Center's study is adopted with regard to these techniques.[2]

[2]Cf. *Lusser, Maria/Parisi, Claudia/Plan, Damien/Rodriguez-Cerezo, Emilio*, New plant breeding techniques – State-of-the-art and prospects for commercial development, Luxembourg 2011, pp. 24-27 (available at http://ftp.jrc.es/EURdoc/JRC63971.pdf).

MIX
Papier aus verantwortungsvollen Quellen
Paper from responsible sources
FSC® C105338

If you have any concerns about our products,
you can contact us on
ProductSafety@springernature.com

In case Publisher is established outside the EU,
the EU authorized representative is:
Springer Nature Customer Service Center GmbH
Europaplatz 3, 69115 Heidelberg, Germany

Printed by Libri Plureos GmbH
in Hamburg, Germany